MUTATION,
DEVELOPMENTAL SELECTION,
AND PLANT EVOLUTION

Mutation, Developmental Selection, and Plant Evolution

EDWARD J. KLEKOWSKI, JR.

COLUMBIA UNIVERSITY PRESS

NEW YORK   1988

Columbia University Press
New York   Guildford, Surrey
Copyright © 1988 Columbia University Press
All rights reserved

Library of Congress Cataloging-in-Publication Data

Klekowski, Edward J.
Mutation, developmental selection, and plant
evolution.

Bibliography: p.
Includes index.
1. Plant genetics. 2. Plant mutation. 3. Plants—
Evolution. 4. Plant morphogenesis. I. Title.
QK981.K54 1988   581.1′5   88-2579
ISBN 0-231-06528-0

Printed in the United States of America

This book is dedicated to my wife Libby, my mother Genevieve,
and my deceased father Edward.

Contents

Prologue   ix

1. Introduction to the Problem   1

2. Uncertainty in Heredity   8

3. Components of Genomic Organization and Stasis   41

4. Unstratified Meristems—General Properties   85

5. Stratified Meristems—General Properties   109

6. Other Plant Meristems   136

7. Phenotypic Responses to Mutation and
   Environmental Mutagens   166

8. Mutation Buffering   182

9. Soft Selection and Life Cycles   219

10. Genetic Load   251

11. Significance of Mutation   293

References   319

Author Index   353

Subject Index   363

Prologue

My interest in the consequences of somatic mutation in plants began with studies of the somatic mutation rates for gametophyte mutations in ferns. In the course of these researches, David W. Bierhorst (University of Massachusetts) convinced me of the significance of apical meristem anatomy and development in influencing both the retention and distribution of mutations in plants. Ferns have relatively simple apical anatomies with regard to somatic mutation since a single tetrahedral apical cell serves as a permanent apical initial. More complicated situations arise in seed plant apical meristems with multiple (and often impermanent) apical initials as well as tunica corpus organizations. In trying to understand the fate of somatic mutations in such complex situations, I was fortunate to enlist the mathematical talents of Nina Kazarinova-Fukshansky (Universität Freiburg); together we carried out a number of heuristic mathematical studies of such meristems. One of the conclusions of these studies was that many apical meristem organizations in higher plants are not very adept at losing deleterious somatic mutations. Thus developed the idea that somatic mutations (especially deleterious mutations and mutational load) may be very significant phenomena in plant evolution. This is the theme of this book.

As a force to be responded to, somatic mutation is primarily destabilizing (i.e., the majority of mutations with phenotypic effect are deleterious to some degree), thus the primary evolutionary responses have been to enhance stability. Such stability enhancement can take many forms, e.g., error rates may be reduced, mutation retention in meristems reduced, developmental homeostasis increased, reproductive systems may have mechanisms to compensate for high mutational loads. The important point to note is that the consequences of somatic mutation must be considered at all levels of biological organization, i.e., molecular, cellular, histological, organismal, and populational. These levels of organization reflect the organization of topics in this book. Many of the topics which certainly relate to the problem of somatic mutation in plants have not yet been studied in plants. I have, therefore, included examples from other organisms (lower eukaryotes, animals) to illustrate these points. I hope the reader will be stimulated to think of other plant characteristics relating to this general theme.

A note should be made about the various symbols used. Because the scope of this book runs across traditional areas of botany and genetics, very often the same symbol or designation has been used for different plant characteristics. I have not changed symbols to maintain uniformity throughout the book but rather have retained the original author's symbols so that the reader can easily refer to the original papers. Thus symbolism is uniform only within chapters. I apologize for this but feel that, in the long run, this method is the more logical policy.

Many people have helped me in the development of the themes in this book. Hans Mohr and Leonid Fukshansky (Universität Freiburg), Steve Robinson, David Mulcahy, Edward Davis, Ronald Beckwith, Leonard Norkin, Mike Marcotrigiano, and Otto Stein (University of Massachusetts), Diana Stein (Mt. Holyoke College), as well as the graduate students who participated in the molecular biology journal club (Botany Department, University of Massachusetts) must all be singled out for special thanks. I would also like to express my appreciation to Valgene L. Dunham (Western Kentucky University), Phil Lintilhac (University of Vermont), Jack Van't Hof (Brookhaven National Laboratory), and Otto and Diana Stein for reading and commenting upon selected chapters and to James Mauseth and Donald Levin (both of the University of Texas at Austin) for reading and criticizing the entire manuscript. Needless to say, the responsibility for any errors that remain in the text is, of course, my own.

I am also grateful to Alena F. Chadwick and Laurence M. Feldman of the Morrill Biology Library (University of Massachusetts) for their tireless efforts on my behalf, Arthur Stern and Sally Klingener (University of Massachusetts), and Frank Woitzik (Universität Freiburg) for their able and patient computer assistance, the U.S. National Science Foundation for supporting some of the researches described in this book, and the Alexander von Humboldt Stiftung (Bonn), which supported my work in Freiburg where this book was conceived and ultimately completed.

Finally, I thank my daughter, Amanda, for her expert translations of German and French articles and my wife, Libby, without whose help this book would never have been completed.

University of Massachusetts, Amherst
and
Universität Freiburg, Freiburg im Breisgau

MUTATION,
DEVELOPMENTAL SELECTION,
AND PLANT EVOLUTION

CHAPTER ONE

Introduction to the Problem

That short, potential stir
That each can make but once,
That bustle so illustrious
'Tis almost consequence,

EMILY DICKINSON
Amherst

The above lines penned by Emily Dickinson in the nineteenth century speak of the mortality of the individual. In biology it is commonly said that although the individual is mortal, the genes are immortal. Mutation, the subject of this book, imposes mortality upon the genes as well.

The theme of this book is that mutation has been primarily a negative force in plant biology. No doubt adaptive mutations do occur and have occurred, but since the majority of mutations are either negative or neutral with regard to the organism, the primary adaptive responses to mutation pressure have been to minimize its consequences. An appreciation of mutation as a disruptive factor begins with an understanding of the origins of heritable molecular changes in the genotype. In chapter 2 the various and diverse molecular mechanisms that can cause mutations are surveyed. In chapter 3 those aspects of genomic organization that can modify mutation rates are considered. Paradoxically, very often the mechanisms promoting genomic stability for some genome components will inadvertently increase instability in other genome characteristics. It is soon apparent that absolute genomic immutability is an impossible evolutionary achievement. In chapters 4, 5, and 6 the fate of somatic mutations is considered. Plants have open systems of growth and lack germlines or immune systems. Thus, whether mutations persist or are lost from meristems is primarily a function of the competitive abilities of the mutant cells and the topologies and growth characteristics of apical and cambial meristems. Because plant organs may occasionally consist of large numbers of mutant cells, the development of an organ may be disrupted (chapter 7) or developmental mechanisms may exist that allow normal organogenesis to occur from mixtures of variously mutant and nonmutant cells. These mechanisms of developmental stasis are considered in chapter 8.

While chapters 2 through 8 consider the problem of mutation at the level of the cell, tissue, organ, and individual, the subsequent chapters analyze the consequences of mutation at the population and species levels. Many reproductive characteristics associated with the formation of spores, gametophytes, zygotes, embryos, and fruits seem almost to have been "designed" to filter effective genotypes from defective genotypes. Such "soft selection sieves" are described in chapter 9. The failure of these sieves in conjunction wth the mechanisms of genomic instability described in earlier chapters results in a high frequency of deleterious alleles in the gene pools of long-lived plants. In chapter 10 the problem of mutational load is considered. Finally, in chapter 11, the problem of mutation and broad-scale evolutionary patterns in plants is discussed. The argument is put forward that the internal factors influencing mutation origin, expression and selection are important constraints on the patterns of adaptive evolution exhibited by various plant groups.

SOMA AND MUTATION

It has long been known that Weismann's doctrine (Weismann 1892) of the separation of soma and germ is invalid in plants. Although plant germ cells ordinarily are produced from undifferentiated cell lineages, these cells are not set aside as in the sex gonads of many mammals (Babcock and Clausen 1927). Despite this fundamental distinction between higher animals and plants, it is only recently that the genetic and evolutionary ramifications of the lack of a germline have been emphasized (Whitham and Slobodchikoff 1981; Buss 1983; Walbot and Cullis 1983; Walbot 1985).

The primary consequence of the lack of a germline in plants is that somatic mutations may occur in cell lines which in turn develop into sporogenous tissues that may give rise to meiocytes. Thus somatic mutations ultimately may be transmitted to the gametes and be passed onto the zygotes of the next generation. The potential for mutation accumulation in organisms lacking a germline may be very great when one considers the number of mitotic generations from zygote to meiocyte in many long-lived plants. Mutation rates are often given per gene and per cell cycle (or per generation in annuals), whereas the accumulation of mutations in a clone is per genome and per a biological time unit that will encompass many cell cycles (or growing periods). The mutation rate per cell cycle is the product of the number of genes in the genome times the mutation rate per gene whereas the accumulated mutation frequency is a function of the number of cell cycles (or growing periods) from zygote to meiocyte. Thus the accumulated mutation frequency in an individual's gametes may be expected to be higher in organisms without a germline.

A direct result of somatic mutation is the development of individuals that are a chimera (i.e., composed of cells with dissimilar genotypes). Tissues and organs that are chimeras may allow competition either between cells within a tissue, between tissues, or between organs in which different genotypes of cells predominate. For example, during the growth of a plant, competition can occur between cells within an apical meristem or, at the other extreme, between buds or branches or ramets; such selection ultimately can alter the frequencies of different mutant and wild type meiocytes and consequently affect the allele frequencies in the offspring.

Competition between cells within meristems has been called diplontic selection (Gaul 1965), whereas competition between organs (or organ systems such as ramets) is one aspect of developmental selection. The concept of developmental selection and its significance in plant biology was discussed originally by Buchholz (1922) and included various life cycle characteristics which promoted competition between meiotic products, gametophytes, gametes, embryos, and gametophytic and sporophytic ramets. Developmental selection was independently reformulated by Whyte (1960, 1964, 1965), who defined it as "The internal selection of mutated genotypes and their consequences, the criterion being compatibility with the coordination of the internal structure and processes of the organism. The restriction of the otherwise possible directions of evolutionary change by organizational factors within the organism" (1964:3). Although Whyte did not specifically consider the problems of mutant cells within chimeric individuals, his ideas are relevant to chimeras. Whyte's concept of developmental selection includes all levels of selection within the organism (sub-cellular, cellular, tissue, and organ) and stresses the loss of mutations as well as the developmental restrictions of future evolutionary change. In this book, all levels of suborganismal or internal selection as well as various life cycle characteristics that may permit game-tophytes, gametes, zygotes, embryos, fruits, or ramets to compete will be considered as aspects of developmental selection *sensu lato*.

Somatic mutations normally would be expected to generate het-erozygous genotypes. Such heterozygotes may convert to homozy-gotes through mechanisms such as mitotic crossing over (Vig 1982) (figure 1.1) or mitotic gene conversion (Zimmermann 1971). Mitotic recombination and conversion mechanisms also generate cell gen-otypes lacking mutant alleles and in this way promote the loss of mutations. Given somatic cells either heterozygous or homozygous for mutant alleles imbedded in a matrix of nonmutant cells, the fate of these mutant cells is dependent upon their phenotypes as well as the growth characteristics of the individual plant. The phenotype of the mutant cell may be expressed as its relative fitness in terms of leaving descendant cells with reference to the nonmutant wild-

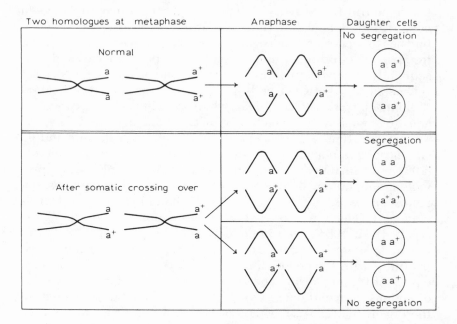

Figure 1.1. Mitotic or somatic crossing over. Such recombinations can generate homozygous genotypes from heterozygotes. (From Sybenga 1972.)

type cells within the chimera. This cell fitness is distinct from the individual organism's fitness when it consists solely of mutant cells. The matrix of these two fitness parameters is shown in figure 1.2. As will be shown in subsequent discussions, the loss or fixation of mutant cells whose fitness (CF) is one (CF = 1) is a function of the stochastic aspects of plant growth whereas the survival or loss of cells whose fitness is other than one (CF < 1 or >1) is more complex, involving various levels of developmental selection. It should be noted that high cell fitness (CF > 1) does not necessarily mean that organisms that are composed of such cells have a high organismal fitness. There are many examples in which the opposite is true, high cell fitness being associated with lowered organism viability (e.g., neoplasms).

As many authors have noted (see Harper 1977 for review), the definition of an individual organism in many plant species is difficult. In an effort to help resolve this difficulty the genet and ramet concept is useful. Following White (1979), a genet is the entire plant, of whatever size or however subdivided or propagated, that is derived from a single zygote or meiospore. In contrast a ramet is a single module or shoot complex of a genet that is conveniently enumerated. All sib ramets have a mitotic origin and, barring somatic mutation or somatic recombination, should have identical genotypes. The organism's fitness (OF) refers to the fitness of the genet (realizing of course it is a function of the individual ramet fitnesses).

CF \ OF	<1	1	>1
<1	Cells and organism handicapped	OF not a function of CF	Cells handicapped, organism advantaged (e.g., perhaps slower growth is adaptive)
1	OF not a function of CF	Fitness of the zygotic genotype	OF not a function of CF
>1	Cells advantaged, organism handicapped (e.g., neoplasms)	OF not a function of CF	Cells and organism advantaged

Figure 1.2. Organism fitness (OF) and cell fitness (CF) need not always be positively correlated. A matrix of the theoretically possible combinations of OF and CF.

When mutant cells occur in sufficient number within a ramet so that its fitness with reference to nonmutant sib ramets is affected, one form of developmental selection may occur. Both Buchholz and Whyte stressed that their respective concepts of developmental selection represented distinct systems of selection which differed from the more "Darwinian" or "Adaptive" selection. In most plants these levels of selection represent a somewhat murky continuum; the emphasis in plant evolutionary thinking, however, has more often been on the organismal end of this continuum rather than considering the levels of selection within the organism.

Developmental selection between the ramets of a genet is evolutionarily meaningful if the ramets are genetically (and phenotypically) variable. Nongenetic phenotypic variation also is important in ramet competition; such competition may cause stochastic changes in allele frequencies for mutations that are fixed in ramets but that do not affect the competitive aspects of the ramet phenotypes. The magnitude of the genetically based phenotypic ramet variance within a genet is a function of mutation rates and, more importantly, the anatomy and the ontogeny of the ramet. Factors such as the number and stability of apical initials, cell pool sizes within meristems, degree of cross-feeding between cells and tissues, degree of apical stratification, and presence of sympodial or monopodial branching patterns all have influence on the magnitude of the genetic-based phenotypic variance among sib ramets.

Another aspect of the expression of genetic-based phenotypic variability in ramets is the topology of the mutant tissue within the context of the growing ramet. The persistence or loss of mutant cells in a growing plant is a function of growth and development of its meristems. Of particular importance is the organization of the shoot apical meristem, since it determines the nature and persistence as well as the patterns of the mutant tissues in ramets. In figure 1.3 the various kinds of chimeras that occur in plants are illustrated. As will be shown in subsequent discussions, ramet variability is at least in part a function of these patterns.

Since the majority of vascular plants (even annuals) are clones to one degree or another with open systems of growth, the accumulation of mutations in such systems is similar to mutation accumulation in a microbial chemostat. Assuming that the majority of mutations are deleterious to some degree, one might suspect the evolution of various strategies to reduce mutation rates and frequencies in plant cells. Characteristics of DNA organization, replication, and segregation at mitosis can contribute to this end (Cairns 1975). Also the

Figure 1.3. The kinds of buds that may originate from a stem chimera of white and green tissues: A: pure white; B: sectorial chimera; C: pure green; D: periclinal chimera; E: mericlinal chimera. (From Kirk and Tilney-Bassett 1978.)

patterns of cell divisions within meristematic cell populations can strongly influence the frequencies of mutations accumulated during the course of plant growth (Kay 1965).

The problem of the maintenance of the genetic integrity of a plant is therefore of considerable importance in any understanding of plant biology (Klekowski 1987). As will be shown, this problem has been responded to in a diversity of ways by various plant groups. It should be noted though that too perfect a resolution of the problem of the maintenance of genetic integrity will invariably lead to a greater problem—extinction.

CHAPTER TWO

Uncertainty in Heredity

NOISE AND NUCLEIC ACIDS

No system of information transmission has perfect fidelity. Errors in the transmission of information constitute *noise* (see Shannon 1949). With regard to nucleic acids, such errors may occur during the storage, copying, or expression of genetic information. Viewed comparatively, RNA and DNA genomes are very different in terms of the noise levels associated with information transmission. Self-replicating viral RNA genomes have an error rate of about 10^{-3} to 10^{-4} substitutions per base per generation, whereas for DNA genomes the corresponding figure is 10^{-9} to 10^{-11} (Pressing and Reanney 1984; Steinhauer and Holland 1986). These five to eight orders of magnitude difference in noise level have considerable consequences.

Eigen and Schuster (1977) have explored the theoretical relationships between information content (in terms of genome size or number of nucleotides), error rates, and the selective advantage of a given sequence. They concluded that the number of nucleotides in a self-reproducible unit is restricted and is inversely proportional to the error rate. The selective advantage of a given self-reproducible unit is not an important parameter in determining the maximum permissable size of the self-reproducible unit. When this threshold in size is exceeded, the information accumulated in the evolutionary process is lost due to an error catastrophe. To quote Eigen and Schuster, "Larger molecules could exist, of course, according to chemical criteria, but they would be of no evolutionary value." In table 2.1 the relationship between error rate and the maximum evolutionarily meaningful genome size is shown.

The deleterious effects of noise can be compensated for to some extent if the damage is dispersed across a number of "information modules." Following the analysis of Reanney, MacPhee, and Pressing (1983), the importance of message redundancy will be demonstrated. If an informational module has L nucleotides and q is the nucleotide copying fidelity, then $(1 - q^L)$ is the probability of an information module with one or more errors. If two copies of a given module are coupled, then the correct information will be absent only when both modules are damaged. Thus for the case of n-fold redundancy, the fraction of correct copies (f) is

$$f = 1 - (1 - q^L)^n.$$

Thus the selective advantage, K, of redundancy is

$$K = f/q^L = [1 - (1 - q^L)^n]/q^L.$$

In figure 2.1, copy number (n) is plotted against the selective advantage of message redundancy. It should be noted that K increases rapidly for small n values. Reanney, MacPhee, and Pressing (1983) have demonstrated that twofold redundancy is the optimal energy efficiency strategy with regard to noise compensation. (See Pressing and Reanney 1984 for further discussions of redundancy.)

Of course the optimal strategy for increasing the fidelity of information transfer is the reduction of noise levels. Comparing the fidelities of RNA and DNA synthesis reveals that noise reduction

Table 2.1. The theoretical limits of information storage for various nucleic acid systems. (Modified from Eigen and Schuster 1977.) These limits were calculated using selection criteria that were more optimistic and sensitive than most biologists would allow (Ninio 1983) and thus they are probably overly generous.

Digit Error Rate	Maximum Digit Content[a]	Molecular Mechanism
5×10^{-2}	$ca.\ 1 \times 10^2$	Expected fidelity of enzyme-free RNA replication
5×10^{-4}	$ca.\ 1 \times 10^4$	Single-stranded RNA replication via specific replicases
1×10^{-6}	$ca.\ 5 \times 10^6$	DNA replication via polymerases including proofreading
1×10^{-9}	$ca.\ 5 \times 10^9$	DNA replication via polymerases including proofreading, repair, and recombination

[a]Assuming high selective superiority of adaptive sequences.

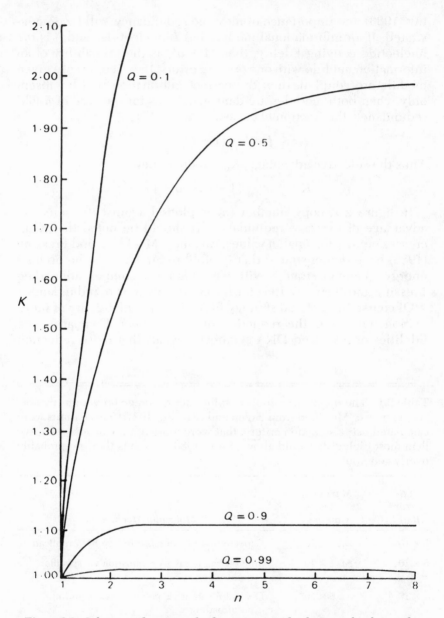

Figure 2.1. Selective advantage of information transfer due to redundancy of information modules, K, as a function of number of copies, n, for selected values of $Q = q^L$. For reasonable values of Q (i.e., those based upon realistic gene lengths and DNA copying fidelity), there is little selective advantage gained beyond $n = 2$. (From Reanney, MacPhee, and Pressing 1983.)

has evolved and is based upon the evolution of a complex series of editing and repair pathways. In table 2.2 the significance of these systems in increasing the fidelity of DNA synthesis is shown. It is clear that without such editing and repair pathways, DNA- and RNA-based genomes would have similar noise levels. Such pathways must be considered as part of an organism's genetic system. Consequently, noise reduction rather than noise enhancement was and still is a very important aspect of fitness (at least at this level of integration) in complex organisms. Fersht, Knill-Jones, and Tsui (1982) have shown that in the bacterium *Escherichia coli,* editing and proofreading are not without costs. Increasing exonuclease activity not only incurs a considerable energy burden but would also slow down DNA replication. (See chapter 3 for a discussion of DNA replication in plants.)

Although editing and repair pathways reduce the error rates from 10^{-4} to much lower levels, these pathways may also generate mutations. In bacteria DNA alterations that result in noncoding or synthesis-blocking lesions are repaired via the SOS pathway that often generates mutations at the site of repair (Radman 1974). Thus lethal lesions are altered to mutant lesions that may or may not be lethal. Repair of apurinic sites during DNA replication may also be an important source of errors. Depurination is probably the most frequent spontaneous alteration in DNA. An apurinic (abasic) site is created when the bond connecting the purine base and the deoxyribose sugar is cleaved, leaving the phosphodiester backbone of the DNA intact. The bypass of these apurinic sites during DNA

Table 2.2. Fidelity accompanying DNA replication in bacteria. (From Hartman 1980.)

Step	Approximate Error Frequencies for Base Substitutions
Base pairing	10^{-2}
DNA polymerase base selection	10^{-3}–10^{-5}
Editing ($3' \rightarrow 5'$ exonuclease) by DNA polymerase	10^{-5}–10^{-7}
Uracil DNA glycosylase repair and new-strand mismatch repair pathways	10^{-8}–10^{-10} [a]

[a] Mutation frequency *in vivo*.

replication may lead to errors during later nontemplate directed (repair) incorporation of nucleotides (Loeb 1985).

Drake, Glickman, and Ripley (1983) have reviewed how specific nucleotide sequences themselves may generate mutational "hot spots" through misalignment mutagenesis. Repair pathways recognize such topological nonconformities in the DNA molecule and generate additions and deletions of nucleotides by attempting "repair." Deletion mutations can arise by misalignments caused by distant repeated sequences as is shown in figure 2.2. Palindromes are also an important source of such mutations since they generate base pair complementarity not only between opposite strands but also within a single strand. Misalignment mutagenesis is a consequence of repair pathways that incorrectly "see" such topological alterations as mutations that must be repaired. Drake, Glickman, and Ripley (1983) also discuss how repair pathways may generate palindromic sequences from imperfect palindromes (quasipalindromes). Thus the repair pathways may inadvertently generate mutational "hot spots" for future misalignment mutagenesis. Viewed in a broader context, while reducing noise levels associated with total information transfer for the whole genome, repair pathways may inadvertently enhance noise levels in specific portions of the genome. Thus we see a pattern that recurs throughout biology—the solution of a problem often (if not always) generates new adaptive problems that must be responded to in some fashion. One could interpret these imperfections as being adaptive (e.g., the editing and repair pathways have been selected to allow low specific levels of noise to occur), or one could take the view that there are no perfect solutions to complex problems.

DNA slippage and mispairing may also cause mutation. Tautz, Trick, and Dover (1986) computer-searched published DNA sequences for direct sequence homologies to 15-nucleotide-long probes, each consisting of an array of all possible mono, di- and trinucleotide motifs. In plants and algae (animals as well), DNA sequences repeating such simple sequence motifs are considerably more frequent than one would expect from chance alone. These results indicate that the frequent occurrence of several directly repeated short sequence motifs in close proximity to each other are important components of eukaryotic genomes and that slippage-like mechanisms are a major source of mutation in all regions of the genome.

Genome size and integrity is, therefore, dependent to a great extent upon a complex, low noise, metabolic machinery that mini-

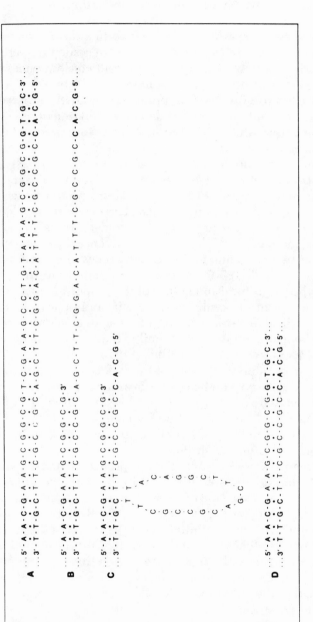

Figure 2.2. The origin of a deletion through misalignment mutagenesis: A: the original DNA sequence in correct alignment; B: replication of the newly synthesized progeny DNA that has reached the end of the first copy of the repeat; C: the second copy of the repeat displaces the first, leaving intervening bases unpaired; D: the resulting mutant has lost one of the repeated sequences and all of the information between the repeats. (From Drake, Glickman, and Ripley 1983.)

mally controls DNA replication, editing, and repair. Since this machinery is itself genetically determined, errors or defects in these
nucleotide sequences will have profound implications in terms of
cell fitness.

Orgel (1963) noted that protein synthesis (primarily translation)
is more error-prone than DNA replication and that such defective
or less-than-perfect proteins may have significant consequences in
terms of long-term viability. Orgel divided the cell's proteins into
two broad classes: those involved in replication of DNA, transcription, and translation; and those involved in other cell functions. The
formation of defective proteins in the first class (information-handling proteins) can result in a kind of feedback, since defective
proteins may heighten the error rate in catalyzing the synthesis of
the subsequent cohort of information-handling proteins. Thus there
may result, with time, a cascade of ever-increasing error frequencies.
Defective proteins would accumulate regardless of whether the cell
divided since the cells resulting from such somatic divisions "in-
herit" or divide the mother cell's cytoplasm. Eventually such cells
would no longer be viable due to the consequences of cytoplasmic
error catastrophe. Orgel (1970) subsequently pointed out that depending upon the magnitude and nature of the proportionality constant between errors in the synthetic apparatus and newly synthesized protein, either an error catastrophe or a steady state error
frequency can be achieved. In more detailed mathematical models
(Hoffmann 1974; Kirkwood and Holliday 1975; Kirkwood 1977,
1980), the degree of catalytic activity of the defective proteins was
shown to be a critical aspect of the error catastrophe hypothesis. For
example, if the catalytic activity is low for defective proteins, then
the error feedback will be trivial. (See Gallant and Prothero 1980
for tests of the various models of error catastrophe as well as Blomberg, Johansson, and Liljenstrom 1985.)

The widespread occurrence of selective proteases in all living
organisms has been taken as evidence of the importance of error
control in protein synthesis. One of the important functions of these
proteases may be to degrade abnormal proteins. Rosenberger and
Kirkwood hypothesize that the universal distribution and conservation of selective proteases are indicative that errors in protein
synthesis, "are a real and ongoing threat to living systems"
(1986:28). With regard to mutation, Holliday and Tarrant (1972) have
discussed how the above cytoplasmic error catastrophe may also
feed back to the DNA by increasing mutability, thus the Orgelian

increase in protein errors may result in more error-prone DNA polymerases and, consequently, mutation rates may be expected to
increase with age (figure 2.3).[1] Kirkwood (1977) hypothesized that
the separation of germline and soma in higher animals is correlated
with an energy-saving strategy of reduced error regulation in somatic cells. Many plants (both primitive and advanced) may live
and propagate vegetatively indefinitely. These organisms lack a
germline, an immune system, and have open systems of growth.
One might expect that such organisms would be very prone to the
error catastrophe scenario. Perhaps the aging effects and ultimate
cessation of growth documented in old plant meristems may be an
expression of either cytoplasmic or nuclear error catastrophe. Ultimately the cell declines associated with error catastrophe must be
interpreted within the dynamics of cell competition within meristems (see chapters 4 to 6).

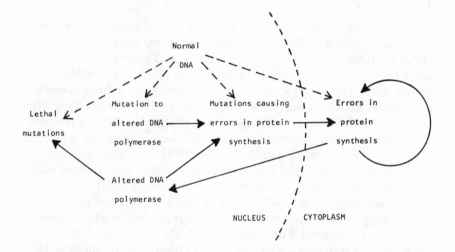

Figure 2.3. Relationship between error catastrophe and mutation rate. Protein
synthesis may become less accurate by either an Orgelian increase in errors in
information-handling proteins or through rare mutations. With time a cell will move
into a metastable condition in which the events marked by the solid arrows occur
with increasing frequency. As these errors feed back to altered DNA polymerases,
the mutation rate would be expected to increase. (From Holliday and Tarrant 1972.)

1. Fungal senescence appears compatible with the error catastrophe theory. Senescence is
accompanied by deterioration of the mitochondrial genetic system. See Benne and Tabak 1986
for review.

GENOMIC FLUX

In addition to the sources of mutation associated with base tauto-mers, depurination, various DNA repair pathways, misalignment, slippage, and the cumulative errors leading to error catastrophe, genomes are also subject to changes through unequal crossing over, gene conversion, transposition, and RNA reverse transcription. These forces of change both singly and in combination may result in levels of genomic restructuring not anticipated in "classical" mutation theory (Dover 1982). Any discussion of the consequences and importance of mutation pressure in plants must consider these novel sources of mutation.

Transposition

Transposition of genetic elements was first documented by Mc-Clintock in the monocot *Zea mays* over thirty years ago (see Mc-Clintock 1984 for review). Using genetic and cytogenetic analysis, what have come to be called transposable elements or mobile ele-ments were shown to cause unstable mutations which were ex-pressed as various kinds of variegations in this species. It is now known that the maize transposable elements are DNA insertions that require some product encoded by the element for insertion (Fedoroff 1983; Döring and Starlinger 1984). In maize such mobile element systems fall into two general categories. One type of trans-poson is autonomously unstable, i.e., the element has the necessary genetic information for both insertion and excision and movement. Another type of transposon system exists that is called a two-element system, one element functions as a stable insert until the other element is introduced into the genome. The second element is itself autonomous but can also cause the first stable insert to become unstable and move. In the two-element system the stable insert is the *receptor;* it becomes unstable in the presence of a second ele-ment called the *regulatory element.*

In *Z. mays* ten such two-element systems have been identified (Peterson 1985). The best known two-element system in maize is the *Dissociation (Ds)* and *Activator (Ac)* pair. *Ds* undergoes excision and movement only if *Ac* is present. *Ds* elements differ from *Ac* by internal deletions and have been designated as "aberrant" *Ac* ele-ments (Döring and Starlinger 1984). In the presence of *Ac*, the *Ds* may be a site for chromosome breakage. In maize lines that are mutationally stable and that at least phenotypically appear to lack mobile elements, the stimulus of genomic shock (e.g., the presence in nuclei of chromosomes with ruptured ends, viral infection, ionizing radiation) can activate previously silent elements to move and insert (McClintock 1984). Mobile elements have been found in dicotyledonous plants *(Antirrhinum majus* and *Glycine max)* as well as monocotyledonous plants and probably occur in most eukaryotes and prokaryotes (Shapiro 1983). Döring and Starlinger (1986) caution that perhaps not all plant species have transposable elements. They note that no genetic evidence for the presence of transposable elements has been reported in the genetically well-studied tomato.

In addition to inactivation of transposable elements by gross structural changes (irreversible changes), transposable elements may undergo reversible changes. For example, in the maize transposable element *Activator (Ac)*, irreversible changes may give rise to nonautonomous *Ds (Dissociation)* elements. Such mutant *Ac* elements may respond to the presence of active *Ac* elements and transpose or cause chromosome breakage. Reversible changes have been termed "changes in phase" (see McClintock 1984 for review); in such situations *Ac* may be active or inactive during different phases of plant development. Reversible changes may also be associated with different genetic backgrounds or whether the element was transmitted through either the male or female parent (Walbot 1986). Chomet, Wessler, and Dellaporta (1987) have shown that when *Ac* elements are in the reversible inactive phase, regions of *Ac* DNA were modified. Such modification involved the methylation of cytosines. Differences in methylation were the only changes that these authors observed between active and inactive *Ac* elements. These results parallel an earlier finding of DNA modification and inactivation in another maize transposable element, *Mutator (Mu)* (Chandler and Walbot 1986).

It is interesting to note that cytosine methylation may have two mutually opposing consequences with regard to genomic stability. On the one hand it may reduce the movement of some transposons and, consequently, promote genomic stability; on the other hand, methylated cytosine residues may act as mutational hot spots (chapter 8).

Most transposable elements have a terminal 10 to 20 base pair sequence showing an inverted repeat pattern (TIR). Integrated plant transposable elements are flanked by a duplication of the target site sequence. Such target site duplications (TSD) may be 3 to 10 base

pairs. The transposable element *Tam3* in *Antirrhinum majus* may
serve as an example of these characteristics. The *Tam3* element
probably is present from 5 to 15 copies per haploid genome (Sommer
et al. 1985). This 3.5 kb transposon causes somatic and germinal
instability at the *nivea* locus, which codes for chalcone synthase
in this species. Sommer et al. (1985) sequenced the junctures of
this insertion and compared it to the wild type and revertant alleles. In figure 2.4 these comparisons are shown. The *Tam3* element generates a 8 bp duplication at the site of integration (TSD =
ATCTCAGC). The TIR consists of a perfect inverted repeat of 12
bp that is followed by 3 additional stretches (8 bp, 12 bp, and 9 bp)
of perfect homology toward the interior of the transposon. In prokaryotes, transposon movement involves movement of a copy of the
element (Shapiro 1979); in plants the evidence indicates that transposition involves excision of the element (Saedler and Nevers 1985).
The *niv* 164 is a revertant caused by such an excision. The excision
and movement of this element was not without consequence; in
addition to the duplicated target sequence, other new base pairs are
also present. Such molecular "footprints" of the visitation of a transposon usually include duplication of the target sequence as well as
other base pair changes. For example, Schwarz-Sommer et al. (1985)
cloned germinal and somatic reversion events induced by the *Enhancer* (*En*) transposable element system at the *wx-8::Spm*-18 allele
of *Zea mays*. The *Spm*-18 insertion is within an exon of the *wx* locus.
Table 2.3 shows the DNA base sequences of the *wx⁺* and *wx-
8::Spm*-18 alleles as well as the sequences generated through the
excision of this element. It is clear that all or part of the target
sequence (GTT) may be duplicated and that additional duplications
or deletions of bases may occur. These sequence alterations can
cause considerable change to the final protein product. Thus, element excisions can generate the kinds of DNA sequence variability
that have the potential for being evolutionarily important. In addition to such base pair changes caused by excisions, insertion may
be associated with a deletion of base pairs from the gene itself
(Taylor and Walbot 1985).

Transposable elements may also have significant phenotypic effects when they are inserted in the nontranscribed regions of a gene.
In *Antirrhinum majus* the *Tam3* transposon may be inserted in the
upstream region of the *pallida* (*pal*) gene promoter. The *pal* gene
encodes a product necessary for the synthesis of red anthocyanin
flower pigment. The *pal^{rec}* allele results in an ivory flower with red
spots; these spots correlate with the somatic excision of the transpo-

son from the *pal* locus (*pal^rec* → → *pal*). Excision events in cells giving rise to meiocytes have resulted in a diversity of mutant alleles that can influence pigment intensity and spatial patterns on the petals depending on the nature of the "molecular footprint" left at the target site after excision (Coen, Carpenter, and Martin 1986; Carpenter, Martin, and Coen 1987). Thus, imprecise excision may

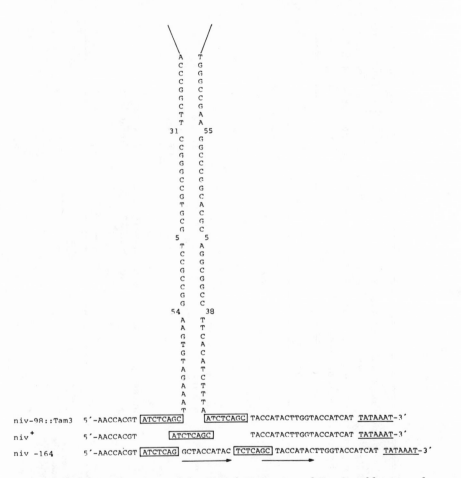

Figure 2.4. DNA sequence of the 5′ and 3′ junction of *Tam*3, wild type and revertant *niv*-164. The 5′ and 3′ termini of *Tam*3 are drawn in a stemloop structure by pairing stretches of homologous sequences. Numbers refer to nonhomologous nucleotides separating homologous regions. The duplicated 8 bp of the target site are boxed in. The arrows below the line indicate the decanucleotide duplicated during excision of *Tam*3. The putative TATA box is underlined. (From Sommer et al. 1985.)

Table 2.3. List of sequences obtained by *En*-induced excision at the *wx* locus in the *wx-m* 8 mutant and their presumptive protein changes. The upper two sequences show the target of the *Spm*-18 insertion (indicated by a bold line) within the wild-type (*wx*⁺) and mutant (*wx-8::Spm*-18) alleles. Capital letters indicate the bases that could be altered by the excision events. (From Schwarz-Sommer et al. 1985.)

Genotype	Events	Structure of Products		Isolates	Protein
wx⁺		GTC AAG TT	C AAC GCG		wild type
wx-8::Spm-18		GTC AAG TT————	G TTC AAC GCG		*
	germinal revertants				
wx⁺-1		GTC AAG TTA	– TTC AAC GCG	1	+ leu
wx⁺-2		GTC AAG T–	CG TTC AAC GCG	1	+ ser
	somatic excisions				
wx-8::Spm-18 + *En*	in leaf DNA	GTC AAG T–	ACG TTc AAC GCG	1	frameshift
		GTC AAG TT	G TTc AAC GCG	2	+ leu
		GTC AAG TTA	– TTc AAC GCG	4	+ leu
		GTC AAG T–	G TTc AAC GCG	2	frameshift
	in endosperm RNA	GTC AAG TT	G TTc AAC GCG	1	+ leu
		GTC AAG TT	– – C AAC GCG	2	wild type
		GTC AAG – –	G TTc AAC GCG	3	frameshift
		GTC AA– – –	– – –C GCG	1	Δ lys, Δ phe

Note: * = Inserted transposon
Δ = Amino acid deleted
+ = Amino acid added

give rise to frameshift mutations, base pair changes and, ultimately, altered protein products when the transposon was originally inserted in a coding region, whereas insertion and imprecise excision near promoters may give rise to alleles with quantitative differences regarding gene product (Coen, Carpenter, and Martin 1986).

Transposons themselves may be subject to genetic variation. Burr and Burr (1982) studied *Ds*-induced insertions into the *Shrunken* locus of maize. Restriction maps of the four *Ds* elements analyzed showed profound differences, even though these elements had a common origin and were separated by only a few generations.

Transposons, therefore, may have a diversity of mutational impacts on the genome: they may cause chromosome aberrations (McClintock 1984; Döring et al. 1985) or, at the gene level, insertions may physically disrupt genes, thus either altering gene regulation or expression; insertions may also occasionally be associated with deletion and excision events which may duplicate target sequences as well as delete or add base pairs. The importance of transposition-like phenomena in restructuring and altering plant genomes may be great indeed. For example, Freeling (1984) suggests that perhaps half of the maize genome consists of transposable or formerly transposable sequences. Nucleotide sequences containing inverted repeats are a characteristic of transposons and are very common constituents of eukaryotic chromosomes (Flavell 1980). The wheat genome possesses over 10^6 sequences consisting of inverted repeats greater than 50 bp in length (Flavell et al. 1985). If such inverted repeat-containing sequences originated through transposition-like phenomena, then these elements will have to be viewed as a major force in mutation and genome change in plants.

RNA Reverse Transcription

In addition to transposition, there are suggestions of other forms of large-scale insertional mutagenesis in plants. In order to more fully appreciate the plant phenomena, a review of retrovirus biology is useful. Retroviruses *per se* are restricted to mammals, although retroviral-like elements are more widespread (Wagner 1986). Retroviruses have genomes of single stranded RNA. Reverse transcriptase is encoded in the genome of the viral particle. This enzyme catalyzes the formation of a complementary DNA strand using the RNA template and degrades the RNA portion of the resulting RNA-DNA hybrid polymer. One or more copies of the double-stranded proviral DNA become integrated into the host genome. The transcription of proviral DNA generates MRNA and RNA genomes for the virons. A consequence of reverse transcription of these retroviral genomes is that proviral DNA is characterized by flanking base sequences (250–1,400 bp) known as long terminal repeats (LTRs) (figure 2.5). Proviral DNA can be inserted at many sites in host genomes. These insertions may physically interrupt or inactivate the host's genes or the regulatory signals in the LTRs may alter the expression of neighboring genes. Retroviruses may contribute not only their genes to the host genome but may also transduce the host's genes (see Varmus 1982 for review).

The characteristics of proviral DNA have been documented in species in which retroviruses are so far unknown. For example, the transposable element *copia* in *Drosophila melanogaster* has such proviral characteristics (Shiba and Saigo 1983; Flavell 1984). In *Zea mays* a heterogeneous family of dispersed repetitive elements known as *Cin* 1 have *copia*-like characteristics (Shepherd et al. 1984). *Cin* 1 insertions have structural similarity to a single retroviral LTR but have never been shown to move in the maize genome (Döring 1985). Also in *Z. mays*, a *copia*-like transposon (*Bs* 1) bounded by 304 bp repeats (LTRs) was inserted into an *Adh* 1 gene after barley strip mosaic infection (Johns, Mottinger, and Freeling 1985). Although barley strip mosaic is an RNA virus, it does not appear to act as a plant retrovirus since the *Bs* 1 insert is not homologous to the viral genome (Mottinger, Johns, and Freeling 1984). In yeast, *Saccharomyces cerevisiae*, the *Ty* transposable elements are 6 kb in length and contain a large central region flanked by direct repeats of 334 bp (LTRs). *Ty* elements constitute 0.04% of the total DNA in yeast. Recent studies by Boeke et al. (1985) indicate that these elements transpose through an RNA intermediate. This DNA → → RNA → → DNA flow of information within the yeast genome involves reverse transcriptase and generates copies of the transposed DNA that contain many new mutations (Boeke et al. 1985). Reverse transcription may also be involved in the replication of cauliflower mosaic virus. This virus infects the Cruciferae. Although this icosahedral virus contains a double-stranded circular DNA, 8 kb in length, Pfeiffer and Hohn (1983) propose that it replicates via an RNA intermediate.

The possibility of the occurrence of the enzyme reverse transcriptase in higher plant genomes may mean that genomes may be occasionally disordered by the reverse transcription of cellular RNAs and the integration of these cDNAs back into the plant genome. In mammals such reverse transcription is the postulated origin of various pseudogenes and "processed pseudogenes" (i.e., sequences lacking introns that are similar to mature mRNA structure) as well as certain highly reiterated short mammalian DNA segments (see Baltimore 1985 for review). For example, in the human genome, *Alu* DNA constitutes the dominant family of short middle-repetitive sequences. Ullu and Tschudi (1984) have shown that *Alu* sequences are related to an abundant cytoplasmic RNA important in the translocation of secretory proteins across the endoplasmic reticulum (7SL RNA). Sequence homologies indicate that *Alu* DNAs were derived from portions of 7SL RNA and are, therefore, processed 7SL RNA genes. The *Alu* sequences probably arose from the reverse transcription of a cellular RNA species, followed by integration of the cDNA into new chromosomal sites in germline DNA.

In higher plants, the origin of genomic DNA through reverse transcription of RNA is not yet documented, although as indicated, indirect evidence suggests it may occur. Recently Schuster and Brennicke (1987) have found an open reading frame with high homology to reverse transcriptase in the mitochondrial genome of *Oenothera*. These authors observe that the majority of plastid and nuclear DNA sequences found in the *Oenothera* mitochondrial genome are probably reverse transcribed RNAs. Schuster and Brennicke hypothesize that the transposition of genetic information between the organelles and the nucleus may involve local reverse transcription and subsequent integration of the newly synthesized DNA into the respective organellar genome.

Considering our state of ignorance about the mechanisms of genomic change in these plants, one would guess that phenomena found in as diverse organisms as humans, fruit flies, and yeasts will probably also be found in plants. If reverse transcription is documented in higher plants, it may prove to be an important mode of genome reorganization. In organisms where reverse transcription has occurred, perhaps as much as 10% of the genomes of these species may have arisen in this manner (Baltimore 1985). What the function of this DNA is to the organisms is unknown. To quote Baltimore: "It remains unclear whether it is an infectious, parasitic invasion or a mechanism with clear biological necessity for higher organisms" (1985:482). (See Weiner, Deininger, and Efstratiadis 1986 for a recent review on genetic material generated by the reverse flow of genetic information.)

With regard to the possible "viral" qualities of transposons, Döring (1985) and Sommer et al. (1985) have pointed out that transposon classification does not follow conventional plant taxonomic patterns.

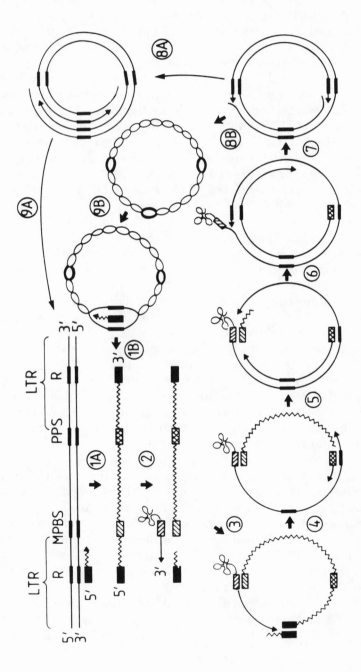

Figure 2.5. Generalized life cycle of retroid elements: RNA with terminal repeats is either produced from an integrated DNA form with long terminal repeats, LTRs (step 1A) or from a supercoiled circular form (step 1B). A primer can bind to a '(−)-strand primer binding site' (MPBS, hatched boxes) of the RNA and initiate (−)-strand DNA synthesis by reverse transcription from the RNA template. Template RNA is consecutively digested by an RNAase H activity (step 2). When the active reverse transcription complex reaches the 5' end of the RNA template, the nascent DNA strand switches to the sister RNA terminal repeat and continues elongation (steps 3, 4). A G-rich RNA sequence known to be relatively stable to RNAase H is spared and can now serve as a primer bound to the (+)-strand primer site (PPS, cross-hatched box) to initiate the synthesis of the (+)-strand DNA complementary to the newly synthesized (−)-strand DNA (steps 4, 5). Nascent (+) and (−) strands will pass the site where (−)-strand DNA synthesis was initiated (step 6). RNA remnants are digested and the hydrogen bonds that held the ends together are now replaced by ones from the newly synthesized DNA sequences (step 6). Short DNA overhangs are produced by peeling off limited stretches of the 3' ends of the two strands, respectively (step 7). Synthesis of (+)- and (−)-strand DNA can continue, in some cases peeling off more and more of the 3' terminal sequences until the origins of (+)- and (−)-strand DNA synthesis are reached again (step 8A). The resulting DNA with long terminal repeats can in some cases circularize or integrate into the host chromosome (step 9A). Alternatively, the form with the short DNA overhangs can be converted into a supercoil with the help of 'repair enzymes' and ligase (step 8B). Either the integrated double-stranded form with the long terminal repeats (step 1A) or the supercoiled form (step 1B) act as a template for the synthesis of the terminally repeated RNA, allowing the replication cycle to continue. (From Hohn, Hohn, and Pfeiffer 1985.)

Two groups of transposons occur in both monocotyledonous and dicotyledonous plants. The *Ac/Ds* system in maize and the *Tam3* element of *Antirrhinum* belong to one group, while the second group comprises the *En* (*Spm*) system in maize, *Tam1* and *Tam2* elements in *Antirrhinum*, and the *Tgm1* sequence in soybean. To quote Döring, "The surprising finding that plants as unrelated as maize and *Antirrhinum majus* bear transposons with similar termini indicates either that these sequences have originated early in plant evolution or that horizontal transmission has occurred between different plant species" (1985:164). (See Finnegan 1983 for the relationship between retroviruses and transposons.)

Gene families have been documented in plant and animal genomes. Such gene families may consist of one or more functional genes and pseudogenes. Pseudogenes are nucleotide sequences structurally related to a functional gene sequence but defective in one way or another. Pseudogenes contain multiple genetic lesions that preclude the translation of any transcript from these sequences to give a polypeptide equivalent to the gene product of the functional gene. Pseudogenes have been found to fall into two categories: those retaining the intron-exon arrangement of their productive counterparts, and those lacking the intervening sequences found in their productive counterparts (Vanin 1984). The second category of pseudogenes (those lacking intervening sequences) have been called processed pseudogenes.

Processed pseudogenes are thought to be the DNA products of reverse transcribed mRNAs that have been inserted into the genome. Processed pseudogenes are scattered in the genome, whereas nonprocessed pseudogenes usually are linked to their productive counterparts. Processed pseudogenes occur primarily in the mammals, whereas nonprocessed pseudogenes occur across a broad phylogenetic spectrum, including plants. Retroviruses are implicated in the origin of processed pseudogenes (Wagner 1986), whereas the origin of nonprocessed pseudogenes is unclear (although unequal crossing over is often postulated as a mechanism).

Nonprocessed pseudogenes occur in legume leghemoglobin genes (Brisson and Verma 1982) and a processed pseudogene was found in the potato actin gene family (Drouin and Dover 1987). In organisms with a germline, the only processed pseudogenes that may readily accumulate through successive generations are for genes expressed in the germline. Since plants lack a germline, processed pseudogenes could be inherited for genes expressed in the apical meristem, meiocytes, spores, and gametophytes.

Other Sources of Genomic Flux

Plant genomes may also be altered by microbial infections. The bacterium *Agrobacterium rhizogenes* induces neoplastic outgrowths by transferring genetic material from the Ri (root inducing) plasmid to the host genome. Furner et al. (1986) detected homology with a portion of the transferred DNA of this plasmid in uninfected species of the genus *Nicotiana*. Comparative studies indicated that this *Agrobacterium* transformation occurred prior to the separation of the subgenera *rustica* and *tabacum*.

Even more surprising is the finding that certain plant pathogenic nucleic acids (e.g., viroids) show nucleotide sequence homologies with a class of plant introns. Viroids are single-stranded RNA molecules (246–375 nucleotides) that replicate through the synthesis of double-stranded replicative intermediates. A number of these plant pathogens have nucleotide sequences that resemble the 16-nucleotide consensus sequence of group I introns (Dinter-Gottlieb 1986; Hadidi 1986). Group I introns are found among mitochondrial messenger and ribosomal RNA genes, chloroplast transfer RNA genes, and nuclear RNA genes (Dinter-Gottlieb 1986); this intron class does not have the GU . . . . . AG termini characteristic of eukaryotic pre-mRNA introns (Hadidi 1986). Whether group I introns evolved from viroids or viroids from group I introns is unknown. Regardless, the viroid and the *Agrobacterium* transformation examples are indicative of levels of interaction between microbes and plant genomes previously unsuspected. From these examples, one might predict that future work will document past and present microbial infections as another component of mutagenesis in plant genomes.

In plants there are numerous examples of the amplification of variously repeated sequences of genomic DNA, although it should be noted that the origin and function of this DNA is still unknown. Within the genus *Secale* there is a 20% interspecies variation of total chromosomal DNA. Bedbrook et al. (1980) reported that telomeric heterochromatin accounts for 12–18% of the DNA of *S. cereale* and less than 9% in *S. sylvestre*, and that four kinds of repeated sequence account for almost all of this difference in telomeric heterochromatin. Flavell (1980, 1985) has reviewed this literature and formulated a number of generalizations concerning repeated DNA sequences in plant genomes. In genomes that exceed 2 pg DNA, more than

75% of the total genome is repeated DNA. The repeated sequences vary in size from a few to thousands of base pairs (bp). Three general patterns of repeated sequences have been documented:

1. repeats in tandem arrays
2. individual members of the repeat dispersed in the genome
3. repeats clustered at a locus but not in tandem.

With regard to these three repeat patterns, a number of characteristics should be noted. Studies of families of tandem repeats in closely related species have shown that evolutionary divergence of these families is very rapid. There appears to be a mechanism to amplify individual mutations in a repeat so that the entire tandem array is characterized by the same mutant repeat, a concerted evolution (Dover 1982). The molecular basis of the origin of tandem repeats and mutant amplification is unclear, but such events are probably the most commonly occurring mutations in plant genomes (Flavell 1985). Dispersed repeats are thought to be related to past transpositional events. It is possible to construct a model based upon some form of replicative transposition to account for the spread of repeats in a species genome (Ohta and Dover 1983). Repeats clustered at a locus but not in tandem are characteristic of various multigene families.

There are numerous reports in the literature documenting dramatic differences in DNA content between congeneric species with identical chromosome numbers. For example, Sims and Price (1985) studied DNA content in diploid *Helianthus* species and reported that DNA values varied over a fourfold range. Such differences are neither exceptionally small nor large compared with other genera (see Sims and Price 1985 for review). It is probable that much of this DNA amplification is due mostly to repetitive DNA changes (Flavell 1985). Flavell (1980) estimated that the minimal genetic information necessary to specify a plant was approximately 15,000 genes per haploid genome or 1.8×10^7 bp of DNA. The angiosperm *Arabidopsis* has 2×10^8 bp and is considered to approach this minimum. The majority of vascular plants (angiosperms, gymnosperms, and pteridophytes) exceed this value by a number of orders of magnitude since 1×10^{10} to 1×10^{11} bp per haploid genome are not uncommon DNA levels in these groups (Sparrow, Price, and Underbrink 1972). If a high proportion of this DNA has originated by the various mechanisms of sequence movement and amplification that have been discussed, then the capacity for genomic change in plants may be very great (Walbot and Cullis 1985). This, coupled with the lack of a germline in these organisms, suggests that mutation pressure *sensu lato* is and has been a very important feature of plant biology.

Perhaps the most thought-provoking discussion of the importance of mutation or noise associated wth the transmission of genetic information is that of Reanney, MacPhee, and Pressing (1983). These authors envision that the biosphere has been shaped by two kinds of selective pressures—extrinsic (environmental) and intrinsic (noise-generated). Botanists have concentrated primarily upon extrinsic selective pressures (e.g., competition, co-evolution, adaptation to specific niches, etc.) in trying to understand the forces involved in plant evolution. In general intrinsic (noise-generated) selectve pressures have been omitted in these considerations but, as will be shown in subsequent chapters, plants have many characteristics that accentuate the problem of noise (e.g., plants have neither germlines nor immune systems but do have open systems of growth). Van Valen (1973) described what has come to be known as the "Red Queen effect," whereby organisms must continue to evolve just to keep abreast of their competitors. Intrinsic noise-generated selective forces interacting with external selective forces may be viewed as generating a similar "Red Queen effect." For example, evolutionary responses to external selection may promote developmental patterns of growth that enhance the retention of mutations that, in turn, again generate intrinsic selective forces. The latter will promote adaptations that either enhance the loss or lessen the effects of these mutant cells. Thus, developmental systems that allow the formation of viable tissues and organs from mixtures of mutant and nonmutant cells will, in effect, reduce diplontic selection within meristems and, consequently, result in higher frequencies of mutants in the meiotic products. This may lead to the selection of reproductive characteristics that enhance the loss of mutant gametes without proportionate losses in reproductive capacity (see chapter 9).

ACTUAL MUTATION RATES

Mutation rate and mutation frequency are often confused. Mutation rate is the probability of a mutation occurring per unit biological time. Units of biological time include generations, growing seasons, cell divisions, DNA replications, etc.; thus mutation rate is the probability of a mutational event per generation or per cell division or per genome replication. In contrast, mutation frequency is simply the frequency of mutants in terms of population size. Mutation frequencies are commonly given as the number of mutants per million gametes, or the frequency of mutants per population of spores, or the number of mutant ramets per genet. Although mutation frequency is relatively easy to measure, it is not as informative as the mutation rate. (See Atwood 1962 for further discussion.) Mutation frequency and rate are analogous to the first derivative (or slope) and the formula that correctly describes the relationship of the dependent and independent variables. Often the derivative is easier to determine for a body of data than the formula for a curve that describes the relations of these data, similarly mutation frequency is easier to measure than mutation rate. To stretch the analogy further, differential calculus allows the correct formula to be deduced from the slope, and a "biological calculus" is necessary to calculate the mutation rate from mutation frequency. In multicellular organisms this "biological calculus" is based upon a thorough understanding of the growth and development of the species being studied.

For example, the frequency of a given kind of mutation per million meiospores may not really indicate a lot about mutability if the estimate is based upon the meiotic products of a single plant. A higher plant often has many mitotic divisions between the zygote and the formation of spore mother cells, thus a high mutation frequency may mean only that a single mutation occurred soon after the zygote was formed, and that this mutation was clonally spread to many of the cell lines giving rise to spore mother cells. Of course if the plant species in our example has a determinate growth pattern, being completed, say, in one growing season, then the average mutation frequency based upon a large number of individual plants can be used to calculate the mutation rate per the period of determinate growth. It should be noted that this calculation is possible only if the same unit of biological time has occurred in each individual studied. This is particularly easy in plants that have finite life spans culminating in flowering, seed maturation, and death (e.g., annuals, biennials). In plants with indeterminate life spans (perennials), calculating mutation rates is more difficult because of the lack of a suitable biological time metric.

Eriksson et al. (1966) reported that 2×10^{-6} pollen grains of the conifer *Larix leptolepis* exhibited the waxy phenotype, 16% of the mature pollen grains were sterile and 1.9% of anaphase II stages of meiosis exhibited chromosome aberrations. Since cloned grafts (branches) growing in pots were sampled rather than large trees, these values are probably fair estimates of the mutation rates for these characteristics per meiosis in this species. Stadler (1942) determined the spontaneous mutation frequency for genes not selected for high mutability in *Zea mays* (an annual). He studied several gene-determined endosperm characteristics and thus could score for mutants by simply scoring seeds. Plants that were homozygous dominant were pollinated by pollen carrying the recessive allele. Nonmutant endosperms exhibit the dominant phenotype and mutant endosperms the recessive, thus the mutational event occurred in the female parent. In table 2.4 Stadler's data are shown. Measurable mutation frequencies for individual genes varies from *ca.* 5×10^{-4} to 1×10^{-6} mutations per gamete. Since *Z. mays* is an annual with a determinate and almost segmental development (see chapter 6), these values can also be used as an estimate of mutation rate per generation. It should be noted, though, that concealed within the term generation are all of those specific developmental parameters that are characteristic of this species (number of apical initials, number of cells in the apical meristem, origin of the flowers, possibilities of megagametophyte competition, etc.). All of these features may influence the estimated mutation rate per generation.

The evolutionary and developmental consequences of somatic mutation are not based upon the mutation rates of single genes but rather on the summed mutation rates for sets of genes. Fitness in any organism is multigenic, thus if the negative effects of mutation are considered, mutation rates must be summed across all loci that

Table 2.4. Mutation frequencies for specific gene loci in maize. (From Stadler 1942.)

Gene Locus	Number of Gametes Tested	Number of Mutations	Frequency
R	554,786	273	4.92×10^{-4}
I	265,391	28	1.06×10^{-4}
Pr	647,102	7	1.1×10^{-5}
Su	1,678,736	4	2×10^{-6}
C	426,923	1	2×10^{-6}
Y	1,745,280	4	2×10^{-6}
Sh	2,469,285	3	1×10^{-6}
Wx	1,503,744	0	0

can mutate to decrease fitness. In *Drosophila melanogaster* ($2n$ = 8) such estimates yield values that are surprisingly high. Wallace (1968) and Simmons and Crow (1977) estimate 0.005 to 0.006 lethals per chromosome per generation for the second and third chromosomes, or an overall mutation rate per generation in the excess of 1% for lethals. Dobzhansky et al. (1977) calculated that the mean rate of all viability mutations is no less than 0.7055 mutations per individual per generation. In *Z. mays* haploids occur in low frequency and may be identified in the seedling stage if suitable genetic markers are present. A small percentage of such haploids may form diploid homozygous sectors. If such sectors include all or portions of the inflorescences (ear and tassel), some seeds may result from self-fertilization. The resulting sib progeny should be genetically identical since they are derived from completely homozygous parental tissues. Sprague, Russell, and Penny (1960) studied such progeny from individual haploids, inbreeding these plants and their offspring through six generations of selfing. These data formed the basis for estimating the mutation frequencies for the set of genes that may mutate to affect agronomically important quantitative attributes. These attributes included plant height, leaf width, number of tassel branches, number of kernel rows, ear length and diameter, weight per 100 kernels, weight of shelled grains per plant, and date of silking. For any given attribute, the conservative estimate of mutability was 4.5 mutations per attribute per 100 gametes tested. As these authors noted, this value is remarkably high. Also in *Z. mays* the mutation rate for that set of genes that can mutate to form the chlorophyll-deficient condition was estimated as 0.002 mutations per gamete per generation (Crumpacker 1967). In barley (*Hordeum vulgare*), the spontaneous mutation rate for chlorophyll deficiencies was measured as 0.0006 mutations per diploid genome per generation (Jørgensen and Jensen 1986).

Johns, Strommer, and Freeling (1983) studied restriction site polymorphism in a 20 kb segment of DNA that included the *Adh* 1 locus in maize. Seven lines of maize carrying different *Adh* 1 alleles were mapped using seven restriction enzymes. Two patterns of DNA variability were observed: within the *Adh* 1 gene (transcriptional unit) no polymorphism was detected, whereas the DNA in the flanking regions was surprisingly variable. Burr et al. (1983) reported considerable restriction fragment length polymorphism in the *Shrunken* locus from two maize strains. Johns, Strommer, and Freeling noted that there is as much variability for restriction site polymorphisms among different lines of corn as had been found among *Drosophila* species that have been separate species for millions of years. The domestication of maize probably occurred 10,000 years ago.

The extrapolation of mutation rate estimates from annuals to perennials is fraught with problems. Genomes are characteristically larger in perennials than in annuals, annuals (and biennials) have a determinate ontogeny often with a specific apical ontogeny (see chapter 6), whereas perennials are more indeterminate in their apical growth. Vegetative growth in perennials may sample different meristems in forming ramets, whereas annuals usually sample only one meristem for sporogenous tissues (typically, the outer tunica layer (LI) of the apical meristem in monocotyledons and the inner tunica layer (LII) in dicotyledons). Finally, the number of mitotic divisions between zygote and spore mother cell is usually greater in perennials. This is not to say that observations of somatic mutations are uncommon in perennials; as discussed by Whitham and Slobodchikoff (1981), somatic mutations are a major source of agronomically significant variations in perennial cultivars. (See Amo Rodriguez and Gomez-Pompa 1976 for examples of within-individual chromosome mosaicism in natural populations.) What has been difficult to devise in these organisms are biologically meaningful measures of mutation rates.

The majority of perennials have indefinite life spans and flower periodically throughout their life spans. Perennials often reproduce vegetatively so that both the physical limits and the life span of individual genets are difficult to determine. One approach to this problem is to calculate the mutation rate on a ramet-doubling generation basis, i.e., the biological time metric is the time necessary for one ramet to give rise to two ramets. Within a single species, one might expect the number of divisions of meristematic initials necessary to double the number of ramets to be as variable as the number of divisions necessary to complete the life cycle of an annual. Thus the ramet-doubling generation is probably as precise a metric in perennials as generation is in annuals. Klekowski (1984) determined the mutation rate in clonal ferns for those genes that could mutate to give rise to mutant gametophytes. As in the previously discussed work on Z. *mays*, the number of genes being monitored for forward mutation in these fern clones is unknown. In table 2.5 are shown the mutation rates per ramet-doubling generation for three species of homosporous ferns. Since these ferns have apical meristems with a single apical initial, the mutation rates in table 2.5 are also for the divisions of a single cell. Mutation rate estimates in other perennials (especially seed plants) are generally unavailable. What is especially important are mutation rate statistics for those genes that can affect fitness in plants with different growth forms and apical meristem organizations.

Recent research in plant tissue culture has demonstrated dramatically the frequency of somatic mutations in plants. Cell and protoplast culture and regeneration have documented high frequencies of variant phenotypes in regenerated plants of various plant species (Shepard, Bidney, and Shahin 1980; Meins 1983). This so-called "somaclonal" variation is often genetic and is an important potential source of mutant germplasm in plant breeding programs (Evans, Sharp, and Medina-Filho 1984). This genetic variation originates in both the cultural and the regenerative process as well as being present within the intact plant (*in planta*, Meins 1983). The *in planta* genetic variability suggests that many plant organs consist of cells heterozygous for postzygotic mutations. Quantification of this *in planta* somaclonal variability may be a useful way of estimating mutation rates in these organisms. If *in planta* somaclonal variation is high in many different taxa, then mutation pressure may be an important intrinsic factor in perennials.

Evolutionary Aspects

As have been described, there are many mutational mechanisms in eukaryotes that alter the genotype. These mechanisms have been discussed under two broad categories—those associated with errors

---

Table 2.5. Somatic mutation rates for three fern species based upon gametophyte mutations. (From Klekowski 1984 and Masuyama and Klekowski, unpublished.)

Species	Clones	Mutant-Free Clones	Chimeras	Mutation Rates[a]
Matteuccia struthiopteris[b]	56	46	10	0.0177
Onoclea sensibilis[c]	37	21	16	0.0341
Woodwardia japonica[d]	29	28	1	0.0339

[a]Mutations per ramet generation: a ramet generation is the time necessary to double the number of ramets or shoot apices.
[b]12.125 shoot apices per clone.
[c]17.623 shoot apices per clone.
[d]3 shoot apices per clone.

or noise that are more clearly a consequence of the second law of thermodynamics (mutations in the classical sense) (see the section Noise and Nucleic Acids), and those changes generated by various kinds of movement mechanisms and nucleotide sequence amplification (see the section Genomic Flux). Of course, both of these categories intergrade and neither is immune from the second law, but modally they do differ in a qualitative sense. Both categories have been included under the general definition of spontaneous mutation, with "Noise" being "less" spontaneous and "Flux" being "more" spontaneous—but not always! With this clear picture in mind, let us consider the evolutionary implications of these phenomena.

The critical question is whether spontaneous mutation causes genome reorganization or disorganization? In other words, have organisms evolved built-in mechanisms to insure a sufficient mutation rate to promote evolutionary change and consequent genomic reorganization, or are mutation rates as low as biologically possible? Obviously this question may have a general answer for all categories of spontaneous mutation, or specific answers for each category (i.e., noise and flux) and, no doubt, different answers for different taxa.

The consequences of spontaneous mutation *sensu lato* have been considered by a number of authors. On theoretical grounds, it is the consensus that the "natural selection of mutation rates has only one possible direction, that of reducing the frequency of mutation to zero" (Williams 1966:139). Mathematical analysis supports the contention that this statement is generally true for all sexual randomly mating organisms (Leigh 1973), whereas in asexual organisms selection may favor an optimum (higher) mutation rate (Leigh 1970). Studies with bacterial chemostat populations have documented selection for increased mutability in populations of haploid asexual organisms (Cox and Gibson 1974). Holsinger and Feldman (1983) demonstrated with a two locus model that in sexual organisms with high levels of selfing there may be an optimal mutation rate even in a constant environment. Modifiers of mutation rate in organisms generally affect all of the genes (or large subsets of the genome) (Cox 1976). Since a greater proportion of mutations is deleterious rather than adaptive, new adaptive mutants may occur within already disadvantaged genotypes. Thus, heuristic studies of mutation rates in sexual organisms should consider the consequences of increasing the mutability of multiple loci.

The view that mutation rates in sexual forms are selected for some optimal value that ensures evolutionary potential has had numerous

adherents in the past (see Williams 1966 for review). Recent research on plant transposable elements by geneticists and molecular
biologists has resurrected this hypothesis. The observation that in
some systems, transposition and consequent mutation may follow
episodes of environmental or biotically induced genomic stress has
prompted speculation that these phenomena are adaptive responses. Thus plants experiencing such "shocks" have been selected to respond by increasing the mutation rates for either all or
portions of their genomes and consequently generating a higher
probability of leaving surviving offspring (McClintock 1984).

These observations may also be interpreted from a viewpoint of
selfish or parasitic DNA. As previously discussed, the primary
means of movement of such plant mobile elements is by excision.
It is interesting to note that such genomic shock-induced excisions
of segments of DNA are also known in bacteria. Prophage induction
in *Escherichia coli* is an excision event that is caused by various
DNA damaging agents (Moreau and Devoret 1977). These authors
suggest that temperate phages recognize when the SOS repair pathway is functioning and that induction is a phage survival mechanism
to escape a genetically damaged bacterium. Similar reasoning may
explain transposition excision and movement in higher plants. Perhaps transposon recognition of genomic stress is not a plant adaptation for evolution but rather a transposon adaptation for transposon
survival. Because plants lack a germline, enhanced transposition in
meristematic somatic tissues may ensure that the spore mother cell
population produced by the plant is chimeric with regard to the
location of the transposon. Thus, haploid cells are formed in which
the location of the transposon varies considerably among the various
linkage groups. Since some linkage groups may be defective genetically (sheltering lethals or other defective alleles), transposon
movement maximizes the probability of its own transmission by
"searching" for viable and competitive genotypes. Of course, an
occasional transposon excision and insertion may be the origin of
evolutionarily interesting mutations. A relevant example in this regard is the report of enhanced response to artificial selection in
Drosophila melanogaster lines with enhanced frequencies of *P* element transposition (MacKay 1985). This response was paralleled by
an increased frequency of recessive deleterious fitness mutations.

Of course the philosophically important question is not whether
mutations (of whatever origin) are sometimes adaptive but whether
plants maintain transposons as part of a genetic system to generate
evolutionary potential. Since such a large proportion of the plant

genome (or practically any eukaryote) may be transposon derived, this is not a trivial question. The view that such DNA is "parasitic" or "selfish DNA" (Doolittle and Sapienza 1980; Orgel and Crick 1980) is countered by the view that at least some of this DNA must have some plant function that is either developmental or evolutionary (Freeling 1984). At present this issue is unresolved.

IS A PLANT A CHEMOSTAT?

A chemostat is a device whereby a microbial population may be maintained in a continuous growth phase for an indefinite period of time. This device is based upon the principle that a cell population growing within a defined and enclosed environment will undergo a period of logarithmic growth followed by a prolonged stationary phase where cell numbers remain relatively constant. The classic growth curve response is manipulated in a chemostat so that the cell population is in continuous logarithmic growth phase. This is accomplished by a continuous flow of fresh nutrient medium into the culture vessel and the simultaneous removal of old medium and cells from the culture vessel.

In many respects the meristems of a vascular plant are analogous to a chemostat. If we consider an apical meristem as an example, the population of apical initials is in a continuous growth phase (at least during the growing season), the excess cells formed are "washed out" to the soma, and finally there is a continuous flow of nutrient medium (water, minerals, photosynthate) into the meristem. Novick and Szilard (1950, 1951) studied the dynamics of spontaneous mutation in bacterial chemostats. The general principles which they developed can be applied to multicellular plants to give a rough appreciation of the impact of continued somatic mutation.

The growth of a newly established cell culture is characterized by a phase of logarithmic growth. The derivative of this part of the growth curve is $dn/dt = \alpha n$, where t is time, n is cell number per unit volume, and α is the growth rate. The reciprocal value, $\gamma = 1/\alpha$, is designated as the generation time. In a chemostat the cell population is maintained in growth phase indefinitely. The consequences of spontaneous mutations in such a system are interesting. If the mutations are neutral (i.e., no selection for or against the mutant with respect to the wild type cell), then the mutations will increase linearly with time. If n^* is the cell number per unit volume of the mutant and if we disregard back mutation then

$$dn^*/dt = (\lambda/\gamma)n$$

where n^* is the mutant, n the wild type, and λ the number of mutations produced per generation per cell. Separating variables and integrating both sides we get

$$\frac{dn^*}{n} = \frac{\lambda}{\gamma}\,dt$$

$$\int \frac{dn^*}{n} = \int \frac{\lambda}{\gamma}\,dt$$

$$\frac{n^*}{n} = \frac{\lambda}{\gamma}\,t + \text{constant}$$

It is clear from the above that the relative abundance of the mutants increases linearly with time if the mutants are selectively neutral. We can apply this equation to the growth of population of n apical initials within the various apical meristems of a vascular plant.

If each gene in the plant has a mutation rate 2×10^{-8} mutations per cell cycle and the total genotype consists of 2×10^4 functional genes, then the overall mutation rate per cell cycle per cell (λ) is 1×10^{-4}. The term t/γ is the number of cell generations or cycles of the initials for the total growth period under consideration. In plants this can be estimated from the following relations; the typical apical initial is thought to divide once every 3 leaf nodes (see chapter 4), if we assume 15 leaf nodes per growing season on the average, then an apical initial may divide 5 times per year (this is a very conservative estimate). Thus

$$t/\gamma = 5 \text{ (the age of the plant in years)}$$

and

$$n^*/n = (1 \times 10^{-4})(5)(\text{age}) + \text{constant.}$$

Disregarding the constant,[2] we can get some impression of the significance of somatic mutation in plants by calculating the relative abundance of mutant apical initials in plants of different ages. In figure 2.6 the results of these calculations are shown. It is clear that the proportion of mutant initials increases linearly with age. The potential magnitude of this somatic mosaicism is obvious if we consider that a century-old plant will have 5% of its apical initials mutant. One could even generalize and say that 5% of the initials in all the meristems are mutant. Considering the number of meri-

---

2. The constant in this equation refers to the mutations inherited from the previous generation.

stematic initials that such an organism may possess, the number of mutant cells and cell lines comprising the plant or genet may be truly astounding!

Of course, the above calculations are very rough, and vascular plants differ greatly from microbial populations. Meristematic initials are not independent of each other but are organized into complex patterns of various kinds of primary and secondary meristems. These meristems are again ordered into patterns to generate the overall form of the plant. Mutations are also not always neutral— diplontic selection may occur. And, finally, the plant species may have other characteristics that enhance the loss or retention of mutations.

In contrast to the chemostat effect in active meristems where mutation frequencies are a function of the number of cell divisions, in dormant structures such as seeds and pollen, mutation frequencies may also increase with time. In many species, seedlings raised from aged seeds exhibit increased mutation (D'Amato and Hoffmann-Ostenhoff 1956; Dourado and Roberts 1984). The conditions of seed storage may influence the magnitude of the mutation in-

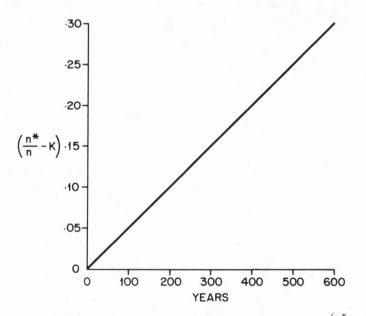

Figure 2.6. The linear increase in the frequency of mutant initials $\left(\dfrac{n^*}{n} - K\right)$ as a function of plant age.

crease. In *Datura stramonium*, seeds stored under laboratory conditions for five to ten years gave 7.9% pollen abortion mutations compared to 0.6% for one-year-old seeds (Carteledge and Blakeslee 1935). Comparisons to old *Datura* seeds collected from more natural conditions gave less clear results, a 22-year-old seed sample from buried soil showed no increase in mutation whereas a 39-year-old sample had a mutation rate of 5.1% (Avery and Blakeslee 1943; Carteledge and Blakeslee 1935).

In *Datura*, seedlings typically develop from three apical initials in the embryo whereas in aged seeds seedlings were derived from fewer or only one embryonic apical initial (Carteledge and Blakeslee 1934). Whether this developmental change reduced competition between mutant and nonmutant and thus was a contributing factor to the increase in mutant seedlings is unknown. It is also possible that the loss of initials was due to their inactivation by mutation. In addition to seed aging, aged *Datura* pollen also exhibited increased mutation (Carteledge, Murray, and Blakeslee 1935).

In conclusion, it appears that mutation frequencies in plant cells, tissues, and organs may be a function of biological time (number of DNA replications and chromosome separations during cell division), and in some dormant structures (seeds, pollen, perhaps buds?) mutation also may be a function of real time (days, months, years).

CHAPTER THREE

Components of
Genomic Organization and Stasis

In addition to the mechanisms of mutation and genomic change already discussed, plants have patterns of cellular organization and metabolic characteristics that can greatly influence genetic integrity. Plants, along with the majority of higher eukaryotes, have large amounts of DNA organized into chromosomes. Chromosome organization, at the minimum, involves further subdivisions of structure and function into telomeres, centromeres, replicons, heterochromatin and euchromatin (Lima-de-Faria 1983). Thus, although chromosomes presumably have evolved to reduce errors in the distribution of genetic material during cell division, chromosomal organization itself may be a source of forces that may destabilize genomic integrity (Schimke, Sherwood, and Hill 1986). Besides the nuclear genome, plants have two cytoplasmic genomes—plastid and mitochondrial. The interaction of these three genomes within a plant cell has not been without important genetic consequence.

In this chapter, the influences on genetic integrity due to the four levels of organizational hierarchy—gene, chromosome, genome, and genomes (nuclear and cytoplasmic)—will be considered.

GENOTYPE DESIGN

Genomic DNA Values

The organization of the genotype also may confer mutation buffering qualities. As previously discussed (chapter 2), the redundancy of genetic information is a very effective way of compensating for the

immediate effects of mutation. (Genetic redundancy also may increase the possibilities for gene repair.) In vascular plants, diploidy
and dominance considerably reduce the immediate consequences
of somatic mutation. Within an individual plant, many cells and
tissues have nuclear DNA values in excess of 4C. Duplication of
genetic material appears to be important in the normal differentiation of many cell and tissue types. In figure 3.1 the two principal
forms of somatic genomic duplication are illustrated. Endomitosis
differs from endoreduplication primarily with regard to the centromeres, in the former the number of centromeres increases (and,
consequently, the number of chromosomes), whereas in the latter
the centromere number remains constant. Genomes that have
undergone one cycle of endoreduplication consist of chromosomes
with four rather than two chromatids per chromosome (such chromosomes are called diplochromosomes). Studies of "karyological
anatomy" in plants have shown that although many plant tissues
consist of mixtures of diploid cells and cells that have undergone
cycles of genomic duplication, the meristematic tissues (pericycle,
procambium, and cambium) almost always remain diploid (Tschermak-Woess 1956; D'Amato 1977). Since dividing cells are more
prone to developmental disturbances from somatic mutations than
nondividing cells (Jones 1935), somatic genomic duplication may
not be the major component of developmental mutation buffering.

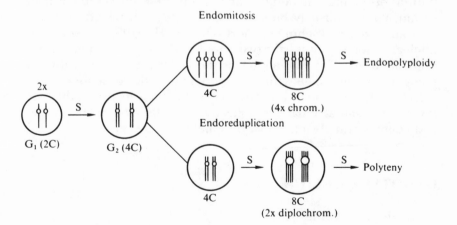

Figure 3.1. Comparison of endomitosis and chromosome endoreduplication starting from an initial diploid (2X) nucleus. Only one pair of chromosomes is shown.
(From D'Amato 1977.)

Plant nuclear genomes generally are characterized by high DNA values (Sparrow, Price, and Underbrink 1972; Price 1976).[1] Genome size can have important consequences on the rate and frequency of somatic mutation. The frequency of many kinds of mutational events is directly proportional to the amount of DNA present in the genome. For example, errors in replication are per base replicated; thus, the mutation rate per genome per replication is a function of genome size in terms of base pairs (see chapter 2). Radiation-induced forward mutation rates per locus are also proportional to the nuclear genome size (Abrahamson et al. 1973). On the other hand, increasing the size of the genome actually may serve as a mutation buffering mechanism.

In plants (as well as eukaryotic genomes in general), a large fraction of the DNA seems to have no strongly sequence-dependent function (Murray, Peters, and Thompson 1981). Following Hinegardner (1976), this component of the genome is described as "secondary DNA" in contrast to selectively constrained sequences, "primary DNA." DNA sequences with no strongly sequence-dependent function are the major component of the majority of eukaryotic genomes. The significance of such sequences is the subject of considerable debate. Such "secondary DNA" may represent "selfish DNA" (Doolittle and Sapienza 1980; Orgel and Crick 1980) or "ignorant DNA" (Dover 1980, 1986) and may play little or no part in organismal adaptation. An opposing view is that selection may operate at the level of DNA mass without reference to base sequence. In plants, many developmental characteristics are correlated with genome size (e.g., seed weight in herbaceous flowering plants, cell cycle length, duration of meiosis, minimum generation time) (Bennett 1976, 1985; Bennett and Smith 1976). Cavalier-Smith (1978, 1985) and Bennett have proposed that DNA mass has an important nucleotypic function in that it strongly influences cell sizes and developmental rates.

Regardless of the causes and origins of secondary DNA, such sequences may lessen the mutational consequences caused by transposable elements. If the number and movement of transposons are not a function of the DNA content of the genome, then decreasing the genomic proportion of primary DNA reduces the probability of a transpositional event inactivating or altering a gene (Wagner 1986).

1. Surprisingly, considerable intraspecific variation in 1C DNA amount has been documented in higher plants (Bennett 1985). In flax such variations may be experimentally induced with relatively simple environmental stimuli; once induced, the DNA changes are heritable (Cullis and Cleary 1985, 1986 for review).

If u is the genomic mutation rate due to transpositions (and similar phenomena), then

Mutations in Primary DNA
$$= [u(\text{Primary DNA})]/(\text{primary DNA} + \text{secondary DNA}).$$

Thus, diluting primary DNA within DNA having no strong sequence-dependent function can reduce the disruptional effects of mobile element mutagenesis. If mobile element movement is totally random within the genome, then the ratio of primary to secondary DNA is more important than the distribution of these two types of DNA within the genome. Thus, whether the genome consisted of two separate blocks of these two DNAs or a number of interspersed segments would not affect mutation buffering capacity.

Movement of transposons within a genome may not be random. The movements of the maize transposable element *Modulator (Mp)* (an autonomous *(Ac)* type of element, see Dellaporta and Chomet 1985) from the *P* locus are biased. The *P* locus is located on maize chromosome 1 and determines red pigment formation in the somatic pericarp and cob tissues of the ear. Greenblatt and Brink (1962) and, more recently, Greenblatt (1984) reported that the majority of new *Mp* locations are linked to the *P* locus on chromosome 1. To quote Greenblatt, "the mechanism of transposition of *Mp* when it is at the *P* locus favors chromosome 1 sites over all other chromosomes combined and, in addition, favors sites immediately distal to the *P* locus over all other locations" (1984:480). In addition to these restricted patterns of transposon movement, transposon duplication was suggested by the occurrence of adjacent sectors on corn ears with co-twin mutations. A model of the pattern of transposition is given in figure 3.2. Greenblatt (1984) has hypothesized that this pattern of transposition is a consequence of the pattern of chromosome replication. Transposition occurs during the replication of a replicon, in this case the replicon initiation site is proximal to *P* and proceeds toward *P*. When the transposon is replicated, it is the replicated *Mp* that transposes to an unreplicated region in the replicon distal to *P* (see figure 3.3). Transposon movement occurs into either unreplicated portions of its own replicon or unreplicated portions of other replicons proximal or distal to *P* or on different chromosomes.

Although this model has not been adequately tested yet, the model is provocative in that it relates transpositional movement to eukaryotic chromosome organization. If transposition and replicon synthetic activity are related and the majority of transpositions are within a replicon, one might intuitively guess that interspersing primary and secondary DNA within replicons would reduce the genetic consequences of transpositions (i.e., reduce the probability of insertions into primary DNA). As is shown in figure 3.4 this is not the case. The probability of insertions into primary DNA is not affected by having individual replicons consist of primary and secondary DNA sequences. Thus, the enhanced buffering capacity against transpositional mutagenesis is primarily a function of dilution of the primary DNA sequences and not the pattern of their distribution in a matrix of secondary DNA. If, on the other hand, transpositions are in some way targeted to certain sequences, then the dispersion of these sequences with secondary DNA sequences will have significance. The possible occurrence of a preferential integration sequence has been documented for the *Tam3* element in *Antirrhinum majus* (Coen, Carpenter, and Martin 1986).

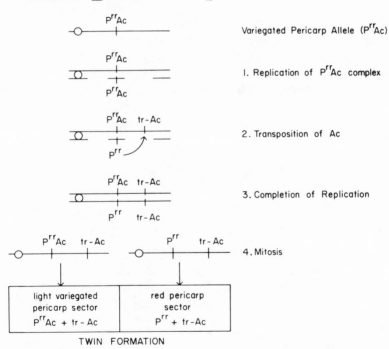

Figure 3.2. Chromosomal model of *Ac* transposition from the *P* locus in maize, which can explain the occurrence of co-twin red and light-variegated pericarp mutations on medium-variegated ears. (From Greenblatt 1984 and Dellaporta and Chomet 1985.)

Centromeres

The late Professor Arnold Sparrow's studies of the cytological characteristics correlated with radiation resistance in plants are important sources of insights on the kinds of cellular characteristics that may confer mutation buffering. Of course, although many of the genetic consequences of ionizing radiation are mediated by physiological interactions (degree of tissue hydration, oxygen effects, etc.), the assumption will be made that characteristics involved in radiation resistance also may contribute mutation buffering for certain kinds of spontaneous mutations. For example, chromosomes in the majority of plants have localized centromeres and, consequently, chromosome breakage often generates acentric fragments that are lost during the subsequent mitosis. Such losses of genetic material often will decrease a cell's viability. The genus *Luzula* in the Juncaceae has a chromosome organization that confers resistance to the effects of chromosome breakage. Members of the genus *Luzula* have diffuse or holocentric centromeres, thus chromosome breakage will generate fragments with centric activity. Consequently, mitotic divisions and breakage are not associated with large losses of genetic

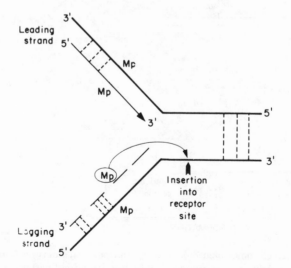

Figure 3.3. Model of a DNA replication fork showing modulator (Mp) transposing. The arrow does not connote active movement. (From Greenblatt 1984.)

material. Evans and Sparrow (1961) compared the radiosensitivity of *Luzula purpurea*, a diploid species with $2n = 6$ and diffuse centromeres, to various other angiosperms with normal centromeres. In figure 3.5 are shown the exponential declines in net dry weight for *Vicia faba*, *Lilium longiflorum*, and *L. purpurea* for various dose rates of chronic gamma irradiation. The mean inhibiting dose rate was 500r/day for *L. purpurea* compared to 45r/day for *Lilium* and

---

I. Primary and secondary DNA intermixed within individual replicons (the frequency of transposition into primary DNA is A).

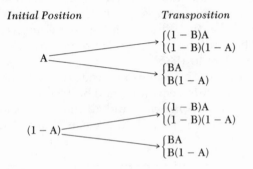

Frequency of transpositions into primary DNA is

$$A^2(1 - B) + A^2B + (1 - B)(1 - A)A + (1 - A)BA = A.$$

II. Primary and secondary DNA occur in different replicons.

Initial Position	*Transposition*
A	$(1 - B)$
	$\begin{cases} BA \\ B(1 - A) \end{cases}$
$(1 - A)$	$(1 - B)$
	$\begin{cases} BA \\ B(1 - A) \end{cases}$

Frequency of transportation into primary DNA is

$$A(1 - B) + A^2B + A(1 - A)B = A.$$

Figure 3.4. Movement of transposons within and between replicons. The proportion of primary DNA is A and the proportion of secondary DNA is $(1 - A)$. The probability of movement to different replicons is B, whereas the probability of movement to another position within the same replicon is $(1 - B)$.

Vicia. In table 3.1 the consequences of diffuse centromeres are more apparent. Four species of plants with normal centromeres are compared to *Luzula*. Regarding the accumulation of chromosome fragments and micronuclei induced by chronic gamma irradiation, it is clear that micronuclei are much more common in species with normal chromosomes. Such micronuclei represent acentric fragments that eventually will be lost with subsequent mitotic divisions. In contrast, *Luzula* has a higher frequency of fragments, presumably with centromeric activity, that will not be lost with subsequent mitotic divisions. Thus, diffuse centromeres may confer mutation resistance against spontaneous chromosome breakage in the same way they promoted radiation resistance. It is an interesting aside to note that *Luzula acuminata* is one of the herbs growing closest to the gamma source at the Whiteshell Nuclear Establishment in Canada, a mixed boreal forest receiving chronic doses of gamma radiation (Amiro and Dugle 1985).

Diffuse (holocentric) chromosomes occur in a limited number of animals such as coccids, scorpions, and butterflies and in some plant groups, i.e., desmids, *Spirogyra* and all the species of the vascular plant families Cyperaceae and Juncaceae (Jones 1978). (A sedge

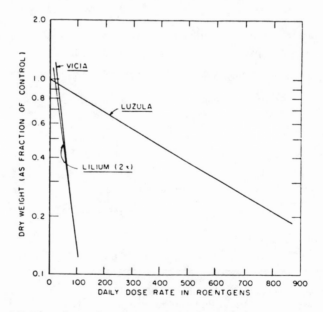

Figure 3.5. The relation between the exponential decline in net dry weight and increasing dose rate in three plant species. (From Evans and Sparrow 1961.)

Table 3.1. Relation between frequency of fragments at metaphase and micronuclei at interphase induced by chronic gamma irradiation. (From Evans and Sparrow 1961.)

Species	Nuclear Volume, μ³	Chromosome No. (2X)	Dose Rate, r/day	Fragments (f) per Cell[a]	Micronuclei (m) per Cell[b]	f/m
Crepis capillaris	105	6	300	0.58	0.06	9.6
Ornithogalum virens	225	6	180	0.39	0.05	7.8
Lilium longiflorum (2X)	1660	24	40	0.44	0.05	8.8
Vicia faba	510	12	40	0.22	0.035	6.3
Luzula purpurea	190	6	375	3.16	0.007	451.4

[a]Mean values from 3 plants (~50 cells) scored per species.
[b]Mean values from 3 plants (~6000 cells) scored per species.

species [Cyperaceae] grew closest to the radiation source at the Brookhaven Gamma Forest, Woodwell 1967). Meiosis in *Luzula* has been well investigated and has been shown to have peculiar characteristics. In this genus two kinds of polyploids occur, true polyploids with multiple genomes and endonuclear polyploids which have resulted from the fragmentation of larger chromosomes. Thus, a single genome is present in an endonuclear polyploid, but it is distributed over many smaller chromosomes. In a classic paper, Nordenskiöld (1961) carried out a kind of tetrad analysis on the meiotic products of diploid X endonuclear polyploid hybrids. Meiosis in organisms with diffuse centromeres is called post-reductional meiosis in contrast to the pre-reductional meiosis characteristic of most plants. In *Luzula*, homologous chromosomes pair and chiasmata occur between nonsister chromatids, chiasmata terminalize, and bivalents are present at metaphase I. At anaphase I (assuming no crossing over) sister chromatids separate (rather than nonsister chromatids). The pairs of nonsister chromatids are separated at anaphase II. In figure 3.6 the consequences of such a meiosis are shown in a diploid X endonuclear hybrid, only one chromosome pair is represented. Pairing and crossing over between normal and fragmented chromosomes results in the formation of genetically unbalanced meiotic products and partial or complete sterility. Based upon figure 3.6, the significance of diffuse centromere chromosome organization as a mutation buffer in plants can be better appraised. As was clearly shown by the work of Evans and Sparrow (1961), diffuse centromeres are an effective buffer against chromosome fragmentation in somatic tissues; consequently, diffuse centromeres reduce diplontic selection within meristems against cells with fragmented chromosomes. Thus, although cell and individual survival is enhanced, this is not without ultimate cost. The reduction in the effectiveness of diplontic selection will result in a progressive sterilization of the individual as meristems are populated with cells having fragmented and unfragmented homologs (figure 3.6). Thus, diffuse centromeres ultimately are not an effective long-term mutation buffer; this may be one of the reasons the majority of plants have localized centromeres. It would be of interest to investigate the species with diffuse centromeres to determine if any compensatory adaptations have evolved.

In table 3.2 the principal nuclear characteristics associated with radiosensitivity are tabulated. In addition to these characteristics, an index known as the interphase chromosome volume (average nuclear volume divided by the chromosome number) was related

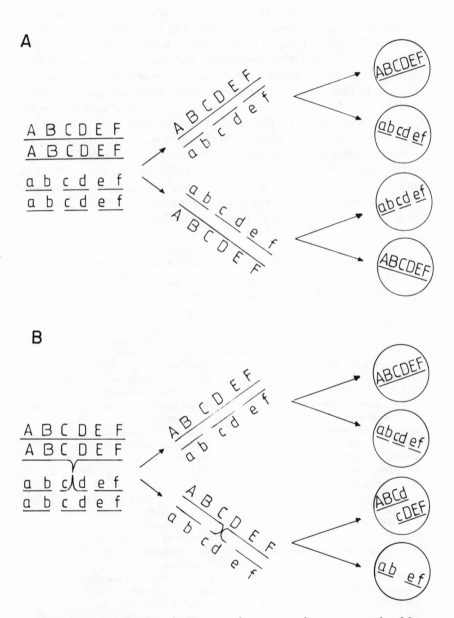

Figure 3.6. Meiosis in *Luzula*. Pairing and segregation between normal and fragmented homologs. A. Without crossing over. B. With crossing over. (Modified from Nordenskiöld 1961.)

to radiosensitivity (Sparrow, Schairer, and Sparrow 1963) (figure 3.7). It was later shown that a similar relationship held for DNA content per chromosome, the greater the DNA content per chromosome the greater the radiosensitivity (for both lethal exposures and heightened mutation rates) (Sparrow, Rogers, and Schwemmer 1968). These correlations have been interpreted as indicating that the major category of mutational damage being screened in these studies was chromosome breakage and consequent loss of genetic material. Thus the more DNA target per chromosome (and thus per centromere), the more likely a "hit" will result in an acentric fragment and consequent loss of genetic material during the following cell divisions. Again, these results probably can be generalized to include spontaneous somatic chromosome breakage and its consequences. The more DNA per chromosome, the higher the probability of an "error" separating a segment of DNA from the centromere.

Table 3.2. The principal nuclear and allied factors influencing plant radiosensitivity. (From Evans and Sparrow 1961.)

Factors Tending Toward High Sensitivity	Factors Tending Toward High Resistance
Large nucleus (high DNA)	Small nucleus (low DNA)
Large nuclear/nucleolar volume ratio	Small nuclear/nucleolar volume ratio
Much heterochromatin	Little heterochromatin
Long chromosome arms (large chromosomes)	Short chromosome arms (small chromosomes)
Acrocentric chromosomes	Metacentric chromosomes
Normal Centromere	Polycentric or diffuse centromere
Uninucleate cells	Multinucleate cells
Low chromosome number	High chromosome number
Diploid or haploid	High polyploid
Sexual reproduction	Asexual reproduction
Slow rate of cell division (long intermitotic time)[a]	Fast rate of cell division (short intermitotic time)
Long dormant period[a]	Short or no dormant period
Meiotic stages present at dormancy[a]	Meiotic stages not present during dormancy
Slow meiosis and premeiosis[a]	Fast meiosis and premeiosis
Low concentration of protective chemical constituents, e.g., ascorbic acid	High concentration of protective chemical constituents

[a]Factors of particular importance only under conditions of chronic irradiation.

Nagl, Pohl, and Radler (1985) make the interesting point that in complex organisms there is probably selection pressure against mitosis. They point out that very often cytokinesis is bypassed in plant tissue differentiation. As discussed earlier, many plant cells undergo varying numbers of endoreduplication cycles during development. Such patterns of development would, of course, greatly reduce the genetic consequences of chromosome breakage by preserving acentric fragments.

Sparrow's correlation of DNA content per chromosome and radiosensitivity may be the key to an important problem in evolutionary theory. Turner (1967) focused attention on this issue with a nice turn of phrase. He asked the question, "Why does the genotype not congeal?" The problem is why do the majority of organisms have their genes dispersed among a number of different chromosomes (or linkage groups) rather than a single large linkage group. Some species have achieved such a congealed condition either through complex translocation heterozygosity, as in some members of the genus *Oenothera*, or with only a single pair of chromosomes ($2n = 2$) but, in general, the majority of species have a number of linkage

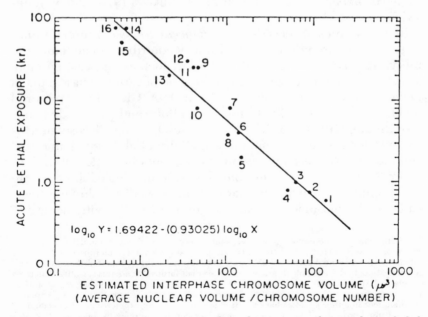

Figure 3.7. Relationship between interphase chromosome volume and acute lethal exposure (kr) for 16 plant species. (From Sparrow, Schairer, and Sparrow 1963.)

groups. Turner (1967) and others (see Bell 1982 for review) have approached this problem from the point of view of recombination, fitness, and selection. The correlation of Sparrow suggests another answer. Increasing the number of linkage groups reduces the amount of DNA per chromosome (all things being equal). Thus, spontaneous chromosome breaks that may result in acentric fragments will have less of an impact on cell viability. The proportion of the genetic material occurring on such acentric fragments is inversely related to the amount of DNA per chromosome (assuming random chromosome breakage). Thus, a totally congealed genotype (one linkage group) will be much more sensitive to the phenotypic effects of chromosome breakage than a genotype consisting of a number of linkage groups. Two examples of such genotypes are known. The nematode *Parascaris equorum univalens* has $2n = 2$; this species also has diffuse or holocentric centromeres (White 1973), which may reduce the consequences of spontaneous chromosome breakage. Recently an ant, *Myrmecia pilosula,* has been reported as $2n = 2$, with localized centromeres (Crosland and Crozier 1986). In vascular plants the lowest chromosome number reported is $2n = 4$ in *Haplopappus gracilis* (Stebbins 1966), and *Zingeria biebersteiniana* (Tsvelev and Zhukova 1974, cited in Jones 1978).[2]

Recent studies of yeast *(Saccharomyces cerevisiae)* chromosomes suggest that chromosome size also may contribute to genomic stability. The DNA component of the kinetochore of yeast chromosomes has been localized to a small region of 220–250 base pairs of DNA. This is referred to as centromere DNA (CEN) and is required for chromosome stability during both mitotic and meiotic divisions (see Bloom, Hill, and Yeh 1986 for review). Using CEN sequences to construct artificial chromosomes, Hieter et al. (1985) found that chromosome size is an important component of mitotic stability. In figure 3.8 the rate of chromosome loss as a function of chromosome size is plotted for both circular and linear artificial chromosomes. Circular chromosomes increased in mitotic stability with increasing

2. In the course of surveying comparative plant radiosensitivities, it was found that, in general, woody plants were twice as sensitive as herbaceous plants for similar interphase chromosome volumes (Sparrow and Sparrow 1965). More recently it has been reported that, as a rough generalization, herbaceous plants are less sensitive than shrubs that are less sensitive than trees (Dugle and Mayoh 1984) and that gymnosperms are more sensitive to ionizing radiation than angiosperms (see Amiro and Dugle 1985 for review). Whether these sensitivity patterns are due to differences in either genetic characteristics, organography, apical meristem organization, or any of a number of developmental differences between these groups is unknown. It is also unknown whether such differences in radiosensitivity can be extrapolated to differentials regarding the accumulation of spontaneous mutations.

size up to about 100kb, circular chromosomes larger than 100kb lost mitotic stability. The instability of large circular chromosomes was attributed to an increasing frequency of sister chromatid exchanges (SCE). In circular chromosomes SCE events may result in dicentric chromosomes and consequent mitotic loss of genetic material due to anaphase bridges. On the other hand, linear chromosome forms allow SCE events to occur without the formation of dicentrics (Hieter et al. 1985). Linear chromosomes smaller than 100kb are less stable than their circular counterparts, at lengths above 100kb linear chromosomes are more stable. Hieter et al. note that the shape of the curve in figure 3.8 suggests that the effect of length on linear chromosome stability has not been saturated; thus, longer chromosomes may show even greater mitotic stabilities.

Murray and Szostak (1985) hypothesized that as genome size increased, one of the major selective forces favoring the evolution of linear chromosomes rather than centromeric circular molecules was sister chromatid exchange. These authors also emphasize that the

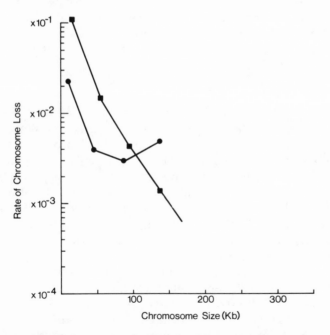

Figure 3.8. Rate of chromosome loss vs. chromosome size for circular and linear artificial chromosomes in yeast. The log rates of loss for circular ($\bigcirc$ – – – $\bigcirc$) and linear ($\square$ – – – $\square$) minichromosome constructions. (From Hieter et al. 1985.)

length of DNA molecules is one of the main determinants of mitotic stability.

Higher eukaryotes have a more complex centromeric organization than yeasts (Jabs, Wolf, and Migeon 1984; Blackburn and Szostak 1984; Lima-de-Faria 1983). Individual chromosome sizes in many plants exceed the total yeast genome (table 3.3). Thus the extrapolation from yeast artificial chromosomes to the chromosomes of higher plants probably is rash at this time. In spite of these caveats, it is tempting to note that if the relationship between chromosome length and mitotic stability is valid in other eukaryotes, then the genome may be subject to two conflicting forces with regard to informational integrity. Longer chromosomes promote mitotic stability and, thus, lessen the frequency of aneuploidy. On the other hand, increasing chromosome length increases the probability of chromosome breakage and the consequent generation of acentric fragments.

Centromere position may also influence mitotic stability. Murray and Szostak (1985) compared the mitotic stability of an artificially created telocentric of chromosome III in yeast with the normal metacentric. The telocentric chromosome was ten times less stable than the normal metacentric. They also reported that the centromere may suppress the amount of recombination of nearby genes.

Telomeres

Although linear chromosomes are more immune to SCEs, linear organization generates problems in DNA replication. When a linear DNA molecule replicates, the excision of RNA primers from the 5' ends of the two daughter strands will leave a gap (figure 3.9). This gap cannot be filled by a DNA polymerase since the gap is terminal and, therefore, a 3'OH template strand unavailable. Continued replications will result in a progressive diminution of chromosome material (Watson 1972; Cavalier-Smith 1974). This problem does not exist for circular chromosomes since they are continuous and, therefore, without ends.

Telomeres are specialized structures found on the ends of eukaryotic chromosomes and are assumed to play a role in the completion of replication of chromosome ends (Blackburn and Szostak 1984). Various mechanisms of telomeric DNA replication have been proposed to fill terminal 5' gaps on eukaryotic chromosomes. These include intrachromosomal mechanisms (Cavalier-Smith 1974; Bateman 1975; Walmsley et al. 1984) as well as interchromosomal mech-

Table 3.3. Genome sizes in vascular plants. (From Bryant 1986.)

Species	HAPLOID GENOME SIZE		HAPLOID CHROMOSOME NO.	MEAN AMOUNT OF DNA PER CHROMOSOME	
	Nucleotide pairs	Meters		Nucleotide pairs	Millimeters
Allium cepa	16.3×10^9	5.62	8	20.4×10^8	703
Anemone virginiana	10×10^9	3.45	8	12.5×10^8	431
Beta vulgaris	1.2×10^9	0.41	9	1.3×10^8	46
Lathyrus latifolium	10.6×10^9	3.64	7	15.1×10^8	520
Lilium pyrenaiecum	31.7×10^9	10.9	12	26.4×10^8	908
Pisum sativum	4.7×10^9	1.62	7	6.7×10^8	231
Vicia faba	14×10^9	4.8	6	23.3×10^8	800
Zea mays	2.3×10^9	0.78	10	2.3×10^8	78

anisms based upon transient telomere-telomere fusions (Dancis and Holmquist 1979).

Telomeres also may lessen the consequences of chromosome breakage. In *Zea mays*, McClintock (1941) found that if chromosomes are broken by various means (e.g., x-rays), the broken ends appear adhesive and tend to fuse with one another (figure 3.10). Such fusions may recur with each mitotic division, resulting in recurrent dicentric bridges and chromosome breakage with consequent deletion or duplication of genetic material. This breakage-fusion-bridge cycle is confined to specific tissues. In *Zea mays*, broken chromosomes will undergo breakage-fusion-bridge cycles in the gametophyte and endosperm tissues whereas broken chromosomes present in the zygote appear to be healed. In subsequent mitotic divisions of the embryo, the previously broken chromosomes undergo normal mitotic chromatid separations. It has been hypothesized that the healing of broken chromosomes may involve the modification of the DNA ends by the telomeric DNA addition enzymes (Shampay, Szostak, and Blackburn 1984; see also Swanson, Merz, and Young 1981). If this hypothesis is correct, then the telomeric system may promote genomic integrity in at least two ways: terminal gap filling and the prevention of breakage-fusion-bridge cycles.

Telomeres also may be a destabilizing force in genomes. They have been implicated as having a significant role in the origin of certain chromosome aberrations (Dancis and Holmquist 1979). Recent molecular studies of telomeric DNA have revealed a very pe-

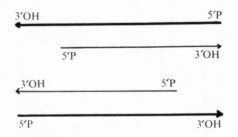

Figure 3.9. The two daughter molecules that result from the replication of a linear DNA molecule, showing the gaps at the 5′ end of the newly synthesized strands left by the removal of the RNA primers. The 3′ end of a polynucleotide is shown by an arrow; newly synthesized DNA by a thin line. (From Cavalier-Smith 1974.)

culiar property of these sequences; they promote their own elongation during replication. In other words, the replication of yeast telomeres is accompanied by regular additions of DNA (see Blackburn and Szostak 1984 for review). The mechanism for this DNA increase is unclear at this time, but if it is a general property of telomeres, then it may have considerable genomic consequences. Blackburn (1984) has noted that if chromosomes are continually being lengthened in this way, then the cell must have a mechanism for limiting the addition to a maximum length. In this regard, it is interesting to note that telomeric heterochromatin often varies considerably between related species. For example, telomeric heterochromatin accounts for 12–18% of the DNA of *Secale cereale* (cultivated rye) and less than 9% in *S. sylvestre* (Bedbrook et al. 1980). In hexaploid triticale (a rye-wheat polyploid hybrid), reduction in the amount of telomeric heterochromatin present on the chromosomes of the rye parent was accomplished in a few years of selection (Bennett 1985). It is possible that such differences in telomeric DNA may reflect alterations in the mechanisms that limit the amount of DNA a telomere can accumulate and maintain.

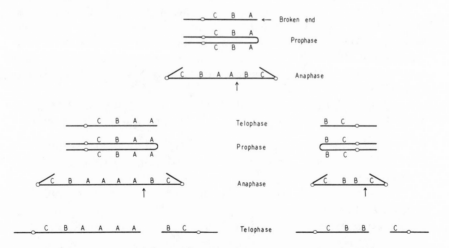

Figure 3.10. Somatic genetic variability produced by the breakage-fusion-bridge cycle. The replication of a chromosome with a broken end may result in the chromatids being interconnected. At mitotic anaphase a bridge forms which may break randomly (arrow). Subsequent cycles of replication and mitosis result in the deletion and duplication of genetic material. (From McClintock 1941).

DNA SYNTHESIS AND REPAIR

Although a lot is known about DNA replication, proofreading, and repair in prokaryotes, this is not the case for eukaryotes (Loeb and Kunkel 1982; Denhardt and Faust 1985). The properties of the DNA polymerases and repair systems in eukaryotes are not well understood and available information is based upon organisms other than higher plants (e.g., mammals, *Drosophila*, fungi) (Bryant 1986). The properties of eukaryotic DNA replication, as have been deduced from these nonplant systems, will serve as points of departure in our discussion of the fidelity of plant DNA synthesis.

A characteristic property of all eukaryotic chromosomes is that the DNA is complexed with histone proteins. This DNA-histone association is called chromatin. Chromatin DNA is wrapped around a histone octomer consisting of two molecules each of histone proteins H2A, H2B, H3 and H4. These structural units have been named nucleosomes. Each nucleosome is associated with a single molecule of histone H1 that is critical for the assembly of nucleosomes into the 300Å filament that is characteristic of chromatin (Watson et al. 1987). A total of approximately 200 base pairs of DNA comprise a nucleosome (including the DNA wrapped around the histone octomer and the linker DNA between octomers) (DePamphilis and Wassarman 1980). Therefore, the replication of eukaryotic chromosomes must involve the duplication of the DNA and nucleosomal units (figure 3.11).

According to Loeb and Kunkel (1982), the fidelity of DNA synthesis is a multicomponent process involving three sequential steps:

1. Discrimination against the insertion of an incorrect nucleotide. This discrimination may be amplified by DNA polymerase and other proteins.
2. Correction during replication at the growing point—proofreading.
3. Postsynthetic correction of mismatches—mismatch repair.

At the present time, quantitative (or even qualitative) information concerning the fidelity enhancing characteristics of higher plant nucleic acid metabolism is almost nonexistent. Existing data are primarily extrapolations from mammalian studies.

Although the spontaneous mutation rate per base pair replicated is probably less than 1×10^{-9} in eukaryotes (Drake 1969; Nagley

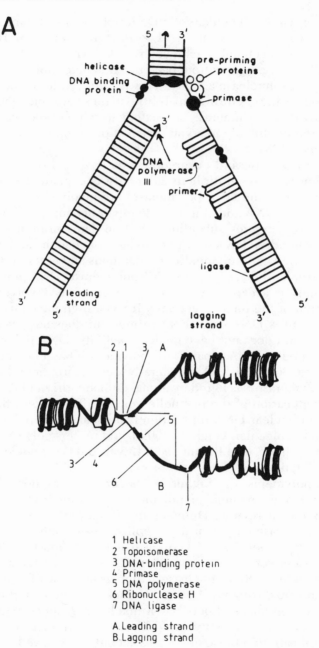

Figure 3.11. Schematic representation of replication forks in (A) prokaryotes and (B) eukaryotes. (From Dunham and Bryant 1985.)

1987 for review), the eukaryotic DNA polymerases themselves exhibit error rates of $1 \times 10^{-3} - 1 \times 10^{-5}$ (Kunkel 1985). How this accuracy enhancement is achieved is unknown.

All DNA polymerases from bacteria and bacteriophage contain a $3' \rightarrow \rightarrow 5'$ exonuclease activity. This is believed to be the basis of the proofreading or editing capabilities of these enzymes and greatly enhances their accuracies. The extent to which proofreading enhances the fidelity of eukaryotic DNA replication is unclear (Loeb and Kunkel 1982).

Both prokaryotic and eukaryotic cells have a diversity of DNA polymerases. This multiplicity generally is explained in terms of function (i.e., some DNA polymerases are involved in the *de novo* replication of DNA and others in the repair of damaged DNA) or, in the case of eukaryotes, subcellular location. Thus, in animals nuclear and mitochondrial DNA polymerases have been found and in plants, chloroplastic DNA polymerase. In mammals and other vertebrates, there are at least three types of DNA polymerase. DNA polymerase-α generally is accepted as the enzyme primarily involved in the replication of the nuclear genome. It has a high molecular weight (>150kd), uses DNA or an RNA primer but does not use an RNA template, and does not have nuclease activity. DNA polymerase-β is implicated in DNA repair. This polymerase has a low molecular weight (ca. 30–40kd) and is generally a chromatin-bound enzyme. DNA polymerase-γ is found in animal mitochondria and is involved in the replication of the organelle genome. The role of this polymerase in nuclear DNA replication is unclear. DNA polymerase-γ has a high molecular weight (>100kd) and prefers poly rA-oligo dT template-primers (see Bryant 1980; Litvak and Castroviejo 1985 for further details).

DNA polymerases in higher plants generally are described and compared to the mammalian polymerases α, β, and γ (Bryant 1980). In their recent review, Dunham and Bryant (1985) indicate that DNA polymerase-α is the primary replicative enzyme for the nuclear genome in higher plants. They also note that most of the DNA polymerase-α activities from higher plants do not have associated nuclease activity. Since nuclease activity is necessary for proofreading or editing during replication, the lack of documented nuclease activity is significant. Loeb and Kunkel (1982) noted that a deficiency in nuclease activity could be due to a number of reasons: DNA replication in these organisms actually may have low fidelity and depend primarily on postreplication repair; high fidelity may be maintained by accessory proteins in the replication complex; a proofreading activity may be removed during polymerase purification.

In a recent review, Goodman and Branscomb (1986) noted that DNA polymerases from animal cells have no measurable exonuclease activities and, therefore, by implication no proofreading or editing ability during DNA replication. These authors suggest that perhaps postreplication mismatch repair may be important. In bacteria, such postreplication error correction can take place only if the newly synthesized DNA strand is distinguishable from the parental DNA strand by its lack of methylated adenines (Glickman and Radman 1980). Thus, mismatched bases are removed from the newly synthesized strand and new bases inserted according to the template of the parental strand. As yet postreplication mismatch repair has not been documented in animals or plants (Goodman and Branscomb 1986).

Although an enzyme strictly similar to animal polymerase-β has not been found in plant cells (Litvak and Castroviejo 1985), a low molecular weight chromatin-bound polymerase which resembles the DNA polymerase-β from mammals has been documented (Dunham and Bryant 1986). The role (if any) this polymerase plays in plant DNA repair is unknown. DNA polymerase-γ is defined generally as an enzyme that recognizes poly rA-oligo dT template-primer more effectively than activated DNA. Chloroplast DNA polymerase has this characteristic (Litvak and Castroviejo 1985; Dunham and Bryant 1986), whereas the classification of the plant mitochondrial DNA polymerase as a γ enzyme is less evident (Litvak and Castroviejo 1985; Litvak et al. 1983). Since the plant DNA polymerase-γ can utilize an RNA template (poly rA), it has been suggested as a potential reverse transcriptase (Dunham and Bryant 1986). Bryant (1986), in a recent review, concluded that the general characteristics of the DNA polymerases of higher plants are still somewhat controversial; the same can be said for the fidelity promoting properties (e.g., proofreading) of these enzymes. The reader is referred to Bryant (1986) for an excellent summary.

The role of the above DNA polymerases (or any other enzymes) in DNA repair in higher plants is also unclear. It was once thought the plants had limited capacities for DNA repair, but more recent research has documented a broad repertoire of repair capabilities (see review by Soyfer 1983). Solar ultraviolet radiation may be a very significant environmental mutagen for green plants. One would predict that for many species somatic and germ cell (pollen) ultraviolet-induced mutagenesis may be a significant adaptive problem (see chapter 7). Of the several effects that ultraviolet radiation produces on DNA, the formation of chemical bonds between adjacent pyrimidine molecules (especially thymine residues) has important biological consequence. Such thymine dimers distort the DNA helix and may have lethal consequences unless repaired. Two major pathways are involved in the repair of ultraviolet radiation-induced DNA damage, photoreactivation and dark repair or excision repair. Photoreactivation involves a photoreactivating enzyme that binds to the dimer and, upon absorption of visible light, converts thymine dimers to thymine monomers. Photoreactivation has been documented in both angiosperm and gymnosperm species (Veleminsky and Gichner 1978).

Excision repair (or dark repair) involves a glycosylase that removes the damaged bases, an endonuclease that makes a single stranded nick in the damaged strand, an exonuclease that digests a portion of the damaged strand, a DNA polymerase that fills in the gap, and a ligase that links the newly synthesized strand to the original strand. Howland (1975) documented dark repair in wild carrot protoplasts but found that dimer excision occurs only at fluences <100 Jm^{-2}. Howland also noted that the efficiency of thymine dimer formation by ultraviolet radiation was 35–40% that observed in bacteria and mammalian cells, presumably due to the shielding effect of plant cytoplasmic organelles. (See Soyfer 1983 for further studies of dark repair in higher plants.) The repair of DNA strand breaks after gamma-irradiation has been documented in protoplasts isolated from wild carrot cells (Howland, Hart, and Yette 1975). The repair of DNA damage caused by chemical mutagens is also known in higher plants (see Veleminsky and Gichner 1978 for review). In contrast to the voluminous literature on mutation induction in plants, the study of DNA repair mechanisms is still in its infancy. The relationship between DNA repair and gene mutation is unclear in higher plants (it should be noted that often mutation is a consequence of some forms of repair). Whether all of the DNA (nuclear and cytoplasmic) is repaired equally efficiently is unknown. Even within a genome there may be differences in repair efficiencies. For example, there is evidence from mammalian studies that the excision of pyrimidine dimers from active genes is much more efficient than in the genome overall (Bohr et al. 1985).

Based upon the results of hybridization experiments using sublethally irradiated pollen, Pandey has suggested that the pollen grains of some angiosperms may either lack or be inefficient in excision repair. As a consequence, pollen grain genomes may sometimes be repaired after fertilization using maternal templates. To quote Pandey,

... during early embryogeny following normal fertilization by sublethally irradiated pollen there may be selection of cells in which radiation-damaged segments of paternal chromosomes have been repaired, through homologous somatic recombination and gene conversion, by segments from the normal maternal chromosomes. (1986:38)

It is unknown whether such a "maternalization effect" is common in angiosperms or if such a repair process is possible for spontaneous genetic lesions in the pollen.

The nuclear genomic DNA is organized into chromosomes; thus, any consideration of DNA replication must refer back to these organelles. Chromosomal DNA is replicated asynchronously at multiple sites along the chromatid. Each of these replication sites or units is called a replicon.[3] Replicons have an origin where replication begins and two divergent bidirectional replication forks.

In plants, replicon size (the origin to origin distance) and the rate of fork movement show little variation, even between species with large differences in genome size (Van't Hof and Bjerknes 1981; Van't Hof 1985). Higher plants have a slower fork movement rate than either bacteria or mammalian cells (Van't Hof and Bjerknes 1979); whether slower fork movement increases fidelity of replication is unknown. Not all replicons replicate simultaneously, but rather there appears to be a chronological hierarchy. A three-tier hierarchy has been proposed starting with the replicon, then a group of replicons on a chromosome which replicate almost simultaneously (or a cluster) and, finally, a family of clusters distributed among the chromosomes of the genome. In *Arabidopsis thaliana,* a species with a

---

3. There is considerable evidence in support of the view that the chromatin fiber in both meiotic and mitotic chromosomes is folded into radially projecting loops and that these loops are tethered along the chromatid axis. These axial components are part of a set of nonhistone chromosomal proteins called the "chromosomal scaffold" (Earnshaw and Laemmli 1983; Earnshaw and Heck 1985). The enzyme topoisomerase II has been identified as a component of the chromosomal scaffold (Gasser et al. 1986; Earnshaw and Heck 1985). DNA topoisomerases are enzymes that catalyze the concerted breakage and rejoining of DNA backbone bonds; the eukaryotic type II topoisomerases catalyze a number of reactions including the relaxation, catenation, decatenation, knotting, and unknotting of closed double-stranded DNA circles (see DiNardo, Voelkel, and Sternglanz 1984 for review). Temperature sensitive topoisomerase II mutants in *Saccharomyces cerevisiae* cannot disjoin sister chromatids at mitosis when grown at restrictive temperature. It has been hypothesized that topoisomerase II may be necessary to cleave the DNA backbone to resolve the intertwinings produced at the sites of convergence of multiple replication forks and, thus, may be required at each replicon (DiNardo, Voelkel, and Sternglanz 1984; Holm et al. 1985).

very small genome (a C-value of DNA of 0.2pg), Van't Hof, Kuniyuki, and Bjerknes (1978) reported that two replicon families were present, one consisting of approximately 687 replicons and the other with 1,888 replicons, and that the two families initiate replication with an interval of 36 min. Replication of an average replicon occurred in little more than 2h or 74% of the S phase. In figure 3.12 the sequential pattern of two clusters of replicons is illustrated. One would assume there to be tight control on the number of initiation events per origin of replication, with each replicon generally "firing" once per cell cycle (Varshavsky 1981).

Recent studies on pea roots indicate that such precision may not be typical for certain cells in this organ. In the last S phase prior to differentiation, these cells replicate 80% of their DNA, accumulate in late S, and then replicate the remaining 20% and differentiate. These cells produce extra chromosomal DNA (exDNA) coincident

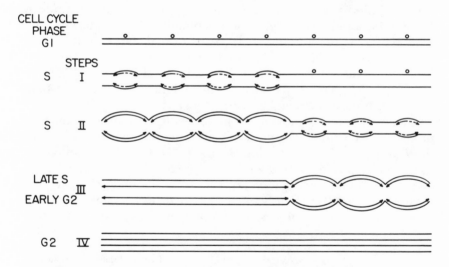

Figure 3.12. A diagram showing the stepwise replication and maturation of chromosomal DNA during S and G2. Top, the parental chromosomal duplex in G1 with Os representing replicon origins; Step I, bidirectional replication by four replicons in a cluster; Step II, the convergence of replication forks of neighboring replicons within the cluster leaving a single-stranded gap between the nascent chains; to the right of the first cluster a second has begun replication; Step III, gaps between nascent chains of replicons of the first cluster are sealed and joined producing a cluster-sized molecule; Step IV, gaps between neighboring clusters and replicons are sealed and joined to give chromosomal-sized DNA. Note that the time between the replication of the first and that of the second cluster may be greater than shown. (From Van't Hof 1985.)

with replicating this later 20%. This exDNA is replicon size, of nuclear origin, linear, and seems to replicate by strand displacement (Kraszewska et al. 1985; Krimer and Van't Hof 1983; Van't Hof, Bjerknes, and Delihas 1983; Van't Hof 1985). The formation of extrachromosomal DNA in mammalian cells is also not uncommon, only these molecules are circular rather than the linear forms so far documented in plants (see Yamagishi 1986 for review). The function of exDNA in plant cell differentiation is unclear. It may be a means of differential gene amplification associated with the differentiation process or exDNA may be a kind of mutational event that is tolerated in these cell lines. Varshavsky (1981) suggested that replicon "misfiring" (additional replication of replicons) may be the origin of exDNAs and that such misfiring should best be viewed as a disruptive or destabilizing force. Varshavsky also suggested that replicon misfiring may even be the origin of the small acentric chromosomes often found in mammalian cell malignant phenotypes. Replicon misfiring may thus represent yet another example of how solutions generate new problems. The imposition of an orderly mechanism of genomic replication (replicons, clusters, and families) generates the opportunity for a unique kind of genomic destabilization— exDNA. Schimke, Sherwood, and Hill (1986) have proposed that occasional overreplication of portions of chromosomes may not be uncommon. Recombination and insertion events involving these free DNA strands with the replicating chromosome may be the origin of a diversity of genomic changes including duplications, inversions, translocations, partial or complete endoreduplication of chromosomes and mitotic recombinants.

Chromosomes may themselves occasionally be a component of genomic instability. In addition to the normal species complement of chromosomes, individuals, clones, or small populations may have supernumerary B-chromosomes. Such B-chromosomes have been documented in numerous plant taxa (gymnosperms and angiosperms) as well as in many different animal phyla (see Jones and Rees 1982 for a complete bibliography). According to these authors, several criteria distinguish B-chromosomes from the rest of the genome. B-chromosomes occur in some individuals and among some populations of a species, yet individuals without B-chromosomes in such populations are normal in growth and reproduction. Typically, B-chromosomes are heterochromatic, smaller, generally lack genes with major effects, and are dispensable to the species. High numbers of B-chromosomes may depress fertility and growth. B-chromosomes often show nodisjunction at mitosis and very often tend to increase in tissues giving rise to reproductive cells (figures 3.13, 3.14).

The suggestion that B-chromosomes are genomic parasites was made over forty years ago (Östergren 1945). Östergren considered B-chromosomes as selfish or parasitic genetic structures which primarily have properties favorable for their own existence and perpetuation in the plant. These properties include:

1. Mechanisms that reduce the phenotypic consequences of B-chromosomes. The loss of B-chromosomes occurs if individuals carrying such chromosomes have lower viability or competitive ability; consequently, mechanisms reducing the phenotypic effects of B-chromosomes should be part of the "parasitic" repertoire. Östergren considered the restriction of B-chromosomes to reproductive organs and tissues a means of minimizing the overall phenotypic effects of the presence of these chromosomes in the plant. B-chromosomes that are genetically inert will have minimum phenotypic effects; thus heterochromatization also may be part of the parasitic repertoire.

2. Mechanisms that increase the frequency of B-chromosomes

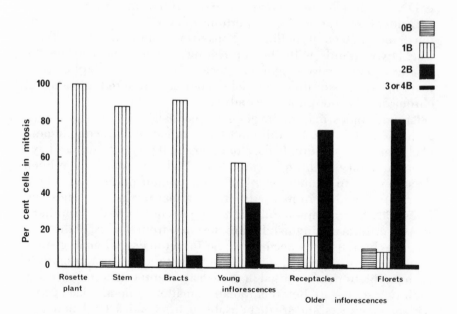

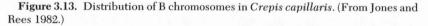

Figure 3.13. Distribution of B chromosomes in *Crepis capillaris*. (From Jones and Rees 1982.)

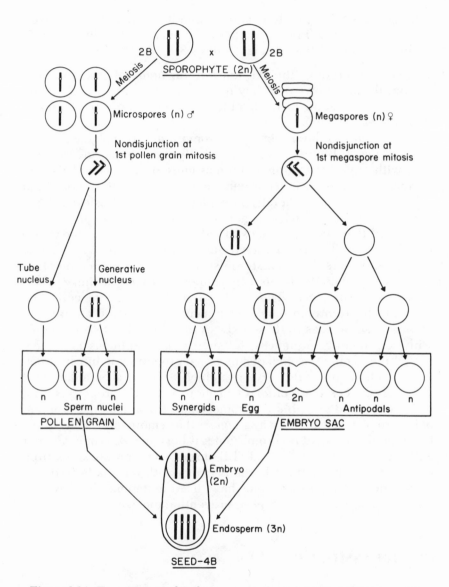

Figure 3.14. Transmission of B chromosomes through male and female gameto-phytes of a rye plant with two B chromosomes. Nondisjunction at the first pollen grain mitosis and at the first megaspore cell mitosis leads to the accumulation of Bs among progenies. Note that the sperm and egg nuclei both contain two B chromo-somes, hence a cross between rye parents, each with two B chromosomes, yields progeny with four B chromosomes. Only the B chromosomes are drawn. (From Jones and Rees 1982.)

in gametes. These include various nondisjunctional mechanisms in gametophytes (mega and micro) that increase the number of B-chromosomes in the gametes (figures 3.13, 3.14).

To quote Östergren, "the existence of fragments (B-chromosomes) in equilibrium shows that they have a use. But it is not necessary that they are useful to the plants. They need only be *useful* to themselves" (p. 163).

With regard to the origin of B-chromosomes, two characteristics are significant; B-chromosomes generally show no pairing homologies with any of the normal chromosomes at meiosis and B-chromosomes seem to lack major gene activity. Although the origin of such chromosomes generally is explained using classical cytogenetic mechanisms which may generate chromosome fragments (see Jones and Rees 1982), perhaps the ideas of Varshavsky (1981) also may be relevant in this context. Such chromosomes may arise *de novo* from rare replicon cluster misfiring which may include DNA sequences that confer limited centromeric activity (Schimke, Sherwood, and Hill 1986). The origin of such exDNAs in cells that normally heal chromosome breaks may result in the addition of telomeric sequences to these exDNA elements. ExDNA elements with centromeric sequences, telomeres, and replicons may have some limited chromosomal disjunctional properties at mitosis. Such exDNA elements could serve as vehicles for the accumulation of DNA sequences that confer even more parasitic properties and, ultimately, to identifiable B-chromosomes. Looking at the problem of B-chromosomes in this way suggests that a more interesting question than the origin of B-chromosomes is how do eukaryotes prevent exDNA elements with potential B-chromosome propensities from occurring and persisting? Perhaps plants with unusually frequent occurrences of B-chromosomes have fixed mutations that have reduced the effectiveness of these prevention mechanisms.

CYTOPLASMIC GENOMES

Green vascular plants have three genomes residing, functioning, and interacting within a single cell. The nuclear genome is the largest (10^8–10^{11}bp), then follows the mitochondrial genome (200–2400kb), and the smallest is the chloroplast genome (120–160kb). Both the nuclear and mitochondrial genomes are the most variable and consist largely of noncoding sequences whereas the chloroplast genome shows the least size variation and consists of primarily coding sequences. The presence of three semi-independent genomes within plant cells has not been without genetic consequences. Since the majority of the structural and functional polypeptides of the mitochondrion and chloroplast are nuclear encoded, it is generally believed that a massive transfer of genetic material must have taken place from these organelles into the nucleus. In addition to the transposition of chloroplast and mitochondrial nucleotide sequences into the nucleus, chloroplast nucleotide sequences have been found in both mitochondrial and nuclear DNAs (see Lonsdale 1985 for review). What has not yet been documented is the transfer of nuclear sequences into these organelles or the transfer of mitochondrial sequences into the chloroplast genome (Lonsdale 1985; Timmis and Scott 1985; Kemble, Gabay-Laughnan, and Laughnan 1985).

Stern and Palmer (1984) documented extensive and widespread homologies between plant mitochondrial and chloroplast DNAs. Restriction endonuclease digestions of these DNAs were studied for cross-homologies by means of Southern hybridization. The pervasive nature of the chloroplast DNA sequences was illustrated by the observation that every ctDNA sequence reacted with one or more mitochondrial DNA (mtDNA) restriction fragments. Stern and Palmer also presented evidence that the transfer of sequences probably occurred from chloroplast to mitochondrion and that such transfers have occurred relatively recently compared to the divergence of the species pairs studied. These authors believe it is very unlikely that these integrated ctDNA sequences play any biological role in the mitochondrion. CtDNA sequence homologies in nuclear DNA (nDNA) have also been documented. Timmis and Scott (1983, 1985) found that virtually the entire chloroplast genome showed significant homology to the nuclear genome, equivalent to five copies of ctDNA per haploid nDNA. The ctDNA is present as relatively short sequences which have integrated at several different chromosomal locations. DNA nucleotide sequence homologies between mtDNA and nDNA in maize has been reported (Kemble et al. 1983).

The frequency of nuclear-organelle transpositions as well as the mechanisms involved are unknown (Palmer 1986). (See Schuster and Brennicke 1987 for the possible involvement of RNA reverse transcription as a mechanism.) What is clear from these studies is that the simultaneous occurrence of three different genomes within a single cell is potentially destabilizing. The possibility of intergenomic invasions by alien sequences is a consequence of multiple genomic organization. Although such intergenomic sequence movements have had evolutionary significance with regard to the transfer of gene function from organelles to the nucleus, one suspects that such "successes" are very rare and that the majority of sequence movements represent a kind of insertional mutagenesis.

Mitochondrial and chloroplast genomes are very dissimilar in size and variation (table 3.4). Mitochondrial genomes of angiosperms are larger and more variable than mitochondrial genomes in other organisms. For example, mitochondria of flowering plants have genome sizes from 200–2400kb whereas 15–18kb characterizes animals, 18–78kb fungi, and 14–47kb protists. In contrast, chloroplast genome sizes are relatively constant in higher plants (Palmer 1985). Even within a single plant family, mitochondrial genomes may vary greatly in size. For example, in the Cucurbitaceae, a seven- to eightfold size variation was found (Ward, Anderson, and Bendich 1981). In spite of the large and variable size, mitochondrial genomes consist of large amounts of nonrepeated, high complexity DNA (Levings 1983). The topology of mtDNA differs greatly from ctDNA. The latter exists as a single circular unit whereas mtDNA exists as a collection of different-sized circular (and some linear) molecules resulting from intragenomic recombination within a variety of repeat sequences (see Palmer 1986 for an excellent review). In addition, various plasmid-like supernumerary DNAs have been documented in plant mitochondria (Sederoff and Levings 1985). Palmer

---

Table 3.4. Comparison of cpDNA and mtDNA evolution in angiosperms. (From Palmer 1986.)

Similarities
  Slow rate of nucleotide sequence evolution
  Constancy of base composition
  Constancy of gene content and operon structure
  Variable (optional) introns
  Slow rate of small deletions/additions

Differences
  Genome size: relatively constant in chloroplasts; highly variable in mitochondria
  Uptake of foreign sequences: rare in chloroplasts; frequent in mitochondria
  Genome conformation: single circular genetic unit in chloroplasts, no plasmids; variable number of circular (rarely linear) genetic units in mitochondria, plasmids common
  Genome arrangement: relatively constant in chloroplasts; highly variable in mitochondria

---

(1986) has pointed out that the fundamental question in angiosperm organelle evolution is, "Why are these genomes (mitochondrial) so large and so variable in size, especially relative to the fairly invariant size of chloroplast genomes?" The answer to this organelle C-value paradox is currently unknown (but see Bendich 1985 for an interesting discussion of possible answers).

Within chloroplast genomes, there is a correlation between genomic content and the degree of structural variation. The two major types of chloroplast genomes in vascular plants are shown in figure 3.15. The majority of vascular plants contain a large, ribosomal DNA-containing inverted repeat (spinach-type in figure 3.15). Species lacking the inverted repeat occur in eight genera (*Pisum, Vicia, Lathyrus, Cicer, Medicago, Trifolium, Melilotus,* and *Wisteria*) of the Fabaceae (pea-type in figure 3.15) (Palmer 1986). Chloroplast DNA rearrangements are correlated with the presence or absence of the inverted repeat.[4] Such rearrangements are generally much more common in species that have lost the inverted repeat sequence than in species in which the inverted repeat is present (Palmer and Thompson 1982). Recently the possibility that small dispersed repeats in ctDNA may promote genomic instability has been suggested. Such repeats may enhance recombination both within and between ctDNA molecules and thus may be a mechanism for generating ctDNAs with duplications, deletions, or inversions (Bowman

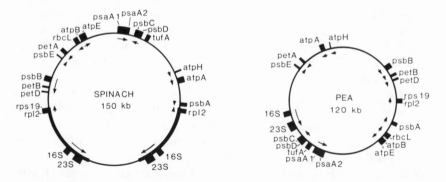

Figure 3.15. Two major chloroplast genome types. Heavy lines centered on the circle indicate the extent of the major inverted repeats in the spinach chloroplast genome. Arrows indicate the direction of transcription. See Palmer (1985) for gene designations. (From Palmer 1985.)

4. The inverted repeat in angiosperm ctDNA can vary from 10 to 76kb in different species. The most common inverted repeat size is 22 to 26kb (Palmer 1985).

and Dyer 1986). In a geranium cultivar *(Pelargonium hortorum)* probably the most extensively altered chloroplast genome so far discovered in vascular plants was reported by Palmer, Nugent, and Herbon (1987). This cultivar has a chloroplast genome size of approximately 217kb (the average land plant has a ctDNA of 150kb), the inverted repeat has spread through adjacent single-copy sequences and tripled in size, gene order has been altered as a result of at least six inversions, and two short sequences are repeated and dispersed to a number of chromosomal locations (most chloroplast genomes lack detectable dispersed repeats). Palmer, Nugent and Herbon (1987) believe that the structural variations (i.e., the inversions) are caused by the small dispersed repeats and, therefore, the mechanisms of chloroplast genomic stasis must in some way involve the suppression or elimination of small dispersed repeats.

Although the molecular mechanisms involved in chloroplast genomic stability are unknown, Palmer (1986) has speculated that the large inverted repeats may be important. Evidence that some kind of correction mechanism occurs within the repeats is based upon two observations: the two repeats are always found to be identical within an individual plant; and all spontaneous or induced mutations within the inverted sequence occur symmetrically in both repeats. Thus, a deletion induced in one inverted repeat is converted to an identical deletion in both repeats. Deletions overlapping a single copy junction can occur asymmetrically. Other characteristics of the inverted repeat include variable size in different taxonomic groups (but always including a complete set of ribosomal genes) and intramolecular recombination within the inverted repeat. Intramolecular recombination has been hypothesized to explain the finding that ctDNAs usually exist as two isomers (figure 3.16) (Palmer 1983).

Mutation in organelle genomes has two aspects: structural rearrangements (including size changes), and base pair changes (substitution deletions or additions of one to a few base pairs). For example, animal mtDNA is highly conserved with regard to large structural rearrangements and size changes, yet the rate of base pair changes is evolutionarily very rapid (Brown 1983). Correlated with the high rate of base pair changes in animal mtDNA are the absence of DNA repair phenomena and replication by γ-polymerase (a more error prone DNA polymerase) (Clayton 1982; Brown 1983). In contrast, plant mtDNA is very variable structurally but is less prone to base pair changes. Chloroplast DNA is similar to animal mtDNA with regard to structural changes but appears to be no more prone to base pair changes than plant mtDNA (Palmer 1986). Considering both structural and base pair changes, angiosperm ctDNA is the most slowly evolving genome known (Palmer 1985). The mechanisms determining stasis and change in plant organelle genomes are as yet unknown.

The physiological interactions between the nucleus and plastids are complex; whether such interactions allow some form of diplontic selection to occur (i.e., cells with mutant plastids are disadvantaged in competition with cells with nonmutant plastids) is unknown. To explore this possibility it may be best to catalogue some of these interactions. There is considerable evidence for the chloroplast control of nuclear gene function (Börner 1986). Mustard (*Sinapis alba* L.) seedlings grown in the presence of herbicides (Difunon, Norflurazon) have carotenoid synthesis inhibited. Seedlings grown in darkness or in continuous far-red light show no developmental effects of the herbicide treatment. Seedlings grown in strong white light experience photooxidative destruction of chloroplasts, the internal structure of the plastid is destroyed and plastid ribosomes and ribosomal RNAs are no longer detectable. The plastid envelope remains intact and the plastid DNA appears unchanged, but plastid

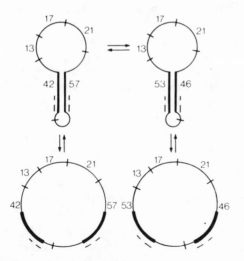

Figure 3.16. Model for chloroplast DNA isomerization via intramolecular recombination between segments of the inverted repeat. *Sal* I sites and fragments (in kb), as well as locations of the inverted repeat (long, heavy lines) and ribosomal RNA genes (short lines at the small single-copy end of the inverted repeat) are given for common bean chloroplast DNA. (From Palmer 1983.)

protein synthesis no longer occurs (Reiss et al. 1983). Using such an experimental system, Oelmüller and Mohr (1986) could show that a molecular signal from the plastid is required to allow the phyto-chrome-mediated appearance of translatable MRNAs from nuclear genes. This signal was independent of the presence or absence of chlorophyll.

The nuclear mutation *"albostrians"* of barley *(Hordeum vulgare)* induces hereditary changes in the chloroplast. Mutant plastids lack plastid ribosomes and, therefore, proteins that are normally synthe-sized on plastid ribosomes are missing (this mutant is similar to *"iojap"* in maize) (Walbot and Coe 1979). Comparing mutant and nonmutant plastids, Bradbeer et al. (1979) documented that many of the plastid polypeptides that are synthesized from nuclear gene transcripts on cytoplasmic ribosomes were absent in the mutants. These researchers postulated that a plastid synthesized RNA (that was absent in the mutant plastids) in some way controlled the syn-thesis of cytoplasmically synthesized plastid proteins. Nitrate re-ductase (a nuclear gene-encoded nonchloroplast protein) requires a chloroplast factor for its accumulation (and probably synthesis) (Börner, Mendel, and Schiemann 1986). Chloroplasts as well as plastids from non-green tissue have been shown to have an impor-tant role in amino acid biosynthesis (Miflin 1974; Miflin and Lea 1977).

Fatty acid synthesis occurs in plastids (Stumpf 1980). Two types of fatty acid synthesis occur in different organisms, Type II occurs in bacteria and is considered characteristic of prokaryotes and Type I occurs in eukaryotes (excluding plants). Plants have the prokar-yotic Type II pattern (Ohlrogge 1982). Although the site of fatty acid synthesis is the chloroplast in leaves and the proplastid in seeds, based upon studies of the above-mentioned barley mutant, Dorne et al. (1982) concluded that all the enzymes involved in fatty acid synthesis are coded by the nuclear genome.

The above examples illustrate the considerable physiological in-tegration of the plastid and nucleus in a plant cell. No doubt the absence of plastids in a plant cell would have dire physiological consequences (disregarding photosynthesis). The critical question is how much genetic damage the plastids of a cell may sustain before that cell becomes physiologically handicapped in competition with other neighboring nonmutant cells.

The path from a mutant organelle genome to mutant organelle to mutant cell to a mutant plant is not a direct one. To illustrate this point, we will restrict our discussion to ctDNA. The first question to be addressed deals with the probability of a somatic mutation in a chloroplast genome becoming ultimately fixed in all the chloroplasts of an organ (or branch or ramet). To appreciate how small this probability is, one must consider the size of the ctDNA populations involved. In table 3.5, the numbers of genome copies per plastid in young and mature leaves are listed for various angiosperms. The minimum number of genome copies per plastid may be about 30 (Boffey 1985), while the number of plastids per cell varies enormously. (Scott and Possingham 1983 report 10–200 plastids per cell in developing spinach leaves.)

Michaelis (1955a, b, 1967) mathematically analyzed the distribution of organelles during cell division using a random model. His model assumes that a cell has n plastids, each of which divides prior to cell division resulting in a population of $2n$ plastids. Each daughter cell receives n plastids randomly sampled from the pre cell division population of $2n$ plastids. Perhaps the easiest way to follow Michaelis' analysis is to use a simple numerical example. A cell has 2 plastids ($n = 2$), where one is mutant and $n - 1$ is nonmutant. Let us follow the mitotic progeny of this cell to determine the fate of these two plastid types. In table 3.6, the various combinations of mutant and wild-type plastids that may be sampled from different cytoplasm pools are listed. The matrix K and its transpose K' are also shown in table 3.6.

The matrix g_0 is

$$g_0 = \begin{bmatrix} a_1 \\ a_2 \\ a_3 \end{bmatrix} = \begin{bmatrix} 0 \\ 1 \\ 0 \end{bmatrix}$$

where a_1 is the frequency of cells without mutant plastids ($s = 0$), a_2 the frequency of cells with one mutant plastid ($s = 1$), and a_3 the

---

Table 3.5. Levels of plastid DNA in various species. (From Boffey 1985.)

| | GENOME COPIES/PLASTID | | |
	Young	Mature	PLASTID DNA/TOTAL DNA
Pea	240	170	8–12%
Spinach	200	30	21%
Wheat	1000	300	17%
Beet	100	30	8–11%

frequency of cells with two mutant plastids ($s = 2$) in generation zero. To calculate the distribution of cell types with regard to mutant and nonmutant plastids, the two matrices are multiplied,

$$\begin{bmatrix} 1 & .167 & 0 \\ 0 & .666 & 0 \\ 0 & .167 & 1 \end{bmatrix} \begin{bmatrix} 0 \\ 1 \\ 0 \end{bmatrix} \quad \begin{aligned} (1)(0) + (.167)(1) + (0)(0) &= .167 \\ = (0)(0) + (.666)(1) + (0)(0) &= .666 \\ (0)(0) + (.167)(1) + (1)(0) &= .167 \end{aligned}$$

$$K'g_0 = g_1.$$

$$g_1 = \begin{bmatrix} a_1 \\ a_2 \\ a_3 \end{bmatrix} = \begin{bmatrix} .167 \\ .666 \\ .167 \end{bmatrix}$$

where g_1 represents the frequencies of cells without, with one, or with two mutant plastids in the first generation. To calculate the distribution in the second generation,

$$g_2 = K'g_1$$

$$g_2 = (K')(K')g_0$$

Table 3.6. Various combinations of mutant (M) and nonmutant (W) plastids which may be randomly from a cell with $2n = 4$ plastids. (From Michaelis 1955a.)

Ratios of (M) to (W) Plastids in Original Cell	Ratios of (M) to (W) Plastids in Daughter Cells		
	M:W = 0:2	M:W = 1:1	M:W = 2:0
M:W = 0:2	100%	0%	0%
M:W = 1:1	16.7%	66.6%	16.7%
M:W = 2:0	0%	0%	100%

or in matrix form[a]

s \ i	0	1	2	
0	1	0	0	
1	.167	.666	.167	= K
2	0	0	1	

[a]Where s is the number of M plastids in the cell prior to plastid division and i is the number of M plastids in daughter cells. Both s and i refer to cells with n plastids.

$$K' = \begin{matrix} 1 & .167 & 0 \\ 0 & .666 & 0 \\ 0 & .167 & 1 \end{matrix}$$

thus for the rth generation

$$g_r = (K')^r g_0.$$

The labor involved in solving this equation with larger numbers of plastids and generations is immense; the reader is referred to Schensted (1958) for more sophisticated methods. The matrix K can be written in a more general form (table 3.7), where p_{si} is the probability of a cell with s mutant plastids giving rise to a daughter cell with i mutant plastids. The individual values for the elements of the K matrix can be calculated from the formula:

$$p_{si} = \frac{\binom{2s}{i}\binom{2n - 2s}{n - i}}{\binom{2n}{n}}.$$

To apply these theoretical formulations to the fate of somatic chloroplast mutations, it is necessary to have an estimate of n, the number of plastids in a meristematic initial. Although there are not a lot of data available, Kirk and Tilney-Bassett (1978) suggest that approximately 10 plastids and 60 mitochondria may be a reasonable number. In table 3.8 the statistics for the sorting out of cells with different numbers of organelles are shown. In each case, the starting situation was a cell with one mutant organelle and $n - 1$ nonmutants. The final frequency of homoplasmic mutant cells is $1/n$ and this value is approximated after $(n)(10)$ division cycles. In figure 3.17 this random sorting of mutant and nonmutant plastids is pictorially represented. In figure 3.17 the first cell, represented as a rectangle at the base of the "leaf," had either $2n = 10$ or $2n = 50$ plastids of which 50% were mutant and 50% were wild type. After 9 cycles of mitosis (2^9 cells), the distribution of homoplasmic cell lines was plotted. Three points should be drawn from these figures: (1) the frequency of homoplasmic mutant cells is related inversely to the

---

Table 3.7. The K matrix. (After Michaelis 1955a.)

$p_{si} =$	p_{00}	p_{01}	p_{02}	$\cdot$ $\cdot$ $\cdot$	p_{0n}
	p_{10}	p_{11}	p_{12}	$\cdot$ $\cdot$ $\cdot$	p_{1n}
	p_{20}	p_{21}	p_{22}	$\cdot$ $\cdot$ $\cdot$	p_{2n}
	p_{n0}	p_{n1}	p_{n2}	$\cdot$ $\cdot$ $\cdot$	p_{nn}

number of organelles; (2) because of the stochastic nature of the selection process, the pattern of homoplasmic sectors within the figure would be very different if the same computer simulations were again recalculated; and (3) for mutations occurring in very young organ primordia, the size of mutant sectors (and therefore the ease of detecting mutant tissues) is inversely related to the number of plastids per cell, i.e., larger mutant sectors are more likely for leaves with small numbers of plastids per cell. This last point is related to the organelle sorting statistics presented in table 3.8, the fewer the organelles the more likely a homoplasmic cell is selected early during the development of the organ. The earlier in development a homoplasmic cell occurs the larger the resultant sector of mutant cells, since the homoplasmic cell and its mitotic derivatives will constitute a greater proportion of the organ.

Using the random model, it can be shown that the rate of appearance of homoplasmic mutant cells is equal to the mutation rate per plastid per cell division (assuming neutral mutations with regard to cell fitness). If each cell has G copies of organelles per cell, if the mutation rate per plastid per cell division is u, then the number of

---

Table 3.8. The consequences of random organelle sorting in cells with n organelles where $n - 1$ is mutant. Organelles are assumed to form a population of $2n$ organelles from which n organelles are randomly selected. (Modified from Michaelis 1955b.)

	$n = 4$	$n = 10$	$n = 25$	$n = 50$	$n = 100$
Fewest number of cell divisions required before a cell with only mutant organelles may be selected.	2	4	5	6	7
Probability of selecting such a cell after the above number of divisions.	0.306%	0.0001%	$1.05 \times 10^{-16}\%$	$1 \times 10^{-\infty}\%$	
Overall frequency of cells with only mutant cells after many cell divisions.	25%	10%	4% $= \dfrac{100}{n}$	2%	1%
Number of cell divisions necessary to approximate the above frequencies.	40	100 $n(10)$	250	500	1000

new mutations per cell generation will be uG. The probability of homoplasmic mutant cells is the product of the mutation rate per organelle per cell division (u) times the number of organelles per cell (G) times the probability of mutant fixation or $(uG)/G = u$. Thus the rate of appearance of homoplasmic mutant cells is equal to u, the mutation rate. Although the overall frequency of homoplasmic mutant cells is not a function of the number of organelles per cell, the organelle number can influence the pattern of sectoring. High numbers of organelles per cell will result in a predominance of many small sectors whereas low numbers of organelles per cell result in large and small sectors (table 3.8; figure 3.17).

At the organism and population level, the mutation rate has also been shown to be the most important factor in governing population polymorphisms for organelle genes (Birky, Maruyama, and Fuerst 1983).

The above formulations are based upon the assumption of randomness regarding plastid distribution into daughter cells. Factors that would lessen the mixing of plastids (e.g., the movement of daughter plastid preferentially into the same daughter cell cytoplasm, insufficient cytoplasm to separate and randomize the daughter plastids) will decrease the number of cell divisions required for the generation of homoplasmic mutant cells. Opposed to this enhancement of plastid sorting, chloroplast mutants ultimately arise as mutations in a chloroplast genome. The minimum number of genome copies per plastid is estimated as 30 (Boffey 1985), thus, a meristematic initial may have at least 300 plastid genomes. The degree of randomness in genome sorting in higher plant plastid genomes during plastid division is unknown.

Two modes of plastid inheritance have been documented in the seed plants, uniparental and biparental. Kirk and Tilney-Bassett (1978), in a literature survey, list 37 genera and 46 species of angiosperms showing uniparental-maternal plastid inheritance and 6 genera and 7 species with biparental inheritance. Some angiosperm species, although predominantly uniparental-maternal, occasionally may exhibit biparental inheritance (e.g., *Antirrhinum majus* Diers 1971). The predominantly maternal plastid inheritance found in the angiosperms is in contrast to the gymnosperms where paternal plastid inheritance appears to be more common (see Kirk and Tilney-Bassett 1978 for documentation). Why maternal is more common in the angiosperms and paternal in the gymnosperms is unknown.

In a recent review Birky (1987) noted that an understanding of

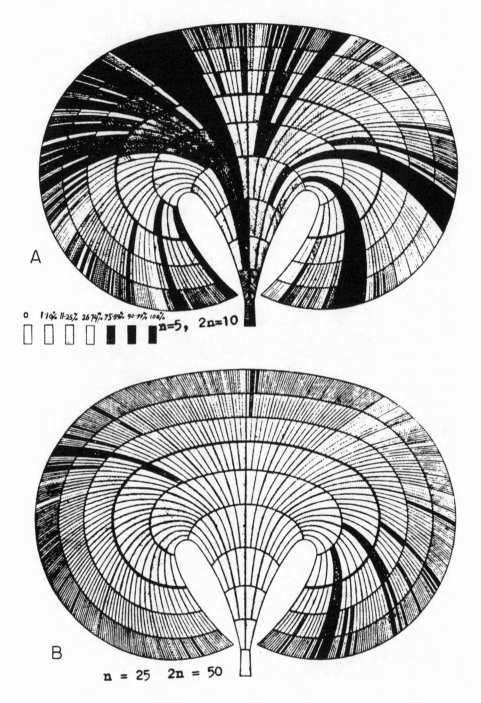

A

0 1·10% 11-25% 26-74% 75·89% 90-99% 100%

n=5, 2n=10

B

n = 25   2n = 50

chloroplast and mitochondrial evolution in plants is still very rudimentary. Quantitative data on chloroplast and mitochondrial genetic diversity in different plant species are rare. (See Birky 1987 for a review of the available studies.) This paucity of empirical data is matched by the lack of a general selection theory for organelles (Birky 1987).

PARTHIAN REMARKS

In chapters 2 and 3 an attempt has been made to discuss the salient and sometimes provocative findings of contemporary plant (and, occasionally, animal) molecular biology that relate to the general problem of genetic stability. This area of biology is one in which new and often revolutionary findings seem to be published almost weekly. Many of the topics covered in these two chapters have concluded with an "it is unknown" statement. The state of knowledge of plant molecular biology with regard to the maintenance of genetic integrity has not yet even generated a satisfactory catalogue of "most significant phenomena." We are still in the "age of exploration," so to speak. Thus, without knowing the kinds of phenomena that exist at the molecular level in plants, generalizations concerning the nature and magnitude of mutational phenomena in plants are, at best, incomplete. Therefore, the meaning of the fantastic range of genomic and chromosomal diversity in plants can only be guessed at, since the molecular basis of this diversity and inherent restrictions of various levels of organization within this diversity are essentially unknown. All we can be certain of is that very likely the current generalizations will need to be revised. Regarding the origin of this diversity, here again the nature of "mutation" *sensu lato* and

Figure 3.17. Pictorial representation of random plastid sorting during organ growth. The basal quadrangle represents the first cell, which consists of 50% mutant and 50% nonmutant plastids. Above this first quadrangle two quadrangles are drawn which represent the two daughter cells after the division of the basal cell. Above these two cells are four cells representing the division products of the two daughter cells, etc., for 9 division cycles. Shading intensity indicates the proportion of mutant plastids in a cell. A: $n = 5$, $2n = 10$ plastids per cell of which 5 are mutant and 5 are nonmutant. B: $n = 25$, $2n = 50$ plastids per cell of which 25 are mutant and 25 are nonmutant. Although the shapes of the figures are leaf-like, leaves do not grow in this manner. The significance of these figures is in demonstrating the different degrees of mutant cell clustering. (From Michaelis 1967.)

the perpetuation of these DNA alterations are important and un-solved problems.

Given that DNA alterations do occur in the nuclear or cytoplasmic genomes, these mutations occur either in somatic cells or in the immediate products of meiosis. If mutations occur in somatic cells, their fate is a function of the growth patterns of these mutant cell lineages within the plant. These patterns are a result of the anatomical and developmental characteristics of primary and secondary plant meristems. In the following chapters, apical meristems, cambia, and branching patterns will be discussed from the viewpoint of the loss or fixation and spread of somatic mutations.

CHAPTER FOUR

Unstratified Plant Meristems—
General Properties

The fate of somatic mutations in plants is a function of the kinds of meristems in which the mutant cells occur. Vascular plants perpetuate themselves through a series of primary and secondary meristems and, depending upon the growth patterns of the plant species, somatic mutations in either of these meristems may be transmitted ultimately to the meiocytes. Although in practically all vascular plant species sporogenous tissues are derived ultimately from apical meristems, many species also form sporogenous tissues that originate from root meristems (root suckers) or from the vascular cambium (adventitious buds and some kinds of cauliflory).

From the perspective of mutation, meristems may be divided into two groups, structured and stochastic. Structured meristems are those with permanent initials. Such initials follow the pattern of cell division illustrated in figure 4.1. The mitotic division of an initial always results in one daughter cell that functions as the subsequent initial (the other daughter cell may divide also and contribute cells to the body of the organ). In contrast, stochastic meristems are based upon impermanent initials. In figure 4.2 are shown the cell division patterns associated with such meristems. The initials in such meristems divide mitotically to give rise to a population of cells from which subsequent initials are selected. Initials may be selected after one or more cycles of division of the previous initials, but the rule that one of the immediate daughter cells of an initial always functions as a subsequent initial is not followed.

The distinction between structured and stochastic meristems is fundamental to any understanding of the fate of somatic mutations in plants. Structured meristems based upon more than a single initial

cell promote the long-term retention of sectorial or mericlinal chimeras as well as lessen the opportunities for competition between
mutant and nonmutant initials. In contrast, stochastic meristems
with impermanent initials may lose mutants through stochastic or
diplontic drift (Balkema 1972) and thus result in transient mericlinal
or sectorial chimeras. Stochastic meristems also maximize the possibility for intercellular competition between mutant and nonmutant cells.

Meristems are ultimately systems of cell renewal or cell multiplication. Kay (1965) considered the patterns of cell division in such
systems from the perspective of reducing the number of cell cycles
between zygote and functional stem cell (although Kay considered
mammalian somatic cell proliferation, his ideas also apply to plant
meristems). If a major source of mutations are events related to cell
division (i.e., disjunctive accidents at mitosis and replication errors
during DNA synthesis), then the heterogeneity of somatic cells will
be, at least in part, a function of the biological age of their stem cells
(initials). Biological age in this case is defined as the number of cell
cycles from zygote. Not only would the frequency of mutations in
initials be expected to be a function of biological age, but if errors
(mutations) accumulate in the DNA replication and repair pathways,

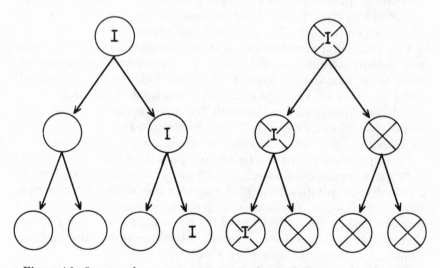

Figure 4.1. Structured meristem concept. Each initial (I) upon division always
gives rise to a daughter cell that also functions as an initial. In this example the
meristem has two initials, one of which is mutant—⊗.

the mutation rates per cell cycle also might increase as cells age (Orgel 1963; Kay 1965; see chapter 2).

Two general patterns of cell multiplication were envisioned by Kay, tangential and logarithmic. In figure 4.3 these cell-multiplication patterns are illustrated. Consider the problem of a single cell giving rise to a population of 8 cells. The logarithmic pattern of cell multiplication results in 8 cells, each of which has a biological age of 3 cell division cycles. In contrast, the tangential pattern of cell multiplication gives rise to a variable population of cells in terms of biological ages (i.e., out of the 8 cells, 2 cells have a biological age of 7 cell division cycles and the remaining 6 cells are 1 through 6 cell divisions in age) with an average biological age of 4.38 cell cycles. It is clear from this simple comparison that meristems with tangential patterns of cell multiplication give rise to cells which are biologically older than meristems with logarithmic patterns. One might argue also that the former meristems give rise to genetically more variable cells than the latter.

If there is a selective advantage in maintaining the genetic integrity of cells within an organism, one would predict a reduction in the rates of division of initials and the concomitant increase in division rates of daughter cells. In this way cell numbers of a tissue or organ could be increased in a predominantly logarithmic mode

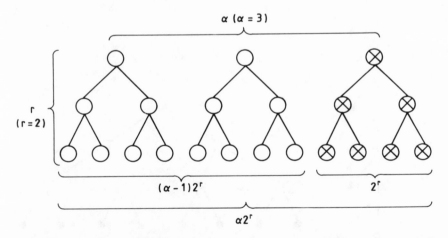

Figure 4.2. Stochastic meristem concept. Initials divide and give rise to a pool of cells from which subsequent initials are randomly selected. In this example the meristem has three initials ($\alpha = 3$), one of which is mutant, $\otimes$. All initials go through two cycles of mitosis ($r = 2$) resulting in a pool of $\alpha 2^r$ cells from which new initials are chosen. (From Klekowski and Kazarinova-Fukshansky 1984.)

and the tangential divisions of initials minimized. Meristems with mitotically quiescent subpopulations of cells which may be later induced to divide and function as initials are not uncommon in plants. It should be noted, however, that mitotic quiescence need not be absolutely qualitative; quantitative differences in division rates for various subpopulations of cells within a meristem also may be effective in reducing mutational divergence. Such cell-multiplication systems have been called skewed asynchronous by Kay (see figure 4.4).

APICAL MERISTEMS

In vascular plants, apical meristems include a great variety of organizations and patterns of cell division ranging from simple meristems with a single apical cell, characteristic of some pteridophytes, to complex tunica-corpus systems (stratified meristems) found in

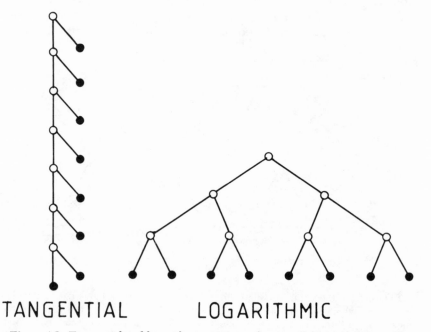

TANGENTIAL   LOGARITHMIC

Figure 4.3. Tangential and logarithmic patterns of stem cell divisions. (From Kay 1965.)

most angiosperms (Popham 1951). Within this spectrum of anatomical and ontogenetic diversity are many features which may either change the frequency of mutations that have accumulated in an apical initial or influence the effectiveness of diplontic selection within the population of cells that constitute the apical meristem. Characteristics such as the degree of permanence of the apical initials (i.e., structured vs. stochastic meristems), the number of apical initials, the size of the cell pool in the apical meristem, patterns of cell division in the meristematic cell pool, presence of a *méristème d'attente* and patterns of DNA strand segregation are among the many facets of apical meristem development that have relevance in the accumulation of mutations in apical meristems. These characteristics and others will be discussed in some detail in this and the following chapters.

UNSTRATIFIED APICAL MERISTEMS

Apical Cell-Based Meristems (Structured Meristems)

Apical cell-based meristems are found in the shoot apices (and other meristems) of many pteridophytes, e.g., the psilopsids, some lycopsids, the sphenopsids (including fossil forms, Melchior and Hall 1961; Good and Taylor 1972), some eusporangiate ferns, and the leptosporangiate ferns (Bierhorst 1971). Although there has been debate as to whether the apical cell is meristematic in such meristems (McAlpin and White 1974), the patterns of cell walls in the

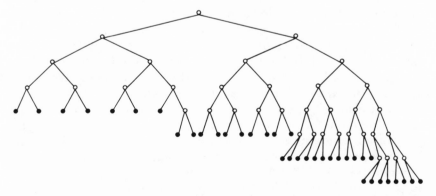

Figure 4.4. Skewed asynchronous patterns of stem cell divisions. (From Kay 1965.)

apex can be best accounted for by assuming that the apical cell is at least occasionally meristematic (Bierhorst 1977; Hébant, Hébant-Mauri, and Barthonnet 1978). In figure 4.5 the apical organization of the sphenopsid *Equisetum* is shown. Cell divisions of the apical cell result in the formation of recognizable cell packets or merophytes. Such multicellular domains show parallel microfibrillar cellulose alignment that reflect successive divisions of the apical cell (Lintilhac and Green 1976) and are perhaps the best evidence for the meristematic role of the apical cell.

In structured unstratified meristems an understanding of the fixation or loss of somatic mutations is relatively straightforward. The fixation of a mutation is assured if an apical initial mutates (of course this assumes that the mutation occurs prior to chromatid duplication and, thus, probably will not segregate during mitosis). If such a mutation occurs in a meristem with a single apical cell, all subsequently derived tissues will consist of cells carrying the mutation. In pteridophytes, spores produced by such sporangia will segregate in 1:1 ratios for mutant and nonmutant alleles (Klekowski 1984). Persistent apical chimeras are impossible in such meristems. The relative developmental rigidity of apical meristems based upon single tetrahedral apical cells suggests that intercellular competition within the meristem is minimized. Thus a mutated apical cell may not be displaced readily by a nonmutant adjacent cell, even if the apical cell is physiologically handicapped.

Although apical cell-based meristems would be expected to retain disadvantageous mutations more readily than other kinds of meristems, perhaps this is in part compensated for by mechanisms that reduce mutation rates. Recent studies by Kuligowski-Andres and Tourte suggest that cell division in apical cells in *Marsilea vestita* embryos has been modified to reduce the mutation accumulation in these cells (Kuligowski-Andres, Tourte, and Faivre-Baron 1979; Kuligowski-Andres and Tourte 1979; Tourte, Kuligowski-Andres, and Barbier-Ramond 1980). Using tritiated thymidine-labeled gametes, two patterns of DNA synthesis in apical cells were noted. When the zygote resulted from the fertilization of a radioactively labeled egg and a nonlabeled spermatozoid, analysis of silver grain counts of autoradiographs of different aged embryos showed a continuous diminution of grains per cell with mitosis. In contrast, when a labeled spermatozoid fertilized a nonlabeled egg, radioactivity was retained selectively by the apical cells up to at least the 5,000-cell stage of the embryo. These researchers have suggested that their data support the hypothesis of nonrandom distribution of paternal

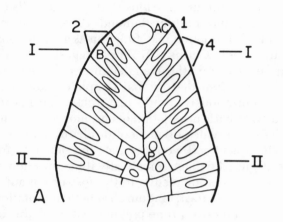

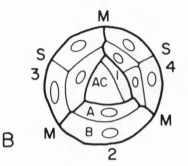

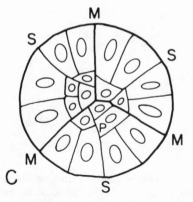

Figure 4.5. A: Median longitudinal section of the shoot apex showing merophytes 1, 2, 4. Merophyte 3 is out of the plane of section. B: Typical transverse section of apex at level I-I. C: Typical transverse section at a lower level (II-II) of shoot apex. A = acropetal cell; AC = apical cell; B = basipetal cell; M = boundaries of merophyte sectors; P = developing pith; S = sextant walls. (From Gifford and Kurth 1983.)

chromatids in the mitotic divisions of the apical cells (Tourte, Ku-ligowski-Andres, and Barbier-Ramond 1980). Thus the paternal copy of genetic information may function as a master copy retained within an apical cell. Cairns (1975) has suggested that during DNA replication the newly synthesized daughter polynucleotide chain is more error-prone than the original template chain and that stem cells which continuously divide may retain the master template chains. The systematic orientation of chromosomes so that the chromatids with the template polynucleotide chain are drawn to the pole that forms the apical cell would reduce the mutation frequency in these cells. Figure 4.6 diagrammatically illustrates how nonrandom orientation of chromatids results in the loss of mutant cells to the somatic tissues. In contrast, random chromatid orientation can result in newly replicated error-prone polynucleotide chains being incor-

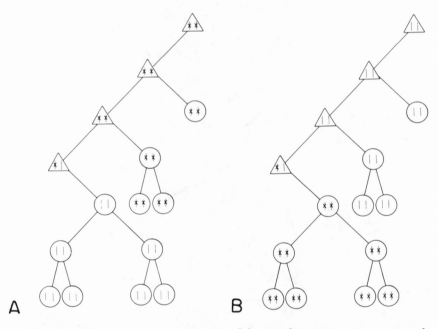

Figure 4.6. Diagrammatic representation of chromatid segregation in a structured meristem based upon an apical cell (△). The (✳) marks the newly synthesized DNA strand. This strand is believed to be more mutation prone. A: Random chromatid segregation, in this case the newly synthesized strand was segregated into the initial during the first mitotic division. Note how all subsequent somatic cells carry copies of the mutation prone strand. B: Nonrandom chromatid segregation, in this case the newly synthesized mutation prone strand is immediately eliminated to the somatic cells.

porated into the apical cell genotype. Once the apical cell has a mutant genotype, all subsequently derived cells and tissues will have the mutant genotype. These results are provocative, to say the least and, if substantiated, will have far-reaching implications. Apical cells are ubiquitous in both embryonic and adult meristems of many pteridophytes. If paternal template conservation occurs in these apical cells, mutations of the maternal genome should be more common than those of the paternal genome within a clone or population of sib ramets.

Other Structured Apical Meristems

Unstratified meristems with permanent apical initials may occur in some pteridophytes. Bierhorst (1971) reports that such apices occur in some members of the Lycopodiaceae, Selaginellaceae, Isoetaceae, and older plants in the Marattiales. As noted by Bierhorst, it is relatively easy to examine median longitudinal sections of shoot tip and point out the initial zone at the crest of the dome, but only in a carefully prepared set of cross-sectional camera lucida drawings can one extrapolate cell lineages back to initials (figure 4.5). In most cases where multiple apical initials have been postulated, whether the initials are permanent (i.e., upon division one daughter cell always remains and functions as an initial) or impermanent is unknown. Unstratified apical meristems also occur in many gymnosperms, although some degree of stratification occurs in some taxa (Popham 1951). Johnson (1951) pointed out that the major difference with respect to apical organization between gymnosperms and angiosperms is that in the former the surface layer not only increases in area but adds periclinal derivatives to the zones within the apex (see figure 4.7). Whether gymnosperm apices have permanent apical initials is open to debate. A consequence of permanent apical initials in such apices would be occurrence of very long and persistent sectorial chimeras. Considering the relatively frequent occurrence of albino somatic mutations in seed plants, the lack of horticultural cultivars with such sectorial patterns in the various cultivated gymnosperms argues against the permanence of apical initials in these plants.

Increasing the number of apical initials within a structured apical meristem can result in increased somatic mutation buffering when mutation rates are high. It seems reasonable to infer that apical meristems consisting of mutant and nonmutant initials are advantaged over apices consisting solely of mutant initials. This advantage

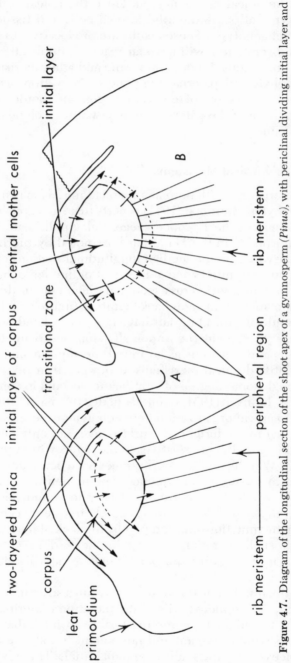

Figure 4.7. Diagram of the longitudinal section of the shoot apex of a gymnosperm (*Pinus*), with periclinal dividing initial layer and central mother-cell zone. Illustrated is the relation of promeristem to derivative regions below it. Arrows indicate direction in which cells are given off from regions within the apical meristem. (From Esau 1960.)

may be an enhanced competitive vigor or a lower frequency of defective gametes. A measure of the selective advantage of apical initial redundancy is the ratio of apices with one or more nonmutant initials (i.e., apices with two or more initials) to mutant-free apices which have single initials. The mathematics of apical initial redundancy follows the pattern outlined by Reanney, MacPhee, and Pressing (1983) for the general problem of message redundancy (chapter 2). Thus, for apices with α initials, where u is the mutation rate per some biological time unit, the frequency of apical meristems with different combinations of mutant and nonmutant initials is given by the binomial

$$[(u) + (1 - u)]^{\alpha}.$$

The frequency (F) of an apical meristem with at least one nonmutant initial is

$$F = 1 - u^{\alpha}.$$

The selective advantage (A) of apical initial redundancy in structured apical meristems is

$$A = F/(1 - u)$$

where $1 - u$ is the frequency of mutant-free single apical initial meristems. In table 4.1 the calculated values of A for various combinations of mutation rates (u) and apical initials (α) are given. It is clear that any selective advantage accruing from apical initial redundancy is totally dependent upon the mutation rate. Mutation rates in these calculations must be based upon sets of genes rather than single genes. In chapter 2 the literature on forward mutation rates for deleterious mutations in plants was reviewed. Mutation rates per ramet generation of 1×10^{-2} and per generation in annuals of 4×10^{-2} to 2×10^{-3} have been reported. For these values of u,

Table 4.1. The effect of apical initial redundancy in structured apical meristems. The values of A are given for various mutation rates (u) and apical initial number (α).

	u =	.01	.1	.2	.5
	2	1.01	1.10	1.2	1.5
α =	3	1.0101	1.11	1.24	1.75
	4	1.010101	1.111	1.248	1.875

apical initial redundancy in structured apical meristems confers very little selective advantage. In fact, if these values of u are realistic (and it should be noted that there is a very small data base), then the occurrence of a single apical initial in the structured apices of most pteridophytes is understandable. As the following discussions will show, increasing the number of apical initials confers mutation buffering only in those apical systems that have evolved more stochastic development.

Stochastic Apical Meristems

Unstratified meristems with impermanent apical initials may occur in some pteridophytes and many gymnosperms. The caveat "may" is used because the determination of whether apical initials are permanent or impermanent is generally difficult for both technical and philosophical reasons. The technical problems associated with studying the stability of apical initials have been outlined already, without the occurrence of permanent sectorial chimeras or detailed studies of cell lineages as outlined by Bierhorst (1971), very little can be said about the stability of apical initials.

On the philosophical side knowledge about apical initials is often hindered because anatomists have adopted a typological apical initial definition. To paraphrase noted anatomists: "The term 'initials' in meristems refers to cells which *always* remain within meristems. When an initial divides, one of the daughter cells continues to fulfill the original function of an initial, whereas the other daughter cells, after several divisions, undergo differentiation and maturation" (Fahn 1974; Bierhorst 1971; Esau 1977; and the Dictionary of Botany, Blackmore and Toothill 1984). Mayr (1963) has clearly analyzed the pitfalls of such a philosophical approach in biology with its emphasis on the idealized concept rather than the vagaries of reality. The typological apical initial definition often leads to an interesting pattern of circular reasoning. For example, in practically all vascular plants, primary apical meristems occur in which a promeristem can be distinguished. By definition the promeristem consists of the apical initials together with cells derived from them. Applying the typological apical initial definition, therefore, almost all vascular plants have permanent apical initials! The typological apical initial definition is often tempered by statements about the possibility of displacements of apical initials during apical growth. If such displacements are common, apical initials cannot be considered permanent. (But see Clowes 1961; Newman 1965; and Soma and Ball 1964 for a more flexible view of apical initials.)

Vascular plant species have evolved a very diverse array of apical meristem types. From the point of view of their respective apical initials, these meristems should be described and understood without the imposition of a typological concept. No doubt some species will be found that have quite permanent initials, while in other species impermanent initials will be the rule. The characterization of these meristems and their associated taxonomic and organographic correlates will further our understanding of the significance of somatic mutation and development in vascular plants.

The opposite of structured meristems with permanent initials is the stochastic meristem concept. The stochastic and structured meristem concepts are two poles of a continuum on which various plant apical meristems, because of organographic and topological constraints, occupy intermediate points (see figure 4.2). Apical initials in such meristems are considered transitory. Such initials divide mitotically to give rise to a pool of daughter cells from which subsequent apical initials are randomly sampled. Two characteristics may vary in such meristems: the number of apical initials (α), and the number of mitotic cycles prior to the random selection of the subsequent set of initials (r). It is important to note three points of symmetry in such a formulation: (1) the number of apical initials is constant; (2) all initials divide the same number of times between cycles of initial selection; and (3) only cells from the resulting cell pool, $\alpha 2^r$, are a source of subsequent initials. These three points of symmetry are primarily a mathematical convenience as in nature one would expect all might vary during the growth of a meristem.

The number of apical initials may change as a plant becomes older. For example, in *Arabidopsis*,[1] mutation studies indicate that the number of apical initials in the seed embryo ultimately contributing their genotypes to the sporogenous tissue in the adult plant is two (Li and Rédei 1969), whereas a nine-day-old seedling has five to eight such initials (Shevchenko et al. 1975a and b as cited in Relichova 1977). Williams (1975) documented many instances where the area of the apical dome increased as much as twentyfold during the embryo/seedling transition. Such size increases probably also reflect changes in α and r. Cell and promeristem displacements during various plastochrons may bring cells into the promeristem from outside of the cell pools shown in figure 4.2. Paolillo and Gifford (1961) noted that in the gymnosperm *Ephedra* (see note 1) the number and identity of the initials may fluctuate during the

---

1. Both *Arabidopsis* and *Ephedra* have a tunica and, therefore, are stratified apical meristems (chapter 5). The apical initials referred to are the subapical initials.

growth of an apex. Therefore the number of mitotic cycles between subsequent initial selections may not be uniform for the population of cells sampled for the subsequent set of initials. Thus it is not hard to visualize exemptions and modifications in the threefold symmetry of the stochastic model, but the results of an analysis of such a "simple" system allows the formulation of a set of generalizations about meristems without permanent initials. How well actual measurements of meristems fit these generalizations may then be an index of the structured or stochastic properties of these meristems.

Klekowski and Kazarinova-Fukshansky (1984a) analyzed such stochastic meristems in terms of the loss or fixation of selectively neutral mutations. The starting condition in each case was a meristem with one mutant and $\alpha - 1$ nonmutant initials. A mutant initial was one that carried a recent somatic mutation in the heterozygous condition, i.e., Aa or $+/-$. The mutant initial mitotically transmitted its heterozygous genotype to both daughter cells. Cells that were heterozygous for the mutation were not selectively advantaged or disadvantaged with respect to nonmutant cells within the context of the meristem. Apical shoot meristems were analyzed in terms of the number of apical initials (α) randomly selected from a pool of potentially meristematic cells resulting from the division of the previous initials. Since r divisions occurred between two subsequent selections the number of potential initials for each selection was equal to $\alpha 2^r$ where 2^r were mutants and $(\alpha - 1)2^r$ were nonmutants. The simplest stochastic meristem to visualize is one in which there are two apical initials ($\alpha = 2$) and new apical initials are selected randomly after one mitotic cycle ($r = 1$) from a pool of four cells. In figure 4.8 such a meristem is diagrammed. From a starting condition of one mutant and one nonmutant initial, after one mitotic cycle three different kinds of meristem configurations may result—the meristem may consist of two nonmutant initials, two mutant initials, or one mutant and one nonmutant initial. The first configuration is termed mutation fixation, the second mutation loss, and the third is the perpetuation of the sectorial chimera. It is obvious that the probability of loss or fixation of a mutation is a function of the number of selection cycles of growth and that the fixation asymptote after many growth cycles is 0.5. Similarly the mutation loss asymptote is also 0.5. The continued growth of a stochastic meristem ultimately results in the loss of the chimeric condition.

The growth of more complex unstratified meristems ($\alpha \geqslant 2$, $r \geqslant 1$) also was modeled (see Klekowski and Kazarinova-Fukshansky

1984a for the mathematical details). Starting with a similar situation, one mutant initial and $\alpha - 1$ nonmutant initials, the fixation asymptote was found to be $1/\alpha$ regardless of the r values (see figure 4.9). Thus if a meristem had one mutant apical initial and two nonmutants ($\alpha = 3$), then with continued growth the probability that the meristem would have only mutant initials is $1/3$ or, conversely, the probability that the meristem would have only nonmutant initials is $2/3$ or $(\alpha - 1)/\alpha$. The value of r, which determines the size of the cell pool from which apical initials are selected, has no effect on the fixation or loss asymptotes.

These fixation asymptotes give an insight on how the number of apical initials influences the frequency of mutations fixed by a growing genet (tree, clone, etc.). The total number of mutations (selectively neutral) fixed in a growing genet is given by the following equation,

$$(R)(\alpha)(\text{Age})(\text{Fixation probability})$$
$$= \text{Total number of mutations fixed per genet}$$

where R is the mutation rate for initials per unit of biological time, and age is the number of biological time units (ramet generations, cell cycles, etc.)

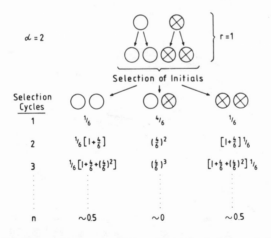

Figure 4.8. Diagrammatic representation of α and r in terms of stochastic apical meristem growth. The beginning state has two initials, one of which is heterozygous for a new mutation ($\otimes$). All initials divide once giving rise to a cell pool of $\alpha 2^r$ cells from which new initials are chosen. The probabilities of two nonmutant initials ($\bigcirc \bigcirc$), two mutant initials ($\otimes \otimes$), and one mutant and one nonmutant ($\otimes \bigcirc$) are given for each cycle. (From Klekowski and Kazarinova-Fukshansky 1984.)

In figure 4.10, a two-dimensional representation of a stem is shown. The frequency of mutations fixed is related to the number of branch meristems that have mutant initials as well as whether the main meristem has fixed or lost the mutation. The number of mutant branch meristems in a sectorial chimera is $1/\alpha$ times the number of branches, assuming that branches are randomly distributed about the stem regardless of the somatic cell genotypes. The probability that the main meristem fixes the mutant is also $1/\alpha$; such a meristem subsequently gives rise to branch meristems that are also mutant. Thus, if we visualize a seed in which one of the three apical initials in the embryo is mutant, then in the subsequent tree (or population of ramets for various other clonal forms) one-third of the branches (or ramets) will be mutant.

If we consider mutations occurring throughout the life of the main stem meristem, then for selectively neutral mutations, the total number of mutations fixed is independent of the number of apical initials in the meristem. This is because increasing the number of apical initials has two opposite effects in terms of mutation: the number of targets for mutagenesis is increased and the fixation asymptotes are decreased. As is shown in the following equation, these factors exactly cancel each other, thus

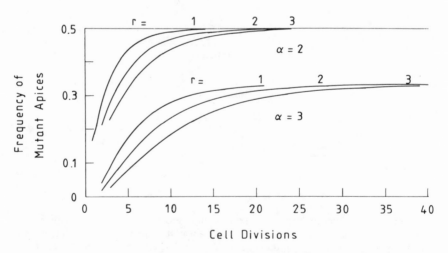

Figure 4.9. The fixation of a mutation in apices with different α and r values as a function of the number of cell divisions after the origin of the mutation. Fixation occurs when all of the initials of a meristem are of the mutant genotype.

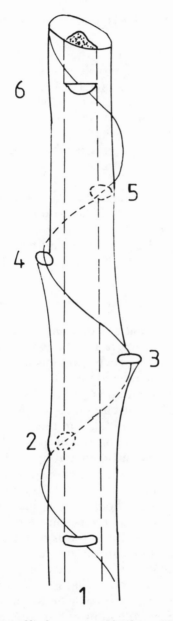

Figure 4.10. Distribution of buds on a stem that has a 1/3 sectorial chimera. The stem has a five-ranked leaf arrangement, the axillary buds are placed over each other every two turns around the stem. Axillary buds 1 and 6 occur in the mutant sector.

$(R)(\alpha)(\text{Age})(1/\alpha)$ = Mutations fixed per main stem apical meristem

$(R)(\text{Age})$ = Mutations fixed per main stem apical meristem.

The above relation is clear for the loss or fixation of mutations in meristems; what may not be so obvious is that for sectorial chimeras a similar relation applies as well. As is shown in figure 4.10, the number of branches having fixed a mutation in a sectorial chimera is proportional to the volume of the stem that is composed of mutant cells. If an apical meristem has α initials, then the volume of the sector of mutant cells (assuming for the moment a structured meristem) is $1/\alpha$ and therefore $1/\alpha$ of the branches will have fixed the mutant.

A mutation, therefore, may be fixed in two ways, either through the loss of nonmutant initials in the main apical meristem or through the origin of branches in the mutant sector on the stem. For either of these cases the probability of mutant fixation is equivalent, $1/\alpha$. Substituting this value into the previous equation gives the following relation,

$R(\text{Age})$ = Number of neutral mutations fixed per genet.

Since the fixation of mutations in branches is related to the persistence of the sectorial chimera in the stem, it is interesting to gain some insight on how long such chimeras may last. The persistence of a sectorial chimera is related to both the number of apical initials (α) and the number of mitotic cycles between selection of initials (r). The actual length of the chimera is also a function of the number of divisions of the apical initials and their derivatives per node. In table 4.2 are given the probabilities of sectorial chimera persistence in different kinds of stochastic unstratified apical meristems.

Selectively neutral somatic mutations are assumed to have viabilities (and fitness values in terms of mitotic progeny) similar to nonmutant cells. Selective neutrality in this case is within the context of the meristem only. There is some evidence that even mutations deleterious to the plant in terms of survival and competition are selectively neutral within the internal environment of a meristem. Langridge (1958) proposed that in the case of deleterious mutations, diplontic selection would be lessened for those genes inactive in the apical meristem as well as for those responsible for diffusible metabolites. Mutations of the latter could be maintained by cross-feeding from nonmutant cells within the meristem. Meinke and Sussex (1979a and b) reported that cells heterozygous for embryo-lethal mutants did not appear to have a growth disadvantage in apices that were a chimera for heterozygous mutant and wild-type initials.

Of course selective neutrality cannot be assumed for all somatic mutations. Competition between mutant and nonmutant cells (or diplontic selection) occurs as well. Two cases may be envisioned— the mutant cells may be either more or less fit than the wild-type cells. As discussed in chapter 1, the fitness of a mutant cell may or may not correlate to total organism or genet fitness. In the following analysis, only cell fitness within a meristem is considered.

Klekowski and Kazarinova-Fukshansky (1984b) analyzed the fate of selectively disadvantageous mutations in apices with stochastic meristems. Starting with one mutant initial and $\alpha - 1$ nonmutant initials where the viability of the mutant initial was less than one ($CF < 1$) and the viability of the nonmutant was one ($CF = 1$), the mutation fixation asymptotes for various combinations of α and r were calculated. In table 4.3 these fixation asymptotes are presented. The total number of disadvantageous mutations fixed in a genet is given by the following familiar relation:

$$(R_b)(\alpha)(\text{Age})(\text{Fixation asymptote}) = \text{Mutations fixed}$$

where R_b is the mutation rate for disadvantageous mutations per apical initial per unit of biological time. A useful index of the propensity of an apical meristem to accumulate disadvantageous mutation is the $(\alpha)(\text{Fixation asymptote})$ product. In table 4.4 these products are given.

It is also possible that cells heterozygous for somatic mutations may have greater cell fitness (CF) values than adjacent wild-type cells within a meristem. Klekowski, Mohr, and Kazarinova-Fukshansky (1986) have analyzed the case where the viability of the mutant cells is one ($CF = 1$) and the nonmutant cells is less than one ($CF < 1$) in an effort to model such a situation. These altered viabilities were substituted into previously cited equations describing stochastic apical meristem growth (Klekowski and Kazarinova-Fukshansky 1984b). If one begins with meristems with $\alpha - 1$ nonmutant initials and one mutant initial, the asymptotes after many cycles of stochastic growth for different values of α and r were calculated. Since the mutant cells have superior viabilities, the nonmutants are the disadvantaged genotypes. In table 4.5 the products of the fixation asymptotes for apices with only mutant initials times number of apical initials, (fixation asymptotes)(α), are given.

Comparing these products for the fixation of disadvantaged and advantaged mutations (tables 4.3–4.5) reveals some interesting patterns with regard to apical development and mutation. If we consider what characteristics promote the loss of mutations, i.e., have lowest (fixation asymptotes)(α) products, then as shown in figure 4.11, there are two opposing trends. Increasing the number of apical initials (α) promotes the loss of disadvantaged mutations ($CF < 1$ for mutant cells) and decreasing the number of apical initials promotes the loss of advantaged mutants ($CF < 1$ for nonmutant cells). A parallel pattern exists for changes in the size of the cell pool from which initials are selected. The major factor determining the size of

Table 4.2. Persistence of mericlinal and sectorial chimeras in stochastic meristems. The probability (or frequency) of a chimera (M) is given in terms of the number of cell generations of the apical initials and immediate derivatives, the number of cycles of apical initial selections, and the number

Cell Generations	Selection Cycles	Nodes	M
$\alpha = 2, r = 1$			
4	4	16	0.198
8	8	32	0.039
12	12	48	0.007
$\alpha = 2, r = 2$			
4	2	16	0.327
8	4	32	0.107
12	6	48	0.035
16	8	64	0.011
20	10	80	0.004
$\alpha = 3, r = 1$			
4	4	16	0.410
8	8	32	0.168
12	12	48	0.068
16	16	64	0.028
20	20	80	0.012
24	24	96	0.004
$\alpha = 3, r = 2$			
4	2	16	0.529
8	4	32	0.279
12	6	48	0.148
16	8	64	0.078
20	10	80	0.042
24	12	96	0.024
28	14	112	0.012
32	16	128	0.006

this pool is r, the number of divisions between periodic selection of cells as subsequent initials. Again, increasing r promotes the loss of disadvantaged mutations, and decreasing r promotes the loss of advantaged mutations (compare the fixation asymptote products in tables 4.3–4.5). It should be stressed again that advantaged mutations are defined as having viabilities greater than wild-type cells within the context of the meristem. It is important to note that it is very likely that such mutations probably do not confer increased fitness to the organism. A more suitable model may be "cancer"-type mutations that result in decreased overall organismal fitness

of nodes of persistence (where four nodes occur every cell generation). The starting condition was $\alpha - 1$ nonmutant initials and one mutant initial and the mutant initials are selectively neutral. (From Klekowski, Mohr, and Kazarinova-Fukshansky 1986.)

	Cell Generations	Selection Cycles	Nodes	M
$\alpha = 4, r = 1$				
	4	4	16	0.479
	8	8	32	0.257
	12	12	48	0.140
	16	16	64	0.075
	20	20	80	0.041
	24	24	96	0.022
	28	28	112	0.013
$\alpha = 4, r = 2$				
	4	2	16	0.659
	8	4	32	0.478
	12	6	48	0.352
	16	8	64	0.263
	20	10	80	0.190
	24	12	96	0.141
	28	14	112	0.102
	32	16	128	0.075
	36	18	144	0.055
	40	20	160	0.041
	44	22	176	0.029
	48	24	192	0.022
	52	26	208	0.016
	56	28	224	0.012
	60	30	240	0.009

(see chapter 7). Thus apical anatomy from the viewpoint of mutation buffering may represent a compromise or conflict between two opposing needs. The anatomical characteristics that promote the loss of mutations with lower viabilities encourage the fixation of mutations with higher viabilities and, conversely, the characteristics that promote the loss of mutations with high viabilities encourage the fixation of mutations with lower viabilities.

Recently there has been speculation about the possibilities of within-individual selection in vascular plants (Whitham and Slobodchikoff 1981). Basically the idea is that if ramets or portions of

Table 4.3. Fixation asymptotes for buds with only mutant initials for the case where cells heterozygous for mutations have viabilities or cell fitness values of less than one ($CF < 1$). (From Klekowski, Mohr, and Kazarinova-Fukshansky 1986.)

a	r	$CF = 0.95$	0.90	0.85	0.80	0.75
2	1	0.449	0.397	0.346	0.296	0.250
	2	0.427	0.357	0.291	0.232	0.181
3	1	0.268	0.207	0.154	0.109	0.075
	2	0.245	0.170	0.111	0.069	0.041
4	1	0.179	0.120	0.075	0.044	0.024
	2	0.157	0.089	0.046	0.022	0.009

NOTE: a = The number of apical initials
   r = The number of mitotic divisions between selections of apical initials
   CF = Cell fitness

Table 4.4. (Fixation asymptotes) α products for the fixation asymptotes in table 4.3. (From Klekowski, Mohr, and Kazarinova-Fukshansky 1986.)

a[a]	r[a]	CF[a] $= 0.95$	0.90	0.85	0.80	0.75
2	1	0.898	0.794	0.692	0.592	0.500
	2	0.854	0.714	0.582	0.464	0.362
3	1	0.804	0.621	0.462	0.327	0.225
	2	0.735	0.510	0.333	0.207	0.123
4	1	0.716	0.480	0.225	0.176	0.096
	2	0.628	0.356	0.184	0.088	0.036

[a]See table 4.3.

ramets are genetically and phenotypically variable and if these differences result in differential growth and mortality among the ramets of a genet, then somatic selection may be of evolutionary significance. These ideas may be extended to the level of diplontic selection within an apical meristem. Assuming that high cell fitness (CF) within the meristem is correlated with an increase in organism fitness of either the ramet or genet, then one can predict apical meristem configurations that might enhance this kind of somatic selection. Increasing the number of apical initials (α) will increase the number of targets for mutation as well as promote the fixation of

Table 4.5. (Fixation asymptotes) α products for the fixation asymptotes for meristems that consist of only mutant initials. The viability of mutant initials is one ($CF = 1$) and the viability of nonmutants is less than one ($CF < 1$). (From Klekowski, Mohr, and Kazarinova-Fukshansky 1986.)

a^a	r^a	$CF^a = 0.95$	0.90	0.85	0.80	0.75
2	1	1.102	1.206	1.308	1.408	1.500
	2	1.146	1.286	1.418	1.536	1.638
3	1	1.212	1.440	1.680	1.911	2.130
	2	1.296	1.608	1.908	2.178	2.403
4	1	1.324	1.696	2.088	2.472	2.824
	2	1.456	1.956	2.448	2.876	3.220

[a]See table 4.3.

MUTANT VIABILITY IS LESS THAN WILD TYPE	MUTANT VIABILITY IS GREATER THAN WILD TYPE
Mutant Loss Is Promoted By:	
1. Increasing the number of apical initials, high α	1. Decreasing the number of apical initials, low α
2. Increasing the cell pool from which initials are selected, high r	2. Decreasing the cell pool from which initials are selected, low r

Figure 4.11. Characteristics which promote the loss of mutant cells in stochastic meristems. Mutants may be either less viable (Mutant CF < wild type CF) or more viable (Mutant CF > wild type CF) than wild-type cells.

advantageous mutations. Increasing the number of divisions (r) between selections of initials will also increase the probability of fixing such advantageous mutations. Thus one would predict that somatic selection would be more effective in stochastic, unstratified apical meristems with large volumes (due to the higher α and r values).

It is important to consider whether some plant species may have evolved meristems with characteristics that enhance somatic selection and evolution. As an untested generalization, it is probable that developmental and ontogenetic constraints are more important factors in determining apical meristem sizes and internal topologies than the consequences of somatic mutation (see, for example, the range of apical meristem sizes found in the Cactaceae,) (Mauseth 1978a, b; Mauseth and Niklas 1979). Given this caveat, it is interesting to note that the angiosperms have evolved apical meristems that almost seem to have been designed to retain somatic mutations. The characteristics of such meristems are the subject of the following chapter.

CHAPTER FIVE

Stratified Meristems—General Properties

STRATIFIED APICAL MERISTEMS

Stratified apical meristems or meristems with a tunica-corpus organization have one or more peripheral layers of cells (the tunica) dividing primarily in the anticlinal plane forming a mantel over the internal corpus. The apical meristem is thus differentiated into two regions distinguished by method of growth—the peripheral tunica in which anticlinal cell divisions predominate, and the internal corpus in which cell divisions occur in various planes. The tunica therefore undergoes surface growth whereas the corpus shows volume growth. Stratified meristems are in reality composed of semi-autonomous component meristems (the individual tunica layers as well as the internal corpus) with limited cell displacements beween component meristems.

Stratified apical meristems occur in a variety of topological configurations depending upon the species. In figure 5.1 the most commonly illustrated configuration is shown. The dome-like shoot apex of *Cassiope lycopodioides* has a two-layered tunica (Lɪ, Lɪɪ) atop a corpus (Lɪɪɪ). The arrangement of leaves is decussate (Hara 1975). In figure 5.2 the flattish shoot apex of *Clethra barbinervis* is shown. In this species a single tunica layer (Lɪ) is atop the corpus (Lɪɪ) (Hara 1971). In both *Cassiope* and *Clethra* the apical meristem maintains a relatively constant shape and position from plastochron to plastochron. In marked contrast to this relative constancy are apical meristems that oscillate in pattern with leaf initiation. In figure 5.3 an example of such an oscillating pattern is shown. In this example *(Michelia fuscata)* vegetative shoots may have either radial or dorsiventral leaf arrangements, in both types of shoots meristem oscillations occur (Tucker 1962). Thus, stratified meristems represent a variety of developmental patterns united by a common feature, the occurrence of semi-autonomous component meristems within a single meristem.

Stratified meristems appear to have evolved independently in a number of different groups of seed plants. In the gymosperms such an apical organization is found in a number of families including the Ephedraceae, Gnetaceae, Araucariaceae (Johnson 1951) and Cupressaceae (Pohlheim 1980). In the angiosperms, stratified meristems are almost universal (Gifford and Corson 1971; Neilson-Jones 1969). Despite widespread occurrence, the adaptive significance of stratified meristems over nonstratified meristems is not clear (Gifford 1954). In the angiosperms, Rouffa and Gunckel (1951) surveyed the Rosaceae and found little correlation between stem morphology and degree of apical stratification. Brown, Heimsch, and Emery (1957) suggested that apical stratification shows taxonomic patterns in the Gramineae but, again, the utility of the various patterns of stratification was unresolved.

Although the adaptive significance of stratified meristems is unresolved, the genetic consequences of such meristems are well understood. Because the component meristems in such apices are almost autonomous in some taxa, somatic mutations often lead to the development of periclinal chimeras. The individual component meristems generally are designated LI, LII, LIII, etc., starting from the outer tunica layer inward to the corpus. In the Cupressaceae many species have stratified apical meristems which consist of a single layered tunica (LI) covering the corpus (LII) (Pohlheim 1980). In figure 5.4 the relationship between these two component meristems in the growth of a *Juniperus* shoot are diagrammed. In this example the shoot is a periclinal chimera, having a chlorophyll deficiency mutation fixed in the LI. Because the epidermis (which

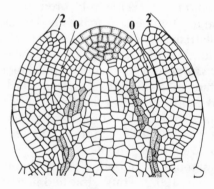

Figure 5.1. Longitudinal view of a shoot apex of *Cassiope lycopodioides* approximately in the maximal area phase of a plastochron. (From Hara 1975.)

is derived from the Lı) does not form chlorophyll, the shoot appears normal. The only evidence that a mutation is fixed in the Lı is from observations of the guard cells of the stomata. These epidermal cells normally develop green chloroplasts. In the periclinal chimera, the guard cells lack chlorophyll. Cell displacement between component meristems may result in buds with different phenotypes. In figure 5.4 cell displacement of Lı cells into the Lıı is thought to occur through periclinal cell divisions. Such displacements may give rise to totally white branches. The opposite may also occur, Lıı cells may displace Lı cells resulting in green branches with green guard cells.

Apical meristems with only two component meristems (Lı and Lıı) have also been documented in some angiosperms. The cultivar *Spiraea bumalda* "Anthony Waterer" is a clone that has fixed a chlorophyll deficient mutant in the Lı. The Lıı is wild type and thus

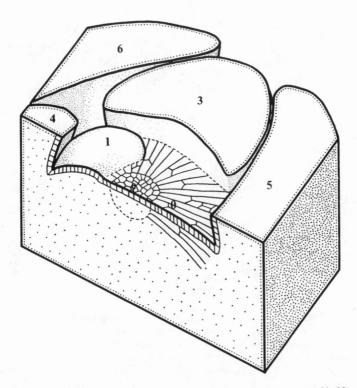

Figure 5.2. Diagram showing a three-dimensional reconstruction of half an apex; the apex is cut through its center *(Clethra barbinervis)*. (From Hara 1971.)

is green. As in the *Juniperus* example, the periclinal chimera is green with guard cells lacking chlorophyll. Cell displacements of the Lɪ into the Lɪɪ result in branches or portions of branches or leaves that are white (Tilney-Bassett 1963). Pohlheim (1971c) doc-

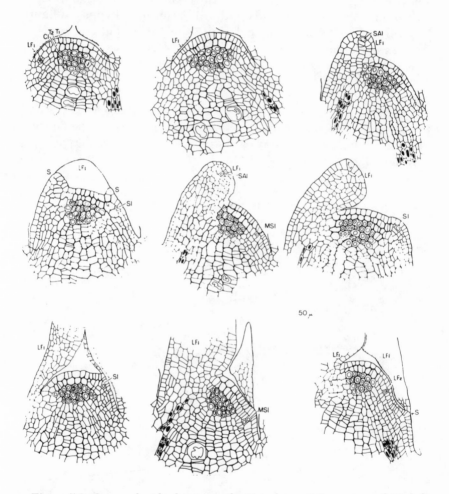

Figure 5.3. Camera lucida drawings of terminal vegetative apices of *Michelia fuscata* in longisection, which illustrate the stages in the plastochron. The youngest leaf (Ff₁) is cut medianly in A, C, E, F, H and I. In the remainder, the plane of section is approximately at right angles to the median of leaf 1, in order to show the stipular margins. Black nuclei indicate procambial cells, and outlined nuclei indicate cells of the central initial zone. cɪ = central initial zone; Lf = leaf; ᴍsɪ = median stipular initial; s = stipule; sᴀɪ = subapical initial; sɪ = stipular initial; ᴛ₁ and ᴛ₂ = tunica layers. (From Tucker 1962.)

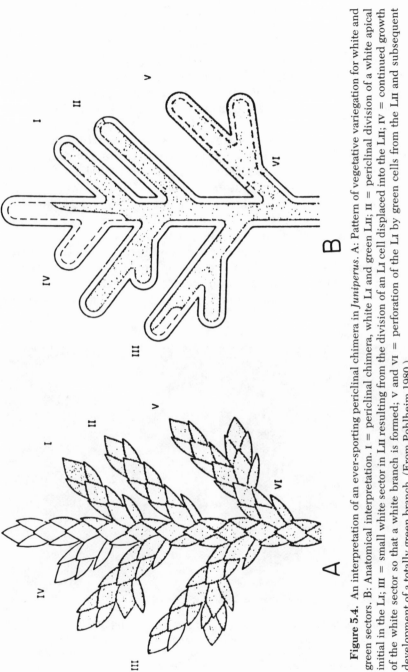

Figure 5.4. An interpretation of an ever-sporting periclinal chimera in *Juniperus*. A: Pattern of vegetative variegation for white and green sectors. B: Anatomical interpretation. I = periclinal chimera, white LI and green LII; II = periclinal division of a white apical initial in the LI; III = small white sector in LII resulting from the division of an LI cell displaced into the LII; IV = continued growth of the white sector so that a white branch is formed; V and VI = perforation of the LI by green cells from the LII and subsequent development of a totally green branch. (From Pohlheim 1980.)

umented a similar situation in *Mentha arvensis* "Variegata" (figure 5.5). The more typical situations in angiosperms are apical meristems that have three (or more) component meristems: LI, LII, LIII (i.e., two tunica layers and a corpus) (Satina, Blakeslee, and Avery 1940). It is a convention to designate these sequential component meristems by their genetic capabilities, thus a GWG periclinal chimera has an LII incapable of forming chlorophyll. This defect may reside in the nuclear or chloroplast genome. The patterns of leaf variegation in periclinal chimeras may vary considerably depending upon which layers are mutant and the contributions of the various layers to the leaf tissues. Bergann and Bergann (1983) have analyzed some of the more complex patterns. In figure 5.6, a GWG periclinal chimera in *Coprosma baueri* is illustrated. The LI is developmentally white, the LII is also white, and LIII is green, resulting in the typical white-bordered leaf. The analysis of the complementary GGW chimera shows an additional complexity. The LI is again developmentally white; the upper portion of the LII forms the hypodermis, which is also developmentally white, whereas the lower portion of the LII forms mesophyll, which is developmentally green

Figure 5.5. *Mentha arvensis* "Variegata," a periclinal chimera with a white LI. Cell displacements from the LI into more internal regions of the apical meristem or stem can give rise to white sectors of various sizes (A). In some cases the entire apical region may become populated by white cells as in B. (From Pohlheim 1971c.)

and therefore expresses the G genotype; the Lɪɪɪ is white. More complex leaf variegation patterns are found in *Elaeagnus pungens*, a species whose leaves form a hypodermis and in which the Lɪɪ and Lɪɪɪ make variable contributions to the internal tissues of the leaf (figure 5.7). These different examples of leaf variegation make the important point that the distribution and to some extent the expression of mutant cells in chimeras is a function of the patterns of organ ontogeny and development.

In clonal plants, depending upon the origin of the ramets, different component meristems may be sampled (see figure 5.8). For example, assuming three component meristems (Lɪ, Lɪɪ, Lɪɪɪ), clonal plants reproduced by modified stems (rhizomes, tubers, corms, or

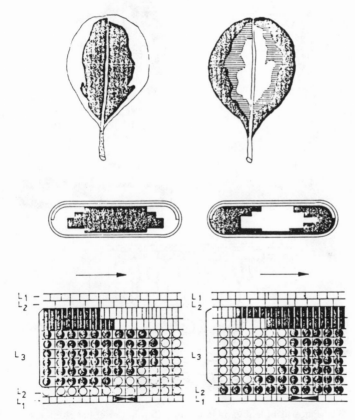

Figure 5.6. Patterns of leaf variegation in different periclinal chimeras of *Coprosma baueri*. The relationship between mature leaf tissues and apical component meristems (Lɪ = L₁, Lɪɪ = L₂, Lɪɪɪ = L₃) is shown. (Bergann and Bergann 1983.)

bulbs) may reproduce the periclinal chimera in the parent plant unless an environmental insult disturbs normal ontogeny and adventitious buds from the L**I** form or if buds are regenerated from mixtures of cells from the component meristems at the wound. Clones that reproduce by modified roots (storage roots, root suckers) will represent primarily the L**III** in the ramets. However, clones in which leaves are capable of vegetative reproduction can give rise to a diversity of offspring since component meristems may vary considerably in the extent to which they contribute to leaf tissues

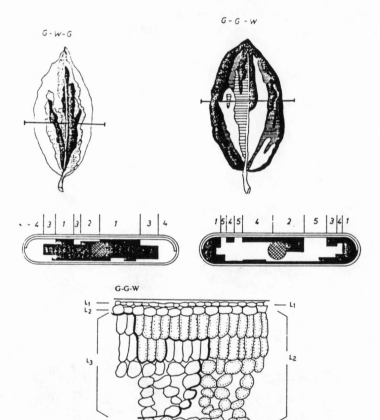

Figure 5.7. Patterns of leaf variegation in different periclinal chimeras of *Elaeagnus pungens*. Note the variable and more erratic contributions of the apical component meristems to the mature leaf as contrasted to figure 5.6. (From Bergann and Bergann 1983.)

(Stewart and Dermen 1975, 1979). Finally, in clones that reproduce via a modification of the sporogenous tissue (as in the dicot apomict *Taraxacum officinale*, Lyman and Ellstrand 1984), the LII genotype will be represented primarily in these ramets (although it should be noted that the LI also may contribute sporogenous tissue in monocots) (Stewart and Dermen 1979). In upright clones, trees, different branches may be composed of cells derived from different component meristems. For example, axillary buds will perpetuate periclinal chimeras, adventitious buds may represent cells from the LI or LII or LIII, and root suckers are derived from the LIII (see Pratt, Ourecky, and Einset 1967) (figure 5.8).

As already indicated, sporogenous tissue generally originates from the LII in dicots and the LI and LII in monocots. Nuclear somatic mutants fixed in the LII component meristem will be transmitted sexually through either male or female gamete (cytoplasmic mutants often are transmitted only through the egg) (Tilney-Bassett 1963; see chapter 3). Mutants fixed in component meristems that do not normally contribute cells to the sporogenous tissue may still be transmitted sexually because of the possibilities of cell displacements between component meristems.

In the past thirty years botanists have realized that shoot apices of many angiosperms with stratified apices are characterized by considerably reduced mitotic activity than would be expected for meristematic tissues. According to a hypothesis formulated by French botanists, the summit of the apex in a vegetative shoot is occupied by a distinctive zone, the *méristème d'attente*, in which there is an absence or near absence of mitotic activity (Buvat 1952). The *méristème d'attente* is surrounded laterally by the *anneau initial* and is subtended by the *méristème medulaire*, in both of which mitosis is common. In more recent interpretations of this hypothesis, the méristème d'attente is characterized by some mitotic activity, but this is much more restricted than in the rest of the apex (see Steeves and Sussex 1972 for discussion). From the viewpoint of mutation theory, the most interesting feature of this hypothesis is that although the *méristème d'attente* is inactive mitotically during vegetative growth, this situation changes radically with the onset of floral or inflorescence development. Abundant mitoses may be detected in the *méristème d'attente* at this time, and the previously inactive region gives rise to most or all of the reproductive structures. Such a pattern of development has been demonstrated clearly in the apical meristems of *Nicotiana* (Sussex and Rosenthal 1973) and *Helianthus* (Langenauer and Davis 1973). Although there is

debate concerning the general applicability of this hypothesis to all seed plants, the average mitotic index of shoot apices is low in many species (Clowes 1961).[1]

From the viewpoint of developmental biology, the function of this population of potentially meristematic cells which "waits" throughout vegetative development is unknown (Sussex and Rosenthal 1973). Although mutation buffering may not be the primary factor, the genetic consequence of such a population of cells destined for reproductive tissue is considerable. Such a pattern of development follows either the asynchronous logarithmic or skewed asynchronous system of cell renewal as described by Kay (1965) (see figure 4.4). Consequently, the number of cell generations between zygote and meiocyte is considerably less in plants with a *méristème d'attente* than in plants lacking such a subpopulation of cells in their apical meristems (assuming that the *méristème d'attente* includes LII initials that give rise to meiocytes). Based upon the previous discussion of tangential systems of cell renewal, the mutation frequency in the meiocytes is in part a function of their biological age (measured in cell generations). If all other factors are similar, the mutational load in species having such populations of nondividing or less frequently dividing cells within their apical meristems should be less than in species lacking this type of apical organization.

The asynchronous logarithmic and skewed asynchronous systems of stem cell renewal described by Kay also can promote the resistance of meristems to damage by physical and chemical mutagens. Grodzinsky and Gudkov (1982) have argued that a cell population that is variable with reference to cell cycle lengths, as well as highly asynchronous in terms of metabolism, will present a broad array of cell sensitivities at any point in time. Thus the exposure of such a population to a genetic insult (ionizing radiation, chemical mutagen, or any other mutationally active environmental influence) will result

Figure 5.8. A: *Pelargonium zonale* var. Mrs. G. Clark. The leaves have a white core (LIII) which is masked by green outer tissue (LI, LII). The white adventitious shoot developed from the root expresses the genotype of the LIII. B: *P. zonale*. This clone has a white LI and green LII and LIII. The epidermis has colorless plastids and occasional white sectors occur on the leaves due to inward cell displacements (presumably due to periclinal divisions in the LI). (From Tilney-Bassett 1963, 1986.)

1. In roots the quiescent center is a parallel situation, see Barlow (1978) for review.

in some cells less damaged than others. Such less-damaged cells may proliferate and restore the meristem and consequently enhance the probability of plant survival.

COMPONENT MERISTEMS—STRUCTURED OR STOCHASTIC?

Whether the component meristems of stratified apical meristems are structured (have permanent initials) or stochastic (have impermanent initials) has been disputed frequently in the botanical literature. Generally botanists have divided into two camps on this issue (although some of the more adroit have maintained a foot in both). The advocates of impermanent apical initials include Soma and Ball (1964) and Newman (1965), whereas Stewart and Dermen (1970) believe permanent apical initials are the rule in most seed plants (see Stewart 1978 for review). Since whether meristems are structured or stochastic greatly influences the dynamics of the loss or fixation of somatic mutations, it is important to consider the evidence in favor of initial permanency.

Stewart and Dermen (1970) have maintained that the critical evidence in favor of permanent apical initials is the occasional occurrence of relatively long mericlinal or sectorial chimeras in various plant species. To quote these authors: "If, as suggested by Soma and Ball (1964), every cell in each layer of the apical meristem had an equal function in the propagation of that layer, there could be no sectors of any length or permanence" (1970:823). Stewart and Dermen then document the occurrence of relatively long sectors caused by the replacement or displacement of apical initials between component meristems of plants that are periclinal chimeras. There is no debate that mericlinal sectors persisting for many nodes occur; the question is whether such phenomena are *critical* evidence for the permanence of apical initials as Stewart and Dermen have maintained.

Klekowski and Kazarinova-Fukshansky (1984a) mathematically modeled stochastic meristems with regard to the dynamics of mutation loss or fixation. Apical meristems with various numbers of apical initials and cell pools were studied with regard to the persistence of sectors composed of selectively neutral cell genotypes. As was shown in table 4.1 mericlinal or sectorial chimeras will persist for various lengths (in terms of nodes) and degrees of permanence in stochastic meristems. Thus the occasional occurrence of long sectors is not evidence for permanent initials, especially when there is a variety of sectors of different lengths with the longest being the least frequent.

The prolongation and stability of mericlinal and periclinal chimeras also is enhanced by increase in size of the apical meristem. The size of the apical meristem of an embryo or an axillary bud at its inception is often much smaller than the apical meristem of a mature stem (see figure 5.9). Such changes in apical meristem volume often are accompanied by increases in the number of apical initials and/or the size of the cell pool from which initials are sampled. For example in *Arabidopsis*, mutation studies indicate the number of initials in the apical meristem of the seed embryo is two, whereas nine-day-old seedlings already had five to eight initials

Linum usitatissimum

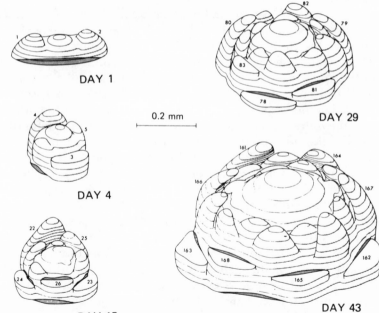

Figure 5.9. Three-dimensional reconstructions of 1-, 4-, 15-, 29-, and 43-day apices of flax drawn to a larger scale and with some primordia "removed" from around the dome. Contour lines are 10 μm apart for day 1 and 20 μm for later stages. (From Williams 1975.)

(Relichova 1977). As will be shown below, such size changes enhance the stability of mericlinal or sectorial chimeras in stochastic meristems.

Using the stochastic apical meristem model described by Klekowski and Kazarinova-Fukshansky (1984a), the consequences of an increase in meristem size may be mathematically analyzed. Consider an apical meristem with two apical initials ($\alpha = 2$, $r = 1$), where one initial is mutant. If this meristem doubles in size so that there are now four apical initials ($\alpha = 4$, $r = 1$) and two initials are mutant, one effect of this size change will be to prolong the retention of selectively neutral mutations as mericlinal or sectorial chimeras. If the meristem size increase is accompanied by a doubling of the cell pool size from which apical initials are randomly selected ($r = 2$), this will further prolong the retention of mericlinal or sectorial chimeras. A similar effect will occur for meristems with three apical initials ($\alpha = 3$, $r = 1$) which increase in size to six apical initials ($\alpha = 6$, $r = 1$ or $r = 2$) (compare the values in table 4.1 with those in table 5.1).

These results are based upon the assumption of random intermixing of all potential initials in the cell pool. This assumption becomes less valid as the number of initials increases because of the topological constraints of a larger promeristematic volume and/or surface within the apical meristem. Thus the meristem acts as if it consists of a series of subdomains within which stochastic processes are occurring. The net result of such topological domains is the prolongation of the mericlinal or sectorial chimera as well as the long-term retention of the original proportions of mutant and nonmutant tissues that existed at the inception of the apical meristem. Thus, if the apical meristem at the time of mutation has two apical initials that increased to eight initials during ontogeny, it is very probable that a mutant sector width of one-half the stem circumference will persist for many nodes even though the apical meristem may have impermanent initials.

It has not been uncommon in the past to deduce the number of apical initials in a mature shoot apex from the proportion of the circumference of the stem occupied by a persistent mericlinal or sectorial chimera (Stewart and Dermen 1970). As shown above, such a procedure is valid only if the apical meristem has remained constant in size since the inception of the chimera. Where somatic mutations are induced in seeds, the size of the chimeric sector indicates only the number of initials in the embryonic shoot apex, nothing can be concluded about the number of apical initials in the

Table 5.1. Persistence of mericlinal and sectorial chimeras in stochastic meristems that increase in size. The probability of a chimera (M) is given in terms of the number of cell generations of the apical initials and their immediate derivatives, the number of cycles of apical initial selections, and the number of nodes of persistence (when four nodes occur every cell generation). The starting condition was $\alpha - 1$ nonmutant initials and one mutant initial; after increasing in size, the meristem consisted of $2(\alpha - 1)$ nonmutant initials and two mutant initials. Thus, for the two cases presented, an apical meristem with two apical initials ($\alpha = 2$) with a single mutant initial became a meristem with four apical initials with two mutant initials or an apical meristem with three apical initials ($\alpha = 3$) with a single mutant initial became a meristem with six apical initials with two mutant initials. In the larger meristems, initials were selected after either one ($r = 1$) or two ($r = 2$) cell divisions. In all cases the mutant initials and their mitotic progeny were selectively neutral within the meristem.

	Cell Generations	Selection Cycles	Nodes	M
$\alpha = 2, r = 1$	4	4	16	0.198
	8	8	32	0.039
	12	12	48	0.007
$\alpha = 4, r = 1$	4	4	16	0.638
	8	8	32	0.342
	12	12	48	0.186
	16	16	64	0.10
	20	20	80	0.054
	24	24	96	0.020
$\alpha = 4, r = 2$	4	2	16	0.756
	8	4	32	0.486
	12	6	48	0.312
	16	8	64	0.200
	20	10	80	0.128
	24	12	96	0.080
$\alpha = 3, r = 1$	4	4	16	0.410
	8	8	32	0.168
	12	12	48	0.068
	16	16	64	0.028
	20	20	80	0.012
	24	24	96	0.004
$\alpha = 6, r = 1$	4	4	16	0.663
	8	8	32	0.407
	12	12	48	0.256
	16	16	64	0.227
	20	20	80	0.106
	24	24	96	.068
$\alpha = 6, r = 2$	4	2	16	0.834
	8	4	32	0.638
	12	6	48	0.483
	16	8	64	0.365
	20	10	80	0.276
	24	12	96	0.209
	28	14	112	0.158
	32	16	128	0.119

mature apex. Similarly, where mutation or cell displacement causes a mericlinal or sectorial chimera to originate in an axillary bud, the proportion of the circumference of the stem occupied by the persistent chimeric sector indicates the number of initials in the bud apical meristem at the time of the origin of the sector.

It is clear from the above considerations that determining whether apical initials within a mature apical meristem are permanent or impermanent is not an easy task. The documentation of occasional relatively persistent mericlinal or sectorial chimeras is not, in itself, critical evidence for the permanence of apical initials as such phenomena would be expected from stochastic meristems as well. Without details on the relative frequencies of different chimera lengths as well as information on the number of cell divisions per node for the cells functioning as initials, little can be concluded about how permanent apical initials are.

Another random process involved in the growth of stratified meristems is the possibility of losing or duplicating component meristems through rearrangements within the apical meristem. Kirk and Tilney-Bassett (1978) have reviewed these kinds of phenomena for two- and three-layered periclinal chimeras. In a two-layered stratified meristem where the L_I is the tunica and the L_{II} the corpus, two kinds of cell displacements are possible. Through one or more periclinal divisions in the L_I, the L_I duplicates and displaces the original L_{II} or the L_{II} undergoes the periclinal division and a cell or cells perforate the L_I to dominate that layer. In three-layered periclinal chimeras, regardless of whether the two or three different genetic types of component meristems occur, only six types of arrangements are possible (figure 5.10). Bud variations in each of the six different arrangements may occur in two general ways:

1. *Displacement.* Periclinal division in the cells of one layer, causing duplication of that layer, leads to displacement of the cells in the adjoining layer. This displacement is either a shifting of one layer farther inward or outward or else an outer layer is perforated by the duplicating layer.

2. *Translocation.* Bergann and Bergann (1962) have suggested the actual layer shifts may occur in trichimeras (e.g., BBC → → BCB, ABC → → ABA, and ABC → → BAC). These shifts may occur in two stages; one of the three layers of the periclinal apical meristem duplicates by periclinal divisions in a sector, thereby making the apical meristem a mericlinal chimera. This shift is followed by one or two layers of the original periclinal sector

moving sideways by anticlinal division to displace the cells of the same layer in the new sector. Kirk and Tilney-Bassett (1978) have noted that displacements probably are much more common than translocations in the origin of component meristem rearrangements within periclinal chimeras.

STRATIFIED STOCHASTIC MERISTEMS

As already indicated, apical meristems that consist of discrete tunica layers (one or more) with an internal corpus are stratified meristems. Such stratified meristems are composed of component meristems, the individual tunica layers, and the internal corpus. The integrity of apical meristems with a tunica-corpus organization is maintained only if the cells of the tunica layer(s) divide primarily in an anticlinal plane. Such apical meristems, therefore, may exhibit both structured and stochastic properties—the former is a function of the anticlinal patterns of cell division in the tunica and whether permanent initials are present in the component meristems. Stratified meristems exhibit stochastic properties if impermanent apical initials occur in the component meristems and if occasional cell displacement occurs between adjacent component meristems. Such cell displacements have been documented in most periclinal chimeras but with very

ONE GROUP			THREE GROUPS								
Three genetic types			*Two genetic types*								
(A	B	C)	(A	B)		(A	C)		(B	C)	
L_I	L_{II}	L_{III}	L_I	L_{II}	L_{III}	L_I	L_{II}	L_{III}	L_I	L_{II}	L_{II}
A	B	C	A	A	B	A	A	C	B	B	C
A	C	B	A	B	A	A	C	A	B	C	B
B	A	C	B	A	A	C	A	A	C	B	B
B	C	A	B	B	A	C	C	A	C	C	B
C	A	B	B	A	B	C	A	C	C	B	C
C	B	A	A	B	B	A	C	C	B	C	C

Figure 5.10. Apical structural arrangements for meristems with three component meristems (L_I, L_{II}, and L_{III}). Regardless of whether the component meristems represent three (A, B, C) or two (A, B; A, C; B, C) genotypes, only six structural rearrangements are possible. (From Kirk and Tilney-Bassett 1978.)

low frequencies (frequencies of from 10^{-2} to 10^{-3} successful displacements per division of the initials or per node have been reported). Such displacements are not simultaneously reciprocal events (Gifford 1954; Bergann and Bergann 1962; Tilney-Bassett 1963; Dermen 1969; and Stewart and Dermen 1970).

The individual component meristems may be either structured (with permanent initials) or stochastic (having impermanent initials). All of the variables that have been discussed previously concerning structured and stochastic unstratified meristems apply equally to the individual component meristems comprising a stratified apical meristem. The mathematical modeling of stratified meristems composed of stochastic component meristems involves a modification of the models of Klekowski and Kazarinova-Fukshansky (1984a, b). Thus, rather than having a single stochastic meristem per apical meristem with distinct α and $\alpha 2^r$ values (numbers of apical initials and size of meristematic cell pools), an apical meristem is assumed to have two adjacent stochastic component meristems with limited leakage of cells between component meristems. Each component meristem has the properties of a semi-independent stochastic meristem; these component meristems may be invaded via cell displacements (e.g., periclinal divisions in a tunica) by cells from the adjacent component meristem.

Recently Klekowski, Kazarinova-Fukshansky, and Mohr (1985) completed a mathematical analysis of the properties of stratified apical meristems. An apical meristem consisting of two adjacent component meristems with the possibility of cell displacement in either direction (this is analogous to the LII and LIII situation in some angiosperm apices) was modeled. The two kinds of component meristems are designated I and II, both having the same number of apical initials ($\alpha_I = \alpha_{II} = \alpha$). The number of divisions between subsequent selections of initials within a component meristem is r where component meristems have identical r values. Cell displacement between component meristems was quantified and included in the model (see figure 5.11). Cell displacement probabilities of 0.01 and 0.001 were selected as these values approximate those reported in the literature (Dermen 1969). If one considers two adjacent component meristems (I and II), the next set of initials for component meristem I is sampled from the cell pool of component meristem I with a probability of p and from the cell pool of meristem II with a probability of q, where $p + q = 1$. Thus, for the 0.001 displacement value, $p = 0.999$ and $q = 0.001$. A similar pattern is used to select initials for component meristem II.

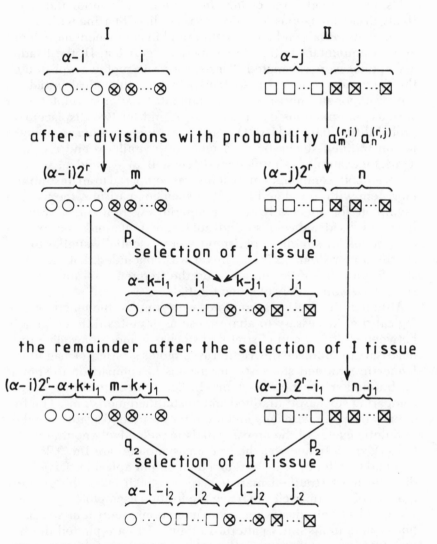

Figure 5.11. General model for a stochastic stratified meristem with two component meristems (I and II). The apical initials for component meristem I are selected from the meristematic cell pool of component meristem I with a probability P and from the meristematic cell pool of component meristem II with a probability q. Similarly, the apical initials for component meristem II are selected from the meristematic cell pool of component meristem II with a probability P and from the meristematic cell pool of component meristem I with a probability q. Mutant cells are designated $\otimes$ or $\boxtimes$ and nonmutant cells are $\bigcirc$ or $\square$. (From Klekowski, Kazarinova-Fukshansky, and Mohr 1985.)

Using this model, the change in frequency of neutral and selectively disadvantageous mutations was studied. Starting with $\alpha_I - 1$ nonmutant initials and one mutant initial in component meristem I and α_{II} nonmutant initials in component meristem II, the fixation asymptotes were calculated. These fixation asymptotes represented the frequency of apical meristems in which the apical initials in both component meristems were mutant. Fixation asymptotes for various combinations of values of α, r, and for two displacement probabilities (0.01 and 0.001) were compared to fixation asymptotes for unstratified stochastic meristems with similar α and r values. Thus, for example, a stratified meristem with $\alpha_I = 2$ and $\alpha_{II} = 2$ and $r = 2$ for both component meristems was compared to an unstratified meristem with $\alpha = 4$ and $r = 2$. In such comparisons, the fixation asymptote for selectively neutral mutations were identical. For selectively disadvantageous or advantageous mutations, the fixation asymptotes for stratified meristems were essentially similar to unstratified meristems ($\pm 5\%$). Thus it was concluded that meristem stratification has little influence on the frequency of apical meristems consisting solely of mutant initials.

Although the fixation asymptotes were similar, the length of biological time necessary to attain these asymptotes differed greatly between stratified and unstratified stochastic meristems. The time necessary to approach the fixation asymptotes is orders of magnitude longer in stratified stochastic meristems in comparison to similar unstratified stochastic meristems. In figure 5.12 are shown curves comparing stochastic stratified and unstratified meristems. The frequency of apices with only mutant initials is plotted against number of mitotic cycles of the apical initials for selectively neutral mutations ($V = 1$). In all cases the fixation asymptotes are $1/\alpha$ (0.25 for $\alpha = 4$ and 0.25 for $\alpha_I = \alpha_{II} = 2$ or a total of four apical initials). Note that the most stratified meristem ($p = 0.999$) takes the greatest number of mitotic cycles to reach the fixation asymptote.

For selectively disadvantageous mutations, meristem stratification results in a similar pattern. In figure 5.13 are plotted the frequency of apices that are a chimera as a function of mitotic cycles of the apical initials for mutant cells having a lowered viability ($V = 0.90$ or $hs = 0.10$). It is obvious that meristem stratification has one important consequence, such meristems promote the retention of those cells having disadvantageous mutant alleles.

Stratified apical meristems, therefore, appear maladaptive from the viewpoint of buffering against disadvantageous mutations. Although such meristems have comparable mutation asymptotes to similar nonstratified meristems, these asymptotes are achieved very slowly. Disadvantageous mutations may persist for long periods as periclinal chimeras within stratified meristems of long-lived perennials and are, therefore, more likely to give rise to sporogenous tissue and ultimately to contribute to mutational load. Evidence for the long-term maintenance of selectively disadvantageous mutations is to be found in the numerous examples of horticultural forms that are periclinal chimeras for nuclear or plastid determined chlorophyll deficiencies (Kirk and Tilney-Bassett 1978). Stewart and Dermen (1979) report finding stable periclinal chimeras in 60 genera within 10 families of monocots. In dicots such forms are even more common (see Hillier 1972). Although such chlorophyll deficiencies are lethal mutations to an autotrophic plant, their persistence is due to the stratified nature of the apical meristems in these plants. In unstratified meristems such mutations would be fixed in a few ramets or buds and be lost eventually.

The computer analysis has shown that even cell genotypes that are selectively disadvantageous within the context of the apical meristem will persist for considerable periods as periclinal chimeras. Dermen (1969) has shown that plant species differ considerably in the frequency of periclinal divisions and cell displace-

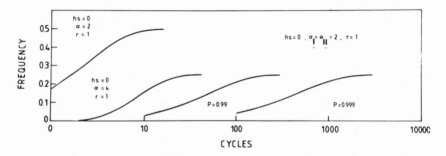

Figure 5.12. Frequency of apical meristems that have only mutant apical initials as a function of cycles of apical initial selections from the meristematic cell pools. Cells heterozygous for mutations have equal viabilities to nonmutant cells; the mutations are assumed to be selectively neutral (hs = 0). The curves for $\alpha = 2$, $r = 1$, and $\alpha = 4$, $r = 1$ are for nonstratified meristems. The remaining curves are for stratified meristems with different frequencies of cell displacements between component meristems. Note that although the asymptotes are similar ($1/\alpha$ or 0.25), the stratified meristems require many selection cycles to attain these asymptotes. In all cases the starting condition was an apical meristem with only one mutant initial and the remaining initials nonmutant (regardless whether in another component meristem). (From Klekowski, Kazarinova-Fukshansky, and Mohr 1986.)

ments between the component meristems. In species with very low frequencies of cell displacements, the component meristems, although in one organism, would be expected to diverge with time due to recurrent mutation. Thus as a long-lived perennial grows, the component meristems could be expected to diverge due to the operation of Muller's Ratchet (Muller 1964; Haigh 1978, see chapter 11) within the relatively small populations of meristematic cells.

Stratified meristems promote the loss or fixation of mutations influencing the frequency of periclinal divisions in the tunica layers of the shoot apex. Mutants that enhance the instability of the layers by promoting periclinal divisions in mutant tunicas will rapidly displace nonmutant cells within adjacent component meristems. The reverse case also is possible, the fixation of mutants that promote periclinal divisions in adjacent nonmutant component meristems. Stratified meristems in which apical initial displacement between

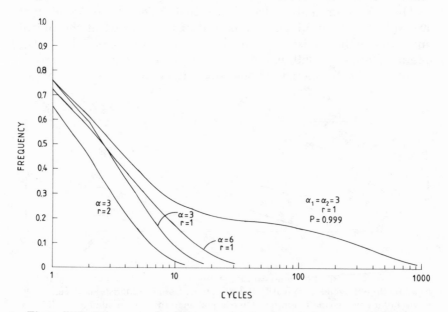

Figure 5.13. Frequency of apical meristems that consist of mutant and nonmutant initials as a function of cycles of apical initial selection (log plot). The cells that are heterozygous for the mutation have lower viabilities (hs = 0.10, V = 0.90) than nonmutant cells (V = 1). The starting condition for all curves was an apical meristem with one mutant initial and the remainder of the initials nonmutant. In all cases the chimera persists longer in stratified meristems. Unstratified meristems with $\alpha = 3$, $r = 1, 2$ or $\alpha = 6$, $r = 1$ are compared to a stratified meristem with $\alpha_I = \alpha_{II} = 3$, $r = 1, P = 0.999$. (From Klekowski, Kazarinova-Fukshansky, and Mohr 1985.)

component meristems is *not* independent of cell genotype are called asymmetric systems. The properties of such systems have been studied by Klekowski, Kazarinova-Fukshansky, and Mohr (1985). Thus, if a somatic mutation becomes fixed in the apical initials of one component meristem, the adjacent component meristem, which has some nonmutant cells, has an enhanced frequency of periclinal divisions. Consequently, there is an increased probability of a mutant apical initial being displaced by a nonmutant cell from an adjacent component meristem. Thus, rather than only competition between mutant and nonmutant cells as in symmetric systems, an additional factor is assumed—component meristem instability related to genotype of constituent cells.

If one begins with a meristem consisting of two component meristems (α_I and α_{II}), where Component Meristem I has $\alpha - 1$ nonmutant cells and one mutant cell and Component Meristem II has α nonmutant cells, then asymmetric displacements result in a rapid loss of mutant cells. In table 5.2, the fixation asymptotes (i.e., frequency of apices consisting of only mutant initials in both component meristems) for various values of α, r, and hs are given. If the cell pools ($\alpha_I 2^r$ and $\alpha_{II} 2^r$) of each component meristem each has at least one nonmutant cell, then the probability of cell displacement from the adjacent component meristem is q or $1 - P$, whereas if all of the cells in a component meristem are mutant, then the proba-

---

Table 5.2. The frequency of apices consisting of only mutant cells for mutations that promote asymmetric displacements between component meristems. For component meristems with at least one nonmutant cell, $P = 0.999$, $q = 0.001$; if a component meristem had only mutant cells then $P' = 0.99$ and $q' = 0.01$. The viability of a mutant cell is $1 - hs$, where the wild-type viability is unity.

			FIXATION ASYMPTOTES		
α[a]	r	$hs =$ 0	0.05	0.10	0.15
2	1	0.045	0.032	0.020	0.010
2	2	0.045	0.029	0.018	0.009
3	1	0.031	0.017	0.008	0.003
3	2	0.031	0.013	0.005	0.002

[a]The number of initials in a component meristem; each meristem has two identical component meristems with regard to α and r.

bility of cell displacement is q' or $1 - P'$. For the fixation asymptotes in table 5.2, P was 0.999 and P' was 0.99. The low values for the fixation asymptotes in table 5.2 indicate that somatic mutations that promote asymmetric displacements are eliminated fairly rapidly from apical meristems.

APICAL MERISTEMS:
SUMMARY OF MUTATIONAL CHARACTERISTICS

Many different characteristics have been shown to affect mutation frequency both within apical meristems and, consequently, within a genet. These characteristics were summarized in outline form in figure 4.12. With regard to those features of apical meristems that promote or retard competition between mutant and nonmutant cells (diplontic selection), three caveats need to be remembered:

    1. Diplontic selection is dependent upon measurable differences in some aspect of cell viability (survival, division rate, etc.) between nonmutant cells and those heterozygous for a somatic mutation. Since mutant and nonmutant cells are in intimate physical contact, nonmutant cells may crossfeed mutant cells and enhance the latter's viability (Langridge 1958).

    2. Somatic mutation may be completely recessive and, thus, have no phenotypic effects in heterozygotes.

    3. Genes not expressed in the apical meristem (e.g., mutants affecting chlorophyll synthesis) will not be influenced by diplontic selection.

Given these limitations, one would predict that most somatic mutations are selectively neutral within the apical meristem and that the primary selective (as opposed to random) loss of such mutations within somatic tissues is through ramet competition. The efficiency of ramet competition is a function of the selective discrimination of genetic-based ramet phenotypes. Such a discrimination is enhanced if the ramet is composed totally of mutant cells. Stratified meristems greatly retard such selective discrimination at the ramet level because mutant and nonmutant cells may be maintained within a single apical meristem as periclinal chimeras for long periods of biological time (nodes). It is likely that a significant (and generally unrecognized) problem facing plants with stratified apical meristems is the accumulation of mutational load during the growth of any long-lived genet.

CELL FITNESS VS. ORGANISM FITNESS

Stratified meristems inadvertently may promote cells with high cell fitness but which decrease the overall fitness of the organism. An example of such selective "cross-purposes" was documented in *Juniperus* (Ruth, Klekowski, and Stein 1985). A juniper clone which had fixed a mutation for albinism in the LI formed white branches when cells in this layer divided periclinally and the white cells invaded the LII. This clone had two component meristems in the shoot apex, a tunica (LI) and corpus (LII). Immediately after the LI cell displacement, the LII was a mixture of white and green cells, and subsequent growth of this shoot apex resulted in a sectorial chimera. Typically, three apical initials in the LII were present at the inception of this chimera, one of the white phenotype and two of the green. Such sectorial chimeras rarely persisted beyond 30 nodes before shifting to either all white or green. If the white and green phenotypes were selectively neutral with regard to apical initial selections, one would predict that one-third of the sectorial chimeras would eventually shift to all white and two-thirds to all green. In actuality almost exactly the opposite occurred, the majority of chimeras shifted to all white. The mechanism generating the higher cell fitness for the white cells appeared to be based on alterations in apical meristem topology due to the formation of axillary buds. Ruth, Klekowski, and Stein hypothesized that bud ontogeny alters the topology of the apical meristem so that cells that will act as new initials will be selected from areas of the apical dome slightly closer to the incipient bud. In the juniper clone studied, sectorial chimeras formed buds more frequently in the white sectors. Thus, the tunica-corpus organization allowed the retention and vegetative spread of a mutant that could invade and dominate the corpus. The fitness of this juniper clone was impaired both with regard to vegetative vigor (30% of the branch tips were albino) and sexual reproduction (sex cells arising from white tissues presumably more frequently formed defective offspring).

Such interactions between apical and shoot ontogeny are not uncommon. Pohlheim (1971b) found that the frequency of periclinal divisions in the LI of the axillary branch is correlated with the age of the axillary relative to the parental apical meristem in the Cupressaceae. In *Chamaecyparis pisifera plumosa argentea*, more white sectors occurred on the lower side than on the upper side of horizontal shoots, suggesting that periclinal divisions in the tunica on the abaxial side were more common (Pohlheim 1971a). These kinds of interactions between apical initial selection or the differential occurrence of periclinal divisions and shoot development are potential "developmental-weak-points" which somatic mutants may exploit to enhance their own fitness without regard to the organism's fitness. Because of the hormonal and source-sink characteristics common in plant development at the tissue and organ levels, one would suspect that there are many "developmental-weak-points" where deleterious mutants (from the viewpoint of the organism) will be preferentially selected.

TUNICA: A SELFISH TISSUE

Apical meristems with one or more tunica layers and an internal corpus are characteristic of some gymnosperms and practically all angiosperms. In spite of the ubiquity of such meristems in seed plants, the adaptive significance of such stratified meristems in comparison to unstratified meristems is unknown. Stratified meristems have very significant genetic properties (as have been enumerated already), but whether these properties singly or in combination are sufficient to account for the evolution of meristem stratification is not clear. In the following discussion, I will make the heretical assumption that "perhaps" meristem stratification is not adaptive at the individual level but, rather, represents a selfish tissue which has evolved repeatedly within those plant groups capable of tolerating it.

Perhaps the best way to explore this possibility is to consider some aspects of the biology of human cancer and then to show that tunica layers may have some of these properties as well. A cancer cell originates as a defect in one cell, allowing it to start multiplying and give rise to an ever increasing population of similarly unrestrained progeny cells. A common characteristic of such cells is that they acquire the ability to spread by way of the blood and lymph systems to form secondary colonies (metastasis) in distant sites (Cairns 1978).

In vascular plants, somatic mutations cannot result in cells with the cancer characteristics of unrestrained division and mobility because of the organization of plant meristems and the lack of a vascular system through which cells may move. Thus the properties of unrestrained mitotic division and mobility are uncoupled in plants.

The consequences of cells that have the property of unrestrained division are well known from studies of plant tumors (see chapter 7), but what types of mutations would result in cell mobility within the plant? The answer is obvious, of course, from our previous discussions—any mutation resulting in a preponderance of anticlinal divisions in the apical meristem. Such a cell genotype will spread throughout all organs formed by the apical meristem with subsequent growth. This selfish anticlinal phenotype will spread even if it is somewhat defective physiologically since it is in a kind of semipermanent sandwich with wild-type cells within the apical meristem. Another aspect of such a "tunica" somatic mutation is that such mutations are more likely to be passed on to the next generation. It already has been noted that sporogenous tissue, both in monocots and dicots, generally originates from one or more of the tunica layers rather than the corpus. Of course these two "selfish" characteristics of tunica are not sufficient to account totally for the evolution of stratified meristems but perhaps they must be included in any consideration of the biological implications of having a stratified meristem.

GENERALIZATIONS AND CAVEATS

Vascular plant apical meristems have been discussed from the perspective of somatic mutations. The great diversity of apical meristem organizations found in the vascular plants was reduced to a number of simple models (stratified, unstratified, stochastic, and structured) to allow analysis and generalization. Of course, the plant world cannot be neatly categorized into these simple theoretical models (see Popham 1951 for a sample of this apical diversity, but even this author was attempting to generalize the total vascular plant kingdom into seven apical types). Perhaps a better view would be to consider a given species as having a constellation of apical characteristics that may vary chronologically (with the age of the individual) and episodically (depending upon development, e.g., flowering vs. vegetative) that influence the fate and frequency of somatic mutations. These varying apical characteristics perhaps can best be understood individually by relating them back to their Platonic ideals, but their total consequence to the organism or species must represent a very complex integration.

In the following chapter some of the properties of "atypical" apical meristems as well as other meristems will be discussed.

CHAPTER SIX

Other Meristems

In chapters 4 and 5 a generalized analysis of apical meristems was developed, stressing the consequences of various anatomical and developmental parameters with regard to somatic mutation. The general thesis presented was that sectorial chimeras in unstratified meristems and mericlinal chimeras in stratified meristems are normally unstable. In the latter, as Clowes (1961) has stressed, since each component meristem is constantly being renewed from its initials, and since those initials may not be constantly the same cells in the same positions, individual component meristems that start by containing cells of two genotypes usually will eliminate one of the genotypes eventually. In the following sections, meristems having either anatomical or developmental characteristics that promote the almost indefinite retention of sectors will be discussed. Such meristems are found in some monocotyledonous taxa and serve as an interesting counterpoint to the gymnosperm and dicotyledonous apical meristems presented in general treatments.

APICAL MERISTEMS IN SELECTED MONOCOTS

In various members of the Commelinaceae, apical meristems have evolved that can perpetuate mericlinal chimeras almost indefinitely (figure 6.1). Such species commonly have distichous or spirodistichous phyllotaxes (i.e., have leaves arranged in two vertical ranks on the stem with each rank separated by *ca.* 180 degrees). Many examples of variegated chlorophyll chimeras are known in which the leaves have white or green stripes running from base to tip. The sectors are regular and stable. The apical meristem of these plants has an outer tunica layer that produces only the epidermis. The remaining tissues are derived from the inner corpus. The corpus can be viewed as a series of component meristems with cells arranged in transverse files running from the region above one leaf insertion to the region above the next leaf insertion in a distichous shoot. Cell divisions in the corpus appear to be confined to planes perpendicular to the line joining one leaf to the next.

In figure 6.2 a diagrammatic representation of this mode of growth is drawn showing the disposition of a mutant cell type as a vertical sheet of mutant cells extending into both the foliar buttresses and the leaves. Variegated plants with this type of apical development may form axillary buds with varying amounts of green or white tissues (or consisting solely of one cell type), since buds sample fewer files of cells than large leaf primordia (see figure 6.2). Such apical meristems have been documented in *Tradescantia albiflora* var. *albovittata* and *Zebrina pendula* and other species in the Commelinaceae (Thielke 1954, 1955; see also Clowes 1961 for a further

Figure 6.1. Variegated branches of *Tradescantia albiflora* var. *albovittata*. (From Thielke 1954.)

discussion). In another member of the Commelinaceae, *Commelina benghalensis,* Thielke (1954) found that leaf variegation is caused by differential proliferation of the LII and LIII during leaf development. For example, in the white-white-green periclinal chimera (LI, LII, LIII), the LI produces the epidermis of the leaf, the LII produces the white layers of the leaf mesophyll, and the LIII gives rise to the green central mesophyll. During leaf development, the white margins are caused by the LI, while the white stripes within the leaf are caused by an abnormal multiplication of the white inner tunica layer, LII, immediately beneath the promeristem. Thus genetically green tissue produced by the corpus, LIII, is invaded by extra growth of the LII and divided into white and green sectors. The leaf primordium is wide enough to include several of these white and green sectors (see figure 6.3).

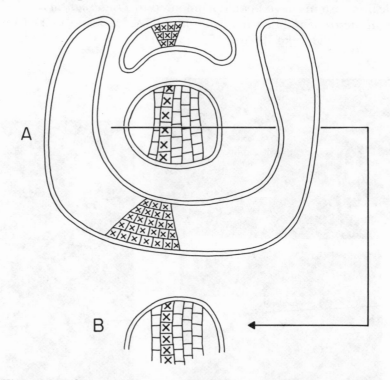

Figure 6.2. A diagrammatic representation of the apical meristem in *Zebrina pendula*. A: The transverse section below the apical dome showing the transverse files of cells in the corpus which make sectoring possible; B: Cross section of the apical dome. The mutant cells are marked with an X. (Adapted loosely from plate 14 in Clowes 1961.)

In the Gramineae, depending upon the species, apical meristems may have one or two tunica layers overlaying a corpus (LI, LII, LIII). Thielke (1951) studied chimera formation and stability in grasses. She showed that recurrent and persistent mutant sectors that occurred in grass leaves were due to differential and invasive growth of tunica-derived tissue into the tissues derived from the corpus. For example, in *Oplismenus imbecillis* var. *variegatus* the single tunica is composed of cells unable to form chlorophyll, whereas the corpus is green. As Thielke has shown, leaf primordia originate from various contributions of the LI and LII resulting in leaves with

PLANT CHIMERAS

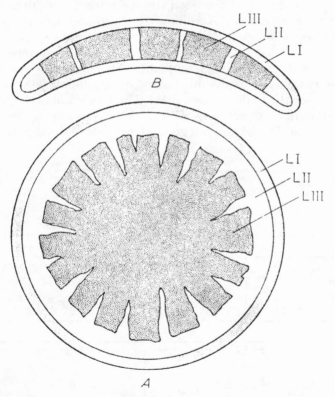

Figure 6.3. *Commelina benghalensis* "foliis variegatus." LI and LII are chlorophyll-deficient, LIII is green. A: Cross section just below the stem shoot apex. B: Leaf cross section. Note how the LII has proliferated into the LIII, in leaves such proliferation results in albino stripes in the leaves. The white leaf borders are due to proliferation of the LI. (From Neilson-Jones 1969.)

sectors of white and green tissue. In figures 6.4 and 6.5 are shown leaves of this species, as well as those of the grass *Phalaris arundinacea* var. *pieta*, having similar patterns of variegation. In contrast, the apical meristem in sugarcane, *Saccharum officinarum*, lacks a tunica, according to Thielke, and is thus more similar to the unstratified apical meristems found in many gymnosperms. Periclinal chimeras are therefore impossible in sugarcane.

The developmental biology and genetics of corn (*Zea mays* L.) are perhaps better understood than in any other grass (or any other vascular plant, for that matter). Using specially constructed genotypes that allow the tracking of tissues derived from somatic mutants, the destinies (in terms of tissues in the mature plant) of different cells in the embryo have been studied intensively (Stein and Steffensen 1959; Coe and Neuffer 1978; Johri and Coe 1983). Corn is grown as an annual, and its development from seed embryo to mature adult is determinate (in contrast to the indeterminate nature of growth in many other seed plants). Somatic mutations occurring in the embryo (or at any other stage of development) will result in mutant sectors in the mature plant. The reciprocal of the fraction of the circumference of the stem represented by a sector is the apparent cell number (ACN) and is a general estimate of the number of cells

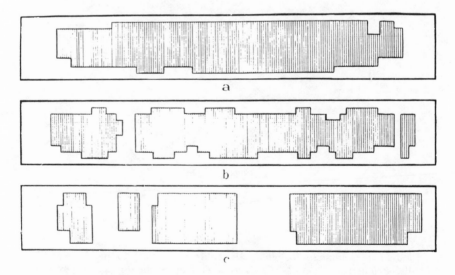

Figure 6.4. Leaf cross sections of variegated grasses showing the differential tunica contributions into the leaf mesophyll. A: *Chlorophytum elatum;* B: *Phalaris arundinacea* var. *pieta;* C: *Oplismenus imbecillis* var. *variegatus.* (From Thielke 1951.)

giving rise to a particular part of a corn plant. In addition to sector width, sector length in terms of nodes is also an important parameter. The majority of sectors start at the base of an internode, continue up through one or more nodes, and terminate in a leaf.

The mature corn plant is approximately 2 to 3 meters tall, with about 15 to 20 nodes. Each node bears a single leaf. The male inflorescence (the tassel) is terminal and the female inflorescences terminate in ear shoots, modified axillary branches. Based upon the distribution of chimeras derived from somatic mutations in the corn embryo within the dry kernel, a type of destiny map can be constructed for the LI cells of the apical meristem (the role of LII cells is still not clear). This map is shown in figure 6.6. It should be noted that within each disk, the individual cells can proliferate to a variable and different extent, but as a group, the proliferation pattern of

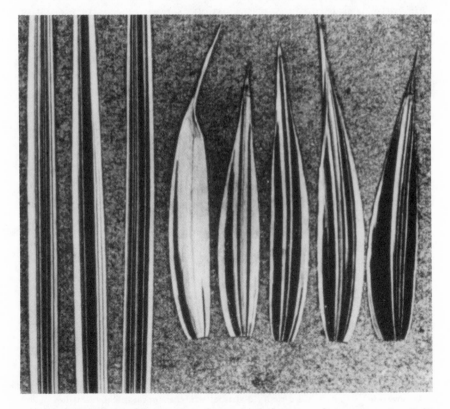

Figure 6.5. Leaves of the variegated grasses *Phalaris arundinacea* var. *pieta* (left) and *Oplismenus imbecillis* var. *variegatus* (right). (From Thielke 1951.)

cells is fixed to some degree and is predictable (Johri and Coe 1982). The number of cells per disk may vary also. For example, the ACN data show that 60% of the tassels develop from 4 ± 1 cells of the dry embryo, while the other 40% come from 1 to 2, or from 6 to 14 cells (Johri and Coe 1983). Even with these developmental caveats, the destinies of cells within the disks are sufficiently predictable to suggest the occurrence of developmental compartments in this plant that are somewhat analogous to compartmentation in the imaginal disks of *Drosophila* (Johri and Coe 1983).

The development of the apical meristem in corn, coupled with the presence of intercalary meristems (see following section), appears to promote the long-term retention of mutants as well as decreasing the possibilities of diplontic selection, although damaged cells in the apex may be replaced by cells situated below (Johri and Coe 1983). Whether such rigorous compartmentation is characteristic only of determinate grasses such as corn or occurs in plants with indeterminate patterns of growth must await future research. It is interesting to note that compartmentation and intercalary growth reduce the number of mitotic divisions between zygote and meiocyte.

INTERCALARY GROWTH

Intercalary meristems are parts of apical meristems that become separated from the apex by permanent tissues as the apical meristem moves forward during the growth of the plant (Hill, Overholts, and Popp 1950). They usually are found at the bases of the uppermost internodes of the stems as well as at the bases of leaves in many

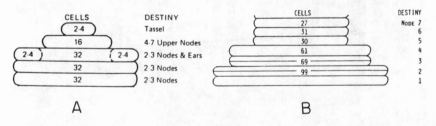

Figure 6.6. Number and density of LI cells in the shoot apical meristem of the corn embryo at the dry kernel stage . A: nodes 8 through 20; B: nodes 1 through 7. (From Johri and Coe 1982.)

monocotyledonous plants. The monocotyledonous apical meristem gives rise to a series of intercalary meristems and leaf initials; the former divides to form internodal tissue and, ultimately, stem segments, whereas the leaf initials proliferate an encircling leaf starting with the tip and continuing until the leaf base is formed. Because of this pattern of development, the basal internodes are oldest, but each internode is youngest at one end, each leaf is youngest at the base (Coe and Neuffer 1978). Intercalary meristems function in increasing the length of the organ in which they occur, but in contrast to other meristems they are wholly transformed into permanent tissues. Internodal intercalary meristems also occur in some species of the Caryophyllaceae, the articulated species of the Chenopodiaceae, as well as in *Equisetum* (Fahn 1974). Such meristems promote the retention of mericlinal or sectorial chimeras throughout the internode or leaf for the same reasons that rapid increase in the size of an apical meristem (in terms of number of initials) promotes the retention of such chimeras (see chapter 5).

BRANCHING

The overall shape of a plant is the result of its branching pattern and how this pattern developmentally responds to environmental variables. Very often plant branching patterns are strongly determined developmentally and therefore have a significant genetic component. Theoretical studies of the factors important in plant architecture or geometry have concentrated primarily on maximizing net productivity with regard to the changing light, water, and nutrient variables associated with competition, evolution, and ecological succession (Horn 1971; Iwasa, Cohen, and Leon 1984; Honda, Tomlinson, and Fisher 1981, 1982; Niklas 1986). Of more interest for the problem of somatic mutation is how patterns of branching affect ramet variation and competition within a growing genet (whether it is a tree or a rhizomatous plant).

Branching in itself is a means of distributing somatic mutational load within a genet. In perennials it is convenient to measure mutation rates on a per apical initial per ramet generation basis (chapter 2). In figure 6.7 a ramet generation is defined in an exponentially bifurcating branch system. In this case, the ramet generation is the biological time unit required to double the number of branches. It also represents a relatively defined number of cell divisions of the apical initials. In terms of actual plant growth, one ramet generation equals two internodes worth of cell divisions (figure 6.7B). The effect of branching pattern on mutation accumulation in growing genets has been studied mathematically (Klekowski, Fukshansky, and Kazarinova-Fukshansky in preparation), but only the qualitative results of the studies will be presented here.

For mutations that are neutral within the apical meristem (Cell Fitness = 1), the majority (if not all) characteristics of apical meristems have no influence on the degree of mutation loading of the apical initials. Thus, whether an apical meristem is structured or stochastic, the number of apical initials, or the size of the cell pools from which initials are selected in stochastic meristems has no consequence on the mathematical expectation of neutral mutations per apical initial at any point in time during the growth of a genet. The most important plant characteristic determining mutation accumulation is the degree of branching. Branching is significant because it is the principal determinant of the biological age of apical initials in terms of the number of cell divisions from zygote. Thus, the more branched an organism becomes, the younger the initials for any given biomass. Comparing different patterns of branching, as a gen-

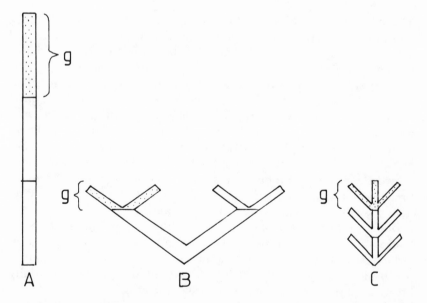

Figure 6.7. Three ramet generations ($g = 3$) worth of growth organized into three idealized branching patterns: columnar (A), deliquescent (B), and excurrent (C).

eral rule the more frequently a stem branches, the younger the apical initials (in terms of cell divisions from zygote) for any given cellular biomass and, consequently, the greater genetic stasis (if DNA replication and chromosome disjunction at mitosis are the major sources of mutation).

The demography of the branches (or ramets) with regard to growth rates and senescence may also influence mutation accumulation in the growing plant (Klekowski, Fukshansky, and Kazarinova-Fukshansky in preparation). Branch systems can vary with respect to the relative growth rates of primary and secondary branches and whether progressive senescence of the oldest branch systems occurs. In figure 6.8A and B a branch system in which the branches and the mainstem grow at equal rates is shown. The degree of mutation loading per apex or per apical initial in such a branch system is not influenced by senescence. In figure 6.8C and D a

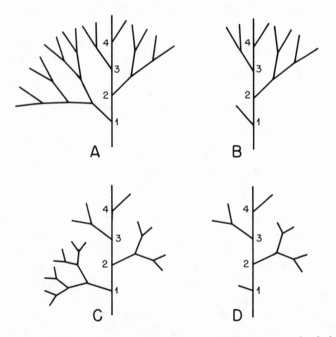

Figure 6.8. Branching patterns and senescence. A: Main stem and side branches grow at identical rates. B: Sequential senescence of the oldest side branches. In both A and B the average number of postzygotic mutants per apical initial is identical. C: Main stem and side branches grow at different rates, the former grows faster than the latter. D: Sequential senescence of the oldest side branch increases the average number of postzygotic mutants per apical initial.

branching system is represented in which the branches grow at slower rates than the mainstem. In such a tree, senescence will increase the average number of mutations per apex or per initial, thus figure 6.8C is less mutation loaded than figure 6.8D. This is because apical initials from the branches originating in the upper portions of the tree are biologically older in terms of cell divisions from zygote than the apical initials of the lower branches. Thus the average number of mutations per apical initial in a growing tree (or genet) is not simply a function of the age of the tree but also is dependent upon the relative growth rates and senescence patterns of the branches. It should be noted that such branch demography is often dependent upon ecological variables. For example, a dense forest in which only the tops of the trees carry living branches would have a higher average number of postzygotic mutations per apical initial than a forest of similar age in which the trees are widely spaced and retain living branches to the tree base (assuming that the lateral branches of the trees in both forests grow at slower rates than the mainstem, e.g., as in excurrent conifers).

The primary component of ramet competition is the discrimination between ramets of different genotypes. If we consider the genetic divergence of sib ramets due to somatic mutation, such mutations often occur as mutant cells imbedded in a matrix of nonmutant cells. Such chimeras are typically sectorial or mericlinal depending upon whether or not the apical meristem is stratified. The development of an axillary bud within the mutant sector will result in an apical meristem composed wholly of mutant cells for nonstratified apical meristems or a periclinal chimera for stratified apical meristems. The reverse is also true: buds forming outside the mutant sector will give rise to apical meristems that lack mutant cells. Thus axillary branching in mericlinal or sectorial chimeras will generate ramets that have fixed or lost the mutant cells.

Whether plants have monopodial or sympodial branching patterns (figure 6.9) may also affect the degree of genetic variation among sib ramets. Sympodial branching occurs when one of the axillary buds near the main shoot apex gives rise to a precocious vigorous side shoot that soon overtops the main shoot in its growth (Foster and Gifford 1959). Sympodial growth may be the result of shoot tip abortion or terminal flowering (Romberger 1963). For example, the shoot tips of many woody angiosperms die and are abscissed each season, and growth proceeds from the uppermost surviving axillary buds (pseudoterminal buds). Such shoot tip abortion is a regular part of the ontogeny and not the result of pathological phenomena

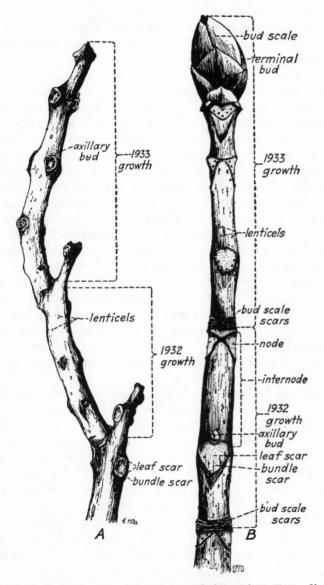

Figure 6.9. A: Twig of catalpa showing sympodial growth. B: Twig of horse chest-
nut showing monopodial growth. Both twigs show characteristic stem markings.
(From Hill, Overholts, and Popp 1950.)

(Romberger 1963). Sympodial branching is common in many temperate taxa, e.g., *Salix, Betula, Carpinus, Corylus, Castanea, Ulmus, Celtis, Platanus, Gleditsia, Robinia,* and *Tilia,* as well as tropical trees (Hallé, Oldeman, and Tomlinson 1978).

Sympodial branching has two important consequences with regard to somatic mutation. The programmed development of subterminal axillary buds may influence the probability that mutant stem sectors are converted into apical meristems that have fixed mutations. The probability of a subterminal axillary bud occurring within a mutant sector is a function of the area of the stem surface that is mutant within the subterminal region as well as the vigor of the buds in the mutant sector.[1] Another aspect of sympodial branching that is not so obvious is that the replacement of a terminal bud by a subterminal axillary bud promotes diplontic selection (i.e., enhances the results of intercellular competition). As was shown in chapter 4, greater numbers of cell divisions between selections of apical initials reduce the probability that cells with disadvantageous mutations are selected as these initials. Since apical initials of axillary buds differentiate from cell populations chronologically separated from the apical initials of the terminal apical meristem by a number of mitotic cycles, this is analogous to increasing the r value in the stochastic meristem model. Axillary buds may arise somewhat distantly from the apical meristem, in partly vacuolated tissue which must dedifferentiate, but more typically from above the second leaf primordium (Esau 1960; Romberger 1963). In either case the r value would be expected to be higher than for stems growing monopodially.

Another aspect of plant architecture that may influence the frequency of somatic mutations is the formation of long and short shoots. Long shoots are responsible for the growth in height and overall framework of a tree, whereas short shoots often have more specialized functions, e.g., photosynthetic units and, commonly, localized sites for sexual reproduction. In terms of apical growth, it might be expected that fewer mitotic divisions would be necessary for the perpetuation of short shoot apical meristems and, thus, one would suspect that their promeristem cell populations are not as mitotically active as those promeristems in long shoots. If one accepts the argument that there is a relationship between the age of a

1. Sympodial branching may promote developmental selection if the probability of a bud breaking is related to the vigor of the cells in a mutant mericlinal sector. Thus, if the buds that form the subsequent year's growth are preferentially selected from the most vigorous cell populations in the twig, sympodial branching may be a very effective means of eliminating mutant cells from subsequent branch (or ramet) generations (Grant Hackett, pers. comm.).

cell in terms of number of mitotic generations from zygote and mutation frequency, then somatic mutations may be less frequent in short shoots in comparison to long shoots. Short shoots are familiar in many gymnosperms (e.g., *Ginkgo* and the Coniferales), but they also occur in many dicotyledonous trees (e.g., *Acer, Betula, Corylus, Fagus*, and many rosaceous fruit trees). In silver birch *(Betula pendula)*, Maillette (1982a) reported that short shoot buds normally formed female catkins, and male catkins formed on the tips of long shoots.

Wilson (1966) has made a detailed quantitative study of short and long shoots in the temperate tree *Acer rubrum* L. Long shoots were defined as branches elongating more than 2cm per year and short shoots less than 2cm per year. In a seventy-year-old shoot system, 85–90% of all branches were short shoots. In figure 6.10 Wilson's nomenclature for branching in *A. rubrum* is shown. Long shoots are related to the main shoots by order numbers, and successive whorls of branches on the main stem of second-order lateral branches are given group number designations proceeding from the youngest to the oldest whorl. In figure 6.11 the distribution of flower buds on successive whorls is plotted for short and long shoots; it is clear that the majority (80–90%) of the flowers are formed on short shoots.

In addition to the anatomical and morphological aspects of branching that promote the fixation or loss of somatic mutations in ramets, branching pattern may strongly influence the degree of ramet competition within a genet. Just as in our discussion of apical meristems, here there is also an inverse relation between order and competition. In stochastic meristems, maximum cell competition was possible, whereas in structured meristems the more rigidly determined developmental destinies of cell lines with regard to initials retarded such competition. Similarly, the development of the vegetative body in vascular plants shows such random and determined components in varying degrees depending upon the species. Tomlinson (1982) proposed that the construction of the plant body is both *deterministic*[2] (i.e., established by genetic processes, or the "design" of the plant) and *opportunistic* (i.e., produced by the interaction of the genotype with the environment, or by "chance" events). As Tomlinson correctly stresses, this is not the same as the contrast between genotype and phenotype, because although the plant's capability for response is influenced by chance events, this capability is itself genetically determined since it differs widely in

2. This should not be confused with the mathematical usage of the same term which refers to random events.

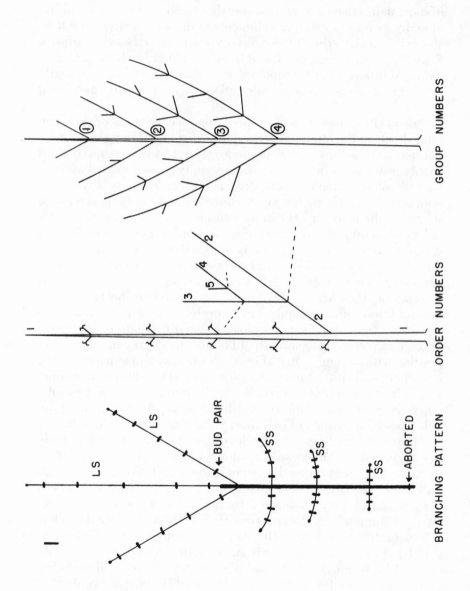

GROUP NUMBERS

ORDER NUMBERS

BRANCHING PATTERN

Figure 6.10. Nomenclature for branching in *Acer rubrum*. BRANCHING PATTERN: A two-year-old long shoot and its laterals. The one-year-old portions are drawn in thin lines with the position of the lateral buds shown. On the two-year-old portion of the parent long-shoot (thick line), the upper pair of buds has developed into lateral long-shoots, the three middle pairs into lateral short-shoots, and the lowest pair has aborted. ORDER NUMBER: Each lateral branch has an order number one greater than its parent branch. GROUP NUMBER: A group of second order branches is analogous to a whorl of branches in a pine tree. The groups are numbered from the top of the tree down. The maximum number of groups of live branches documented by Wilson was 29. (From Wilson 1966.)

different plant groups. For example, single-stemmed palms cannot respond to crown damage because they lack lateral meristems, whereas most trees have such meristems and thus can respond and survive. If we expand the nature of the environmental chance events to include the internal environment (i.e., the occurrence of apical meristems having fixed somatic mutations), then one would predict that deterministic growth retards and opportunistic growth promotes ramet competition.

In trees, deterministic and opportunistic growth may be a consequence of the form of branching. Buds may develop into elongating orthotropic or plagiotropic branches or remain dormant or form short spurs. Two general modes of tree growth have been recognized—excurrent and deliquescent (Gray 1887).[3] The former has a single

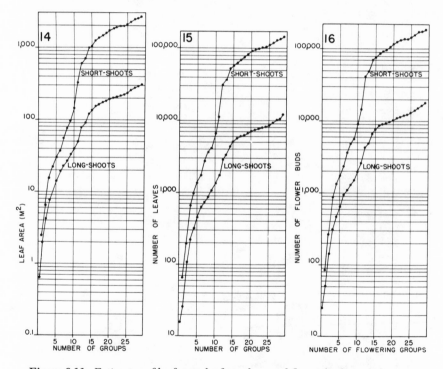

Figure 6.11. Estimates of leaf area, leaf number, and flower bud number on trees of *Acer rubrum* with 1 to 29 groups of second order long-shoots. (From Wilson 1966.)

3. Tree architecture is, of course, more complex and its nomenclature even more complex (see Hallé, Oldeman, and Tomlinson 1978, and Guédes 1982 for conflicting views). The relative possibilities of opportunistic and deterministic growth are different for practically all of the tree models described by these authors.

well-marked leader shoot (as in many conifers), whereas the latter has a large number of branches of roughly equal status. There are all gradations between these two modes. Branching pattern is often strongly genetically determined (many horticultural mutants with unusual branch conformations are known). Although intuitively one would suspect that the difference between excurrent and deliquescent growth forms is due to greater apical dominance in the former, Brown, McAlpine, and Kormanik (1967) have shown that this explanation is too simplistic. In deliquescent species such as oaks, hickories, and maples, the lateral buds on the current year's twigs are completely inhibited, whereas in excurrent conifers and angiosperms many lateral buds on the current year's twigs elongate to varying degrees. Thus the twigs of deliquescent species exhibit stronger apical dominance than those of excurrent species! In spite of this difference in apical dominance, the mainstem leader outgrows the laterals in excurrent species (unless the leader is damaged). Brown, McAlpine, and Kormanik (1967) have coined the term *apical control* to describe this relationship between leader and laterals. The stronger the *apical control* the more deterministic is growth, with the consequent lessened possibilities for competition and displacement of the leader by a more genetically and physiologically vigorous lateral. For example, the mainstem apical meristem in a conifer could become populated with mutant cells of lower vigor and still not relinquish its position as the primary meristem because of the developmental suppression of the lateral nonmutant branches. Of course if the mainstem apical meristem becomes moribund, lateral branches will overtop it and take over its role. The important question relating to ramet competition is how much of a decline in vigor in this terminal meristem can occur before competition with the subadjacent lateral branches is initiated. In contrast, in deliquescent forms one would expect that competition between potential terminal mainstem branches (as well as laterals) would occur continually.

In contrast, competition among lateral branches is more complex to analyze. With regard to lateral bud development, excurrent species with weak apical dominance may allow maximum competition among the developing lateral branches. On the other hand, deliquescent species exhibiting strong apical dominance manifested in lateral bud suppression may allow only the most vigorous of buds to later develop into branches. In the former the competitive emphasis is on developing and interacting lateral branches, whereas the latter is based upon bud vigor (of course branch competition must occur as well). Since we have little information on the quantitative aspects of these various levels of branch or ramet competition, only qualitative distinctions can be made.

External environmental factors also can contribute to altering the amount of branch competition in trees. Abiotic factors influencing the number of branches produced or the growth form will influence branch competition. Biotic factors such as herbivory have similar consequences. For example, in the pinyon pine *(Pinus edulis)*, herbivory by the moth *Dioryctria abovitella* may considerably alter tree architecture, converting trees to shrub-like forms (Whitham and Mopper 1985). Repeated attacks by the white pine weevil will convert the strongly excurrent white pine *(Pinus strobus)* to the almost deliquescent cabbage pine so common in abandoned pastures in northeastern America.

Demographic studies of bud populations within deliquescent tree crowns have documented differential patterns of bud survival with regard to bud position. Maillette (1982a, b) has shown that buds upon branches of different orders show very different survival and growth properties.[4] These bud populations were studied as populations with demographic properties as summarized in the following equation:

$$N_{t+1} = N_t + B - D$$

where N_t and N_{t+1} are the numbers of buds in the population at times t and $t + 1$, B is the number of new buds, and D the number of buds lost during the time interval. Maillette studied different trees of silver birch, calculating bud production for a winter-to-winter cycle and presenting it in a matrix format. In figure 6.12 this matrix formulation is shown. The transition matrix **T** multiplies a column vector V_t, whose elements are numbers of buds in each branch order at time t. The product **T** $\times$ V_t is another vector V_{t+1} which gives us the distribution of buds in the various branch orders at time $t + 1$. For subsequent times, the same transition matrix is used for all multiplications, thus,

$$\mathbf{T} \times V_t = V_{t+1}; \quad \mathbf{T} \times V_{t+1} = V_{t+2}; \quad \ldots \quad \mathbf{T} \times V_{t+n} = V_{t+(n+1)}.$$

In figure 6.13 the transition matrixes of bud production for three silver birch trees are shown. The non-zero elements in the upper right hand corners of these matrixes are always the largest numerical

---

4. The tree trunk was designated order number 0; branches growing directly from the trunk were ordered 1; branches growing on order 1 branches were of order 2, etc. The reader is referred to chapter 3 for a description of matrix notation and usage.

values in the matrix; this indicates that the majority of bud production is occurring on the branches of orders 0 and 1. This concentration of bud production in the top of the tree is seen clearly when the matrix for all buds is split into separate matrixes for the top and the bottom of the trees (figure 6.14). It is clear that the matrixes for the tops of the trees have much larger values than the bottom matrixes. Maillette has noted that changing the values within these matrixes will alter tree shape and form. These studies show that even within deliquescent growth forms, various kinds of developmental orders and priorities prevail. Models of branch competition with regard to somatic mutations will be greatly constrained by these nonrandom or developmentally structured aspects of bud and branch development.

$$
\begin{bmatrix}
P_0 & 0 & 0 & 0 & \cdots & 0 & \cdots & 0 & 0 & 0 & 0 \\
P_{0+1} & P_1 & 0 & 0 & \cdots & 0 & \cdots & 0 & 0 & 0 & 0 \\
P_{0+2} & P_{1+1} & P_2 & 0 & \cdots & 0 & \cdots & 0 & 0 & 0 & 0 \\
0 & P_{1+2} & P_{2+1} & \cdot & \cdots & \cdot & \cdots & 0 & 0 & 0 & 0 \\
0 & 0 & P_{2+2} & \cdot & \cdots & \cdot & \cdots & 0 & 0 & 0 & 0 \\
\cdot & \cdot & \cdot & \cdot & \cdots & \cdots & \cdot & \cdot & \cdot & \cdot \\
\cdot & \cdot & \cdot & \cdot & \cdots & \cdots & \cdot & \cdot & \cdot & \cdot \\
\cdot & \cdot & \cdot & \cdot & \cdots & \cdots & \cdot & \cdot & \cdot & \cdot \\
0 & 0 & 0 & \cdot & \cdots & \cdots & \cdot & 0 & 0 & 0 \\
\cdot & \cdot & \cdot & \cdot & \cdots & \cdots & \cdot & \cdot & \cdot & \cdot \\
\cdot & \cdot & \cdot & \cdot & \cdots & \cdots & \cdot & \cdot & \cdot & \cdot \\
\cdot & \cdot & \cdot & \cdot & \cdots & \cdots & \cdot & \cdot & \cdot & \cdot \\
0 & 0 & 0 & 0 & \cdots & \cdot & \cdots & \cdot & P_m & 0 & 0 \\
0 & 0 & 0 & 0 & \cdots & 0 & \cdots & \cdot & P_{m+1} & P_n & 0 \\
0 & 0 & 0 & 0 & \cdots & 0 & \cdots & 0 & P_{m+2} & P_{n+1} & P_p
\end{bmatrix}
\times
\begin{bmatrix}
n_{t,0} \\ n_{t,1} \\ n_{t,2} \\ n_{t,3} \\ n_{t,4} \\ \cdot \\ \cdot \\ \cdot \\ \cdot \\ \cdot \\ \cdot \\ \cdot \\ n_{t,m} \\ n_{t,n} \\ n_{t,p}
\end{bmatrix}
=
\begin{bmatrix}
n_{t+1,0} \\ n_{t+1,1} \\ n_{t+1,2} \\ n_{t+1,3} \\ n_{t+1,4} \\ \cdot \\ \cdot \\ \cdot \\ \cdot \\ \cdot \\ \cdot \\ \cdot \\ n_{t+1,m} \\ n_{t+1,n} \\ n_{t+1,p}
\end{bmatrix}
$$

P_x = Production rate of buds of order x by buds of order x
P_{x+1} = Production rate of buds of order $x + 1$ by buds of order x
*P_{x+2} = Production rate of buds of order $x + 2$ by buds of order x
$n_{t,x}$ = Number of buds of order x at time t

$$P_x = \frac{\text{Number of buds of order } x \text{ at time } t}{\text{Number of buds of order } x \text{ at time } t - 1}$$

* No production rate P_{x+n} was ever found where $n > 2$.

Figure 6.12. Theoretical matrix model of bud production in silver birch. (From Maillette 1982b.)

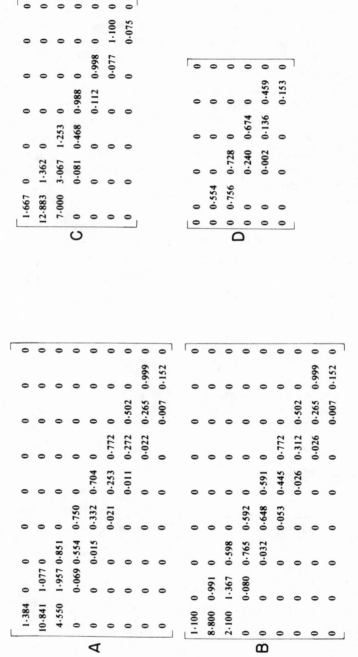

Figure 6.13. Total transition matrixes of bud production of silver birch trees growing in the open at Treborth gardens. Each element is a weighted average of bud production rates of (a) trees B1, B2, and B3; (b) tree B1; (c) tree B2; (d) tree B3. (From Maillette 1982b.)

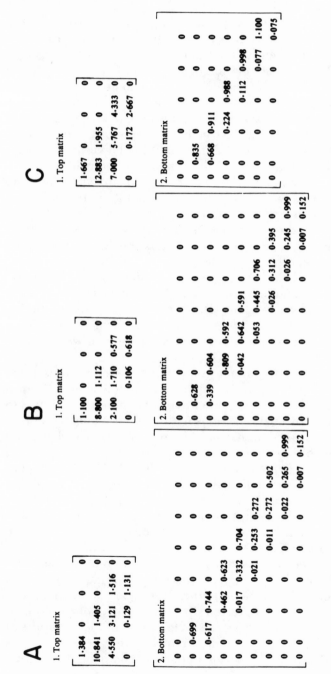

Figure 6.14. Transition matrixes for bud production on top and bottom branches of silver birch trees. (a) Weighted average of trees B1, B2 and B3; (b) tree B1; (c) tree B2. Each element is a weighted average of the bud production rates of the relevant branches of trees B1, B2 and B3. (From Maillette 1982b.)

Rhizomatous plants have all of the branching characteristics described for shrubs and trees (monopodial and sympodial growth,
long and short shoots) as well as rhizome branching patterns which
may or may not enhance sib-ramet competition. Bell and Tomlinson
(1980) noted that the architectural pattern of a rhizomatous plant is
based upon three simple parameters of the rhizome system: a linear
component, a component of divergence from linearity, and a spacing
component by means of which meristems or appendages are separated in time and space. An instructive example of these architectural parameters is the pattern of growth of the flowering herb *Medeola virginiana* L. In figure 6.15 the generalized branching pattern
of a population of sib ramets is depicted in an idealized manner (the
vagaries of environmental factors have been omitted). It is clear that
ramets are positioned with a minimum of direct overlaps; this is a
consequence of the three rhizome parameters already discussed and
an additional factor, ramet longevity. In this species, an individual
ramet has a one-year life span; what survives the following year are
its daughter ramets. Ramet competition in *Medeola* is reduced by
the nature of the branching system which minimizes the overlap of
ramets in time and space. Contrast this branching pattern with that
of the sand dune plant *Carex arenaria* L., for which there is no
obvious adaptive strategy to minimize competition between ramets
(figure 6.16).

Probably the most important component of rhizomatous plant
architecture influencing somatic mutation and ramet competition is
ramet longevity. The longer a ramet lives, the more difficult it is to
displace in competitive interactions with younger ramets. An older
ramet may more easily dominate a site and not be displaced by more
genetically and physiologically vigorous younger ramets. For example in the ferns, *Thyrsopteris elegans,* the monotypic genus endemic to the Juan Fernandes Islands, is an example of long-lived
ramets; to quote Bower, "It is a fern with an upright axis, sometimes
as thick as a man's thigh, and three to five feet high. It spreads by
runners, from which shoots come up at some distance from the
parent plant" (1926:260). The fern *Matteuccia struthiopteris* has a
similar clone structure consisting of long-lived erect rhizomes interconnected by stolons. In angiosperms long-lived ramets are found
in many taxa (e.g., *Fagus* and *Populus*). If we compare the possibilities of ramet competition between sib ramets in *Medeola virginiana*
where the individual ramets last a single year with the possibilities
of a young ramet displacing an older ramet (a tree) in *Populus,* it is
clear that ramet longevity is an important component of ramet
competition.

SECONDARY GROWTH
(VASCULAR AND CORK CAMBIA)

The capacity to form secondary vascular tissue occurs very early in land plant evolution. By the end of the Devonian, secondary growth was present in representatives of all major plant groups then in existence (Barghoorn 1964). Studies of Carboniferous fossils indicate that the woody pteridophytes extant at that time had evolved a vascular cambium with very distinctive properties when compared to modern seed plants. Recent studies of Carboniferous arborescent lycopods and sphenopsids indicate that the vascular cambium in these widely divergent cryptogamic groups was a determinate meristem. Fusiform initials in the vascular cambium did not undergo anticlinal divisions. Thus, the number of fusiform initials remained constant as the girth of the vascular cylinder increased, the increased circumference was accommodated by an increase in fusiform initial size (figure 6.17). Since these initials were not capable of increasing in size infinitely, the duration of cambial activity was circumscribed by the limits of growth of the fusiform initials (Cichan 1985a, b, 1986).

In the monocotyledons, the tree-like habit occurs in three groups, the screw pines, the palms, and the Agavaceae (see Philipson, Ward, and Butterfield 1971 for review). In the screw pines *(Pandanus)*, the stems are incapable of secondary thickening, the weight of the crown is supported partially by stilt and prop roots. In the palms, cell divisions immediately behind the apical meristem are very frequent and result in a rapid increase in diameter of the stem. (See Stevenson 1980 for an analogous pattern of development in the cycads.) Further increase in bulk of the palm stem is due to cell division and cell enlargement remaining prevalent throughout the tissues of the stem (diffuse secondary thickening) (Tomlinson 1961). In the Agavaceae a cambium is present, but it is unlike cambia found in the dicotyledons (Cheadle 1937; Stevenson and Fisher 1980).

Secondary growth in gymnospermous and dicotyledonous plants is based upon mitotic divisions in the lateral meristem known as the vascular cambium. The cambial fusiform initials normally divide to give rise to phloem tissues to the outside and xylem tissues to the inside of a stem as well as have the capacity to increase in number as the vascular cylinder increases in girth. The fusiform initials have the capacity for anticlinal divisions; thus, the cambium theoretically

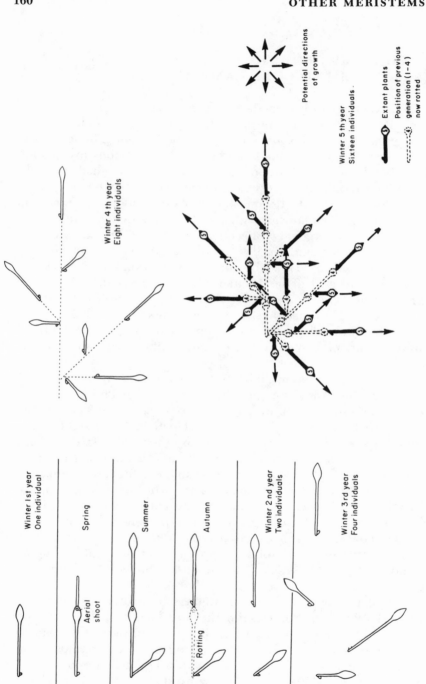

Potential directions
of growth

Winter 5 th year
Sixteen individuals

Extant plants
Position of previous
generation (1–4)
now rotted

Winter 4 th year
Eight individuals

Winter 1st year
One individual

Spring

Aerial
shoot

Summer

Autumn

Rotting

Winter 2 nd year
Two individuals

Winter 3 rd year
Four individuals

Figure 6.15. Rhizome branching and longevity in *Medeola virginiana*. Each ramet is represented by a rhizome segment which in winter consists of a tuberous distal portion and a proximal stoloniferous portion with a basal bud. In spring this produces an erect leafy shoot distally (not in the plane of the diagram) and by sympodial branching a distal renewal shoot growing in the direction of the parental segment, forming by autumn the next dormant segment. The basal bud produces a similar segment diverging at an angle of 45° from the direction of the parent segment. Increase in ramet number is exponential, i.e., two ramets after the second year, then 4, 8, 16 after four years of growth (fifth winter). Since ramets last one year, the pattern of branching minimizes superposition of ramets. (From Bell 1974; Bell and Tomlinson 1980.)

is capable of indefinite growth. Somatic mutations occurring in the cambium may be passed on to more permanent and biologically significant tissues through the formation of adventitious buds and/or cauliflory. Adventitious buds may give rise to branches that may flower and thus pass somatic mutations to the gametes. Cauliflory, the development of flowers on older branches or the trunk of a tree, may or may not represent cell lines derived from the vascular cambium. In some species, inflorescences or flowers originate in leaf axils on young shoots from primary meristems that have remained dormant for extended periods before flowering (e.g., cocoa), whereas in other species the flower buds are truly adventitious (see Hallé, Oldeman, and Tomlinson 1978 for examples). It is only in the latter that cambium derived cell lines may contribute to the sporogenous tissue.

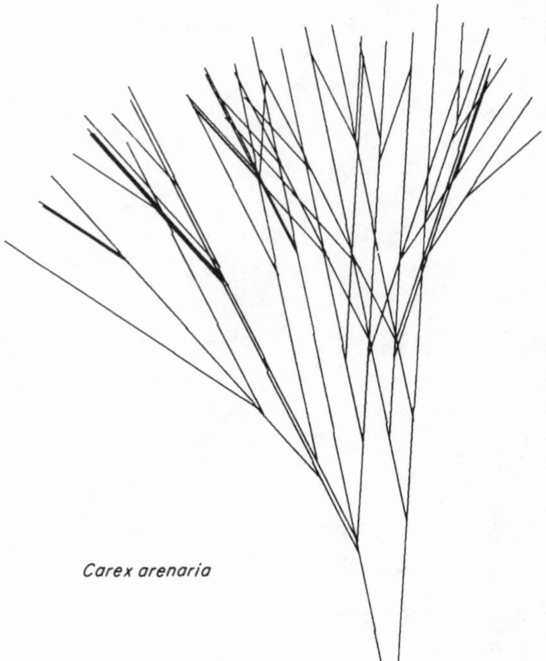

Carex arenaria

Figure 6.16. *Carex arenaria* (Cyperaceae). Plan view of an eight-year-old rhizome generated by computer on the basis of observed examples. (From Bell and Tomlinson 1980.)

The cambium is a meristem that has large numbers of fusiform initials; e.g., a 10m-tall pine tree may have 10^7 initials on the stem, branches, and roots (B.F. Wilson, pers. comm.). These large populations of mitotically active cells almost make it a certainty that mutant cells must be relatively common in this meristem. Many woody plants have cambia with stochastic characteristics that may, at least indirectly, promote cell competition. Cambial behavior in both gymnosperms and woody dicotyledons is characterized by both high rates of production of initials and almost equally high rates of loss (see review by Philipson, Ward, and Butterfield 1971).[5] In figure

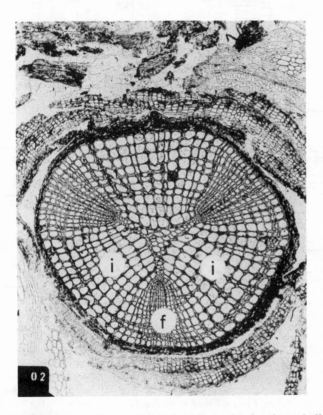

Figure 6.17. Secondary xylem in the Carboniferous fossil *Sphenophyllum pluri-foliatum*. Cross section of the stem showing triarch primary xylem surrounded by fasicular (f) and interfasicular (i) zones of secondary xylem. The pattern of tracheid enlargement from inner to outer wood may be indicative that cambial fusiform initials increased in size during secondary growth. (From Cichan 1985b.)

5. In some herbaceous dicotyledons, the loss of fusiform initials is minimal (Cumbie 1963, 1969).

6.18 is shown a diagrammatic representation of the loss of fusiform initials in the conifer *Chamaecyparis*. Bannan (1950) observed that over 1,100 anticlinal divisions of the fusiform initials in *Chamaecyparis* were required for a net gain of only 162 functional initials, suggesting that fusiform initials are under intense competition in

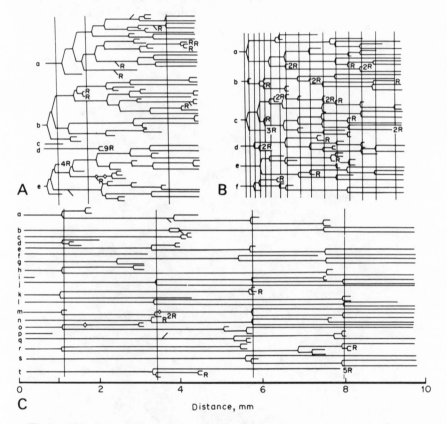

Figure 6.18. Diagrammatic representation of the succession of tracheid rows in selected lineal series. The diagrams are to be interpreted as showing the relative time and frequency of anticlinal divisions involved in the production of new fusiform initials, the disappearance of fusiform initials from the cambium and the origin of rays. A: young, vigorous stem, *Chamaecyparis lawsoniana* (A. Murr.) Parl. B: young, slow-growing stem, *C. thyoides* (L.) B.S.P. C: old vigorous stem 22 cm in dia., *C. thyoides*. Equal forkings of the horizontal lines indicate centered pseudotransverse division, unequal forkings, eccentric pseudotransverse division, and side branches, lateral divisions. The termination of horizontal lines indicates the disappearance of fusiform initials from the cambium, and the letter R signifies the establishment of one or such a number of rays as indicated by the preceding digit. The vertical lines represent the boundaries of annual rings. (Summarized from Bannan, Figure from Philipson, Ward, and Butterfield 1971.)

an overcrowded environment. Bannan and Bayly (1956) point out that the overproduction accompanied by extensive cell loss is a characteristic of conifer cambium. Typically the largest fusiform initials survive and repeat the cycle of elongation and division, thus there is a continued selection of the longest newly formed fusiform initials. Other than contributing to the maintenance of cell length in the secondary vascular tissues, how this competition influences the spectrum of cell genotypes proliferating in the cambium and, in turn, the genotypes present in adventitious buds and reproductive organs is unknown.

Although no information is available regarding error rates and DNA synthesis in plant meristems, one might speculate that DNA fidelity would be maintained at different levels in apical meristems and in the vascular and cork cambia. This is based upon the supposition that errors incur a greater biological cost in the former than in the latter. Cells derived from apical meristems are more likely to contribute their genotypes to the next generation than cells derived from the cambia. Daughter cells derived from the cambia are destined for what could be termed genetically inert tissues (composed either of dead cells or cells with defined life spans and fairly limited metabolic activities), whereas cells from the apical meristems give rise to cells that must continue to be active mitotically. Finally, one would guess that the opportunities for cell competition are greater in these secondary meristems.

PARTHIAN REMARKS

In chapters 4, 5, and 6 the above-ground primary and secondary meristems have been discussed from the viewpoint of the loss or fixation of somatic mutations. The long-term consequences of mutation loss or fixation were stressed, i.e., how meristem topologies and architectures influenced the probability of the transmission of mutations to the meiocytes and, ultimately, the next generation. In the next chapter, the short-term or immediate consequences of somatic mutation will be considered. As our discussions of meristems have shown, plant tissues and organs may occasionally become populated with high frequencies of mutant cells. The problem that is of immediate importance (from the viewpoint of the plant!) is the development of a relatively normal and functional plant structure from a mixture of variously mutant and nonmutant constituent cells. In chapters 7 and 8 this problem will be considered.

CHAPTER SEVEN

Phenotypic Responses to Mutation and Environmental Mutagens

ABNORMAL AND DEVIANT GROWTH

Growth and development of vascular plants is ultimately based upon cell divisions in various kinds of primary and secondary meristems. The total number of meristematic cells populating these meristems is so large it is almost axiomatic that any plant (other than perhaps small annuals) must be to some degree a chimera for postzygotic mutations. For example, considering only cambial initials, the number of initials in even a moderate sized woody plant (0.1m stem diameter) exceeds 1×10^6 initials per meter stem length.

Because each leaf subtends an axillary bud which may develop into a short or long shoot (at least in dicotyledonous plants), the potential number of apical meristems and, consequently, the total number of apical initials per genet are also very large. Values of 10^5 to 10^6 buds per tree are not unreasonable. Maillette (1982b) estimated that 2×10^5 buds were present on a silver birch (*Betula pendula*) tree 10m in height. If three apical initials are present in each apical meristem, the number of apical initials in this tree exceeds half a million. If we add to these initials the dividing cells in leaf primordia and other developing organs, then the total population of dividing cells per genet is very large indeed. In both corn and tobacco, leaf primordia arise from approximately 100–200 cells in the shoot apical meristem (Poethig 1984a, b; Poethig and Sussex 1985a, b). These numbers of apical and primordium "initials" are a gross underestimate of the population of dividing cells in a plant, since such "initials" give rise to daughter cell lines which continue dividing. Thus, given reasonable mutation rates per cell cycle (see chapter 2), almost all genets must be genetic mosaics to some degree.

Although there is not a lot of information about the magnitude of somatic mutation rates (or frequencies) (see chapter 2), the available data indicate that somatic mutation rates may vary between individuals as well as between the tissues and organs of a single individual. Unfortunately, the majority of data documenting differential mutation in various plant meristems and organisms are estimates of mutation frequencies rather than rates. Mutation frequency within an organ is a function of both the mutation rate per cell cycle and the number of meristematic cells during different stages of organ development. Since the number of meristematic cells giving rise to an organ is, in most cases, unknown, the most reliable estimates of differential mutation rates are based upon comparisons of the same organ in different individuals.

Estimates of chromosome breakage and the consequent generation of chromosome aberrations for different tissues of *Trillium grandiflorum* indicated that some tissues may be very mutable. Rutishauser (1956) and Rutishauser and La Cour (1956a) documented a very high frequency of chromosome aberrations in endosperm cells (0.158 chromosome breaks per nucleus) whereas root tissues were essentially aberration free. Giles (1940) studied spontaneous chromosome aberrations in species of *Tradescantia* as well as in their F_1 and F_2 hybrids. Pollen tube mitosis was scored for chromosome aberrations resulting from one or two chromosome breaks. The frequency of aberrations varied considerably between hybrid and nonhybrid plants, often being higher in the hybrids (chromosome aberrations in hybrid endosperm tissues are also more frequent) (Rutishauser and La Cour 1956b). Giles calculated the aberration frequency on a per chromosome basis. For nonhybrids the spontaneous frequency was 4×10^{-4} chromosome aberrations per chromosome. Since these plants had $2n = 12$, the frequency of aberrations per mitosis is approximately 4×10^{-3}. This value is similar to the frequency of mitotic figures with aberrations for root tissue in *Vicia faba* ($2n = 12$) (Davidson 1960). Based upon these studies, a conservative estimate for the frequency of cells wth chromosome aberrations may be 1×10^{-3}. If chromosome aberration rates per mitotic cycle are a function of the number of chromosomes per cell, then the rates of chromosome aberrations may exceed 1×10^{-3} in genomes with higher chromosome numbers. Since even a small annual consists of approximately 1×10^8 cells and the majority of plants exceed this value by many orders of magnitude (Dyer 1976), the actual number of cells of a genet that have chromosome aberrations may be quite high.

Probably one of the most important mechanisms for generating genotypically and phenotypically different somatic cells is mitotic crossing over (see figure 1.1). This process will generate somatic cells homozygous for recessive alleles inherited from the previous generation as well as postzygotic mutations. Based upon studies with marker genotypes that generate twin spots when the appropriate somatic crossover and mitotic orientation occurs, the mitotic crossover frequency has been estimated as 1×10^{-5} to 1×10^{-4} per cell (this estimate is probably the rate per cell division as well) in a number of different plant species (Evans and Paddock 1979). Since this estimate is for a pair of homologous chromosome arms and the number of pairs of such homologous chromosome arms is equal to the diploid chromosome number (assuming meta or submetachromosomes), the mitotic crossover frequency per genome per cell may be much higher. In angiosperms, for example, the average chromosome number is $2n = 32$ and many species exceed this value (Grant 1963). Thus the mitotic crossover frequency for the entire genome may be as high as 1×10^{-4} to 1×10^{-3} mitotic crossovers per cell.

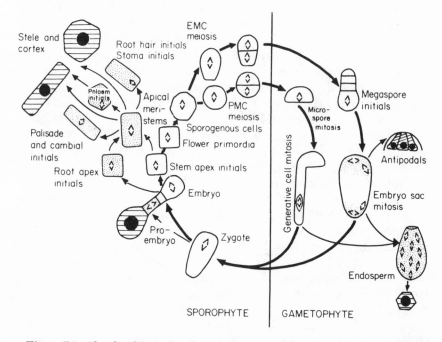

Figure 7.1. The dividing cells of the angiosperm life cycle. The cycle of broad arrows represents the direct cell lineage from one generation to the next. (From Dyer 1976.)

It should be noted that although it has often been stressed that plants lack a germline and, therefore, that somatic mutations may be transmitted mitotically to germinal tissues, the majority of meristems typically do not do so. Many meristems give rise to tissues and organs that are genetic dead-ends (figure 7.1). For example, the cabium forms secondary xylem and phloem, tissues seldom giving rise to reproductive organs (except for adventitious branches). Other meristems give rise to organs that are shed periodically (e.g., leaves, fruits, flower parts, even branches), therefore mutations accumulated in the cells of these organs will have little long-term genetic impact on the gene pool. Hardwick (1986) has pointed out that such a programmed senescence of meristems or ramets may lessen the long-term consequences of "selfish" mutations (i.e., those with high cell fitness but low organismal fitness).

Of more importance with regard to the individual genet are the short-term effects of somatic mutations. What are the developmental consequences of somatic mutations? Can tissues and organs develop and function normally from mixtures of mutant and nonmutant cells? Are there categories of somatic mutations that are life-threatening to a genet or ramet? In animals, these kinds of questions understandably have attracted a great deal of critical attention. Considering somatic mutation in its broadest sense (see chapter 2), in animals such genetic changes in somatic cells may result in cell variants with increased fitness, that is, they can multiply faster than normal cells or can displace nearby cells whenever space is limited. Such variant cells can form tumors of various sizes; if the abnormal cells remain within their normal territory, benign tumors result, whereas if the abnormal cells have the ability to spread to new sites (metastasis) they are called malignant tumors or cancers (Cairns 1978). The latter are generally more life-threatening than the former (see Hiatt, Watson, and Winsten 1977 for general review). In plants such phenomena appear to be much less important. The diminished importance of cancer-like phenomena is in part related to the fundamental differences between plant and higher animal development and organization.

In table 7.1 some of the major differences between plant and higher animals with regard to development and organization are listed. With regard to susceptibility to cancer-like phenomena (presumably arising from somatic mutations), probably the most significant factor is the impossibility of metastasis in vascular plants. Plant somatic cells have permanent domains since they are encased within cell walls which strongly adhere to the cell walls of adjacent cells. At the organismal level, vascular plants lack a circulatory system through which cells can move to other locations within the organism. These two plant characteristics (cell walls and lack of a circulatory system) confer almost total immobility to plant cells within the organism (but see chapter 5 for a discussion of the tunica as a selfish tissue); consequently malignancy is not possible (Jones 1935).

Although malignant tumors are lacking in plants, localized aberrant growths are not uncommon. Such abnormal growths may take two forms, depending upon whether the atypical patterns occur at the histological (amorphous changes or neoplasms) or organ (teratomata) level (Bloch 1954; Sinnott 1960). Amorphous changes include tumor-like growths similar to wound callus, crown gall, and the genetic tumors occurring in certain interspecific hybrids (Smith 1972; Schieder 1980) or resulting from specific mutations as, for example, the pea pod neoplasms caused by the dominant gene Np

Table 7.1. Some characteristics that distinguish plant and animal organization and development. (In part summarized from Trewavas 1982 and Gottlieb 1984.)

PLANT (TRACHEOPHYTE)	ANIMAL (MAMMAL)
ca. 15–20 cell types	ca. 100 cell types
Majority of cell types are only distinguishable because of unusual cell wall structure; secretion is a critical event in cell differentiation.	Majority of cell types have specific cytoplasmic characteristics, numerous cell specific proteins (e.g., keratin, haemoglobin).
Cell movement almost nonexistent.	Surface cell contact and cell movement are critical in development.
Plastic development, many cells in a mature organ are totipotent.	Highly canalized development, once a developmental choice is initiated, cells and tissues remain irreversibly fixed in their chosen pathway.
Steps in development are repeated many times in many places since similar organ and tissue types differentiate in separate places and at separate times.	Most organs are differentiated only during embryogenesis according to highly regulated developmental schedules.
Immune system absent.	Immune system present.
Germline absent.	Germline present.

in *Pisum* (Snoad and Matthews 1969). Tumors occur spontaneously in natural plant populations and can be induced by ionizing radiation treatments; many are thought to represent somatic mutations (Kehr 1965). Tumors may also be induced to form in response to viral, bacterial, or fungal infection as well as insect predation and environmental treatments (Bloch 1965).

Of the pathologically induced tumors, the most well known is the induction of crown gall by the soil bacterium *Agrobacterium tumefaciens*. Crown gall tumors occur in a diverse array of seed plants (many dicotyledons and, to a more limited extent, in monocotyledons and gymnosperms). The cause of the disease is the transfer of genetic information (T-DNA element) from bacterium to plant via a tumor-inducing (Ti) plasmid. The T-DNA element is a portion of the Ti-plasmid DNA which is integrated into the plant genome (Stachel, Timmerman, and Zambryski 1986; Lichtenstein 1986 for literature). It has long been known that although normal plant cells require the phytohormones auxin and cytokinin to grow *in vitro*, cultures of crown gall cells do not have such requirements (White and Braun 1941). The phytohormone-independent characteristic of crown gall tissues is due to genes on the Ti plasmid that code for enzymes that can convert common plant metabolites to critical phytohormones (Thomashow et al. 1986). These phytohormones have key roles in controlling normal development, and tumor cell proliferation is, at least in part, due to the abnormal synthesis of phytohormones (Sachs 1975).

Abnormal growths at the histological level are teratomata and are well known in vascular plants (Masters 1869). The most common type of teratoma is fasciation, a term derived from the Latin *fascia* meaning band, and most fasciations are characterized by a band-shaped stem (Gorter 1965). Two forms of fasciation have been described depending on the cause of the altered morphology. *True fasciations* are characterized by a change in the structure of one growing point or apical meristem; *connations* arise from the fusion of two or several growing points. The distinction between these two forms is often difficult.

Morphologically, fasciation is a change from the normal round or polygonal stem to a stem that becomes flat, banded, or ribbon-shaped. In some cases the apical meristem becomes linear or comb-like and may develop numerous growing points. Fasciated stems may terminate in a coiled, snail-like helix or may become a distorted tangle of coils and partly broken stem. If the apical meristem develops numerous growing points, clusters of branches will produce a witch's broom (White 1948) (see figure 7.2). The most significant characteristics of fasciated growth are the increase in weight and volume of these tissues, in comparison to the normal tissues of the same taxa, and the loss of control of the growth area, which results in irregular and unpredictable growth. Excepting cell mobility, White (1948) correctly notes the parallels of this type of growth to animal cancer.

In conifers, witch's broom has been utilized as a source of new horticultural forms. Two general kinds of witch's broom occur, those caused by pathogens and those not associated with any known pathogen. The latter appear to be caused by somatic mutations. Waxman (1975) studied the progeny from cones produced by witch's brooms that appeared to be caused by somatic mutations. Typically in *Picea* and *Pinus* witch's broom forms only ovulate cones; thus pollen must originate from cones formed on wild-type branches (of the same or different trees). In table 7.2 the phenotypes of witch's broom progeny are tabulated. It is clear that in *Picea* and *Pinus* the dwarf trait (presumably witch's broom) is inherited as a dominant whereas in *Tsuga* the trait shows recessive inheritance. In addition to the dwarf and normal overall phenotypes, the progeny from witch's brooms are highly variable for other traits. Witch's brooms generally produce more cones than normal branches (Waxman 1975).

As with plant tumors, fasciations may result from a diversity of causes, including pathological responses to viral, bacterial, and fungal infections, predation by insects and environmental factors (including ionizing radiation) (Gunckel and Sparrow 1954). Of more importance to the present discussion is the evidence that such teratomata also may result from genetic causes. Fasciation was one of the seven traits in the garden pea that Mendel studied. He found that the fasciated phenotype was due to a recessive allele. More complex patterns of inheritance have been documented for fasciation in another legume, *Glycine max* (Albertsen et al. 1983). In many cultivated plants, fasciated forms have been selected because of their larger fruit size (e.g., pineapple, mulberry, strawberry, and tomato). (See Gorter 1965 and White 1948 for other examples of genetically based fasciation.) The primary cause of true fasciations is a change in the shape of the shoot apical meristem; this occurs at a certain point in the development of the plant. The usual pattern is for the apical meristem to change from the normal conical shape to a ridge-like form. Subsequent growth of this abnormally shaped apex results in the various aberrant morphologies associated with the fasciated syndrome.

Figure 7.2. Witch's broom in *Pinus silvestris* (= *P. sylvestris*) growing in Kazakh-stan. Open pollinated seed was collected from cones formed on the witch's broom (A). Seedlings grown from these seeds segregated into two phenotypes; 40% had the witch's broom phenotype (B) and 60% had the normal phenotype (C). (From Shul'Ga 1979.)

In the beginning of this section, the question was posed as to what kinds of somatic mutations might be the most developmentally disruptive and/or life-threatening to a vascular plant. Disregarding the obvious consequences of mutants that disrupt photosynthesis, which, of course, have the most dire consequences to an autotroph, probably the most developmentally disruptive mutants are those that alter either apical meristem shape or phytohormone metabolism. The former may result in fasciations and the latter in tumors. Comparatively, the susceptibility of plants to fasciations (and other teratomata) and tumors varies greatly. The phenotypic effects of exposing plants to ionizing radiation have been extensively catalogued. Recognizing, of course, that in addition to inducing mutation, ionizing radiation has a variety of biochemical and physiological effects in plants (Gunckel and Sparrow 1961), a comparison of these aberrant phenotypes is instructive.

The most common stem responses to irradiation are dwarfing and fasciation. Common leaf abnormalities include dwarfing, thickening, changes in form and texture, puckering of blade, abnormal curvatures, curling of leaf margins, distorted venation, fusions, and mosaic-like color changes. The most frequent modifications in floral development are in the form and number of floral parts, particularly of petals and stamens (Gunckel and Sparrow 1954, 1961). What is interesting in these surveys is the observation that plant tumors are not the most common phenotypic response to irradiation in most species (Stein, Sparrow, and Schairer 1959; Smith 1972). Spontaneous tumor formation is rare in most plant species and species hybrids. Even in the genus *Nicotiana* where spontaneous tumor formation occurs in interspecific hybrids, tumor-forming hybrids are

Table 7.2. Segregation of normal and dwarf seedlings from the seeds formed by ovulate cones borne by individual conifer witch's brooms. (From Waxman 1975.)

Species	Dwarf	Normal	Approximate Ratio
Picea abies	33	40	1:1
Pinus resinosa	30	42	1:1
P. strobus, No. 1	104	98	1:1
P. strobus, No. 2	373	368	1:1
Tsuga canadensis	621	1952	1:3

unusual. Among the 64 species recognized in *Nicotiana*, more than 300 interspecific hybrids have been studied, yet only about 30 of these produced tumors (Smith 1972). To quote Smith:

Considering the intricacy of the interrelated gene-controlled metabolic pathways that are finely coordinated in normal plant development as a result of selection during evolution for systems best adapted for survival, and considering the vast numbers of plants that have been hybridized and genetically altered by experimenters, it is in a sense surprising that so few reports can be found in the literature of spontaneous undifferentiated plant growth. (1972:147)

Plants appear, therefore, to be much less susceptible to tumorigenesis than higher animals. How plant developmental pathways are buffered against tumor proliferation is an important and often unrecognized question.

PHENOTYPIC RESPONSES
TO ENVIRONMENTAL MUTAGENS

As noted by Schmalhausen (1949), adaptive responses or modifications to environmental disturbances are primarily processes with a long antecedent history, that is, the disturbances must have occurred sufficiently often in the evolutionary past for the adaptive response to have evolved. With regard to phenotypic adaptations that may protect a plant from natural environmental mutagens, the responses to certain genotoxic chemicals and sunlight are probably the most important.

Plant characteristics that promote either sensitivity or resistance to chemical or physical mutagens have been mentioned incidentally in our discussion of mutation buffering. For example, meristem characteristics such as quiescent centers in roots have been shown to act as a reserve of cells that can repopulate meristems that have been damaged with ionizing radiation (Barlow 1978). Grodzinsky and Gudkov (1982) have suggested that meristems that are populated by a cell population that is variable with reference to cell cycle lengths as well as is highly asynchronous in terms of metabolism will present a broad array of cell sensitivities at the time of mutagen exposure. Thus, less sensitive cells may survive undamaged and repopulate the meristem. In the following discussions, other biochemical and anatomical characteristics that may affect the plant's survival with regard to environmental mutagens will be considered.

Xenobiotics

Plants are commonly exposed to a diversity of potentially mutagenic chemicals. These include various agrochemicals, soil, water, and air pollutants, as well as chemicals constituting the "more natural" components of their environments (e.g., heavy metals in certain soils, pyrolysis products after a fire, mycotoxins, plant decomposition products) (Clark 1982; Tazima 1982). Many of these chemicals may act directly as mutagens, whereas others require chemical modification by the target organism's metabolism before they are mutagenic (promutagen activation). In higher animals, the liver is the main organ involved in xenobiotic detoxification and often, incidentally, promutagen activation as a by-product of these detoxification and excretion pathways (see Hiatt, Watson, and Winsten 1977 for review). In plants, the cells have greater biochemical autonomy; therefore detoxification and excretion processes can take place within a single cell (Sandermann 1982) in contrast to mammalian systems where several organs with different cell types have to cooperate in order to metabolize, process and, finally, excrete environmental chemicals. Sandermann (1982) and Callen (1982) have reviewed xenobiotic metabolism in plants from an environmental mutagen perspective. A great diversity of transformation reactions have been documented in plants (including cytochrome P-450-dependent reactions, peroxygenase reactions, phenolase reactions, reductase reactions, hydrolytic enzyme reactions).

Probably the most versatile and widely distributed are the peroxidase reactions. Peroxidase enzymes occur in soluble as well as membrane-associated and cell wall-associated forms (Sandermann 1982). Plant cells are capable of various conjugation reactions, e.g., β-D-glucosides are important glucosides in plants whereas β-D-glucuronides are the major type of carbohydrate conjugation in the liver. In figure 7.3 the metabolism and excretion by plant cells is summarized. Two sites of "local excretion" occur, the deposition of conjugates into plant vacuoles and the copolymerization of foreign chemicals and their metabolites into lignin. The latter, being the second most abundant natural polymer, may be a very important sink for environmental chemicals (Sandermann 1982). Thus, in addition to enhancing cell wall rigidity, lignification may have a role in the plant cell's defenses against chemical mutagens.

Ultraviolet Shielding

Ultraviolet radiation exposure is almost an unavoidable environmental mutagen for land organisms relying upon sunlight for an energy source. The ultraviolet spectrum is generally classified into three biological divisions—UV-A, UV-B, and UV-C (table 7.3). Although UV-C, especially wavelengths of approximately 260nm, gives maximum effectiveness for inducing mutation, the atmosphere around the earth effectively eliminates radiation less than 280–290nm from reaching the earth's surface. Most of the UV-B band does reach the earth's surface, but the dose varies considerably in different environments. In lower latitudes (tropical environments), UV-B flux is enhanced because solar angles are closer to the zenith and the total stratospheric ozone column is thinner (figure 7.4).[1] In addition to latitudinal effects, snow cover, and cloud cover, altitude may also considerably alter UV-B flux (Caldwell 1979; Caldwell, Robberecht, and Billings 1980). There is substantial UV-B absorbance by nucleic acids (and proteins) and evidence for mutagenicity (Hannan, Calkins, and Lasswell 1980). The degree to which plants have evolved adaptations to lessen the impact of UV-B flux has been the subject of considerable speculation.

Plants have long been thought to possess adaptations that lessen the biological impact of UV-B radiation. (See Lockhart and Brodführer Franzgrote 1961 for a review of this early literature.) In addition to photo repair and dark repair, plant organs have properties

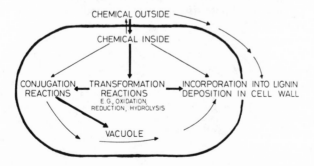

Figure 7.3. Proposed scheme for the metabolism and "local excretion" of environmental chemicals in plant cells. (From Scheel and Sandermann 1981.)

1. Ozone absorbs strongly in the UV-B portion of the spectrum.

that may either reflect UV-B or have outer tissues with a high degree of UV-B absorbance and thus act as a UV screen for the inner tissues. In many species the leaf epidermis has a low UV-B transmission (high absorbance or reflectance) and very often the upper epidermis has a lower UV-B transmission than the lower epidermis (Caldwell 1968; Robberecht, Caldwell, and Billings 1980). Soluble cell constituents such as carotenoids, flavones, flavonones, flavanols, and anthocyanins have absorption maxima in the UV-B band and have been hypothesized to be part of a plant's ultraviolet screen (Lockhart and Brodführer Franzgrote 1961; Caldwell 1971). Special features of plant surfaces such as leaf hairs, wax-layers, and flavone powder may all reduce UV-B transmission.

UV epidermal transmission is also responsive to the environment. Caldwell (1968) showed that as stems of the alpine species *Ranunculus adoneus* emerged from a rapidly melting snow bank, UV absorbance of the outer tissues increased markedly in parallel with the increase in UV-B radiation the stems received. Pate (1983) has postulated that the function of Δ^9-tetrahydrocannabnol in *Cannabis* is UV-B screening. Pollen commonly contains carotenes and flavonoid derivatives. Asbeck (1954) hypothesized that the presence of such yellowish compounds in pollen and spore exine is related to their function as components of an ultraviolet shield.[2] He noted that a yellow filter has the complementary chromatic absorption to blue-

---

Table 7.3. General biological divisions of the ultraviolet spectrum and common synonyms used in referring to the various regions. The wavelength limits are only approximate. (From Lockhart and Brodführer Franzgrote 1961.)

400–320 nm	320–280 nm	280–200 nm	Less than 200 nm
Long-wavelength UV	Middle-wavelength UV	Short-wavelength UV	Extreme UV
Near UV	Dorno region	UV-C	Vacuum UV
UV-A	UV-B	Germicidal region	Schumann region
"Black light"	Erythymal region		

2. Johnson and Critchfield (1974) reported that occasionally individual pine trees that form pollen without flavonoids are found. Such genotypes could be used in experiments to test Asbeck's hypothesis.

violet. (See also Stanley and Linskens 1974 for further discussion of pollen and spore UV-B shielding.)

The majority of UV-B effects on plants are destructive. Experimental studies with soybean cultivars have shown variability with regard to reduction in leaf area, dry weight, height, stunting, leaf chlorosis, and loss of apical dominance in response to UV-B irradiance (Biggs, Kossuth, and Teramura 1981; Teramura 1983). Growth reductions may also occur that are not a consequence of cellular damage which may be part of normal patterns of morphogenesis (Steinmetz and Wellmann 1986; Lindoo and Caldwell 1978). Whether such morphogenetic responses are part of a plant's defense against UV-B damage is unclear.

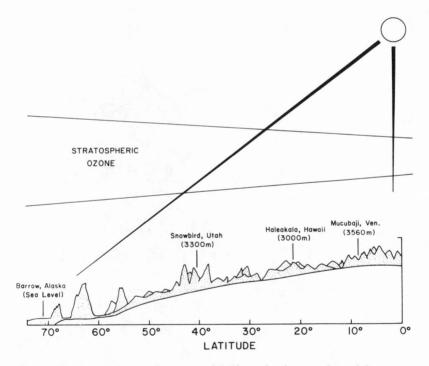

Figure 7.4. Diagrammatic depiction of the latitudinal arctic-alpine life zone gradient. The stippled area represents this life zone that progresses from sea level in the Arctic to high elevations at lower latitudes. The change in effective solar UV-B irradiance is due partly to a latitudinal gradient in thickness of the atmospheric ozone column, changes in prevailing solar angles that results in different pathlengths through the atmosphere, and elevation above sea level. Most of the atmospheric ozone is in the stratosphere. (From Caldwell, Robberecht, and Billings 1980.)

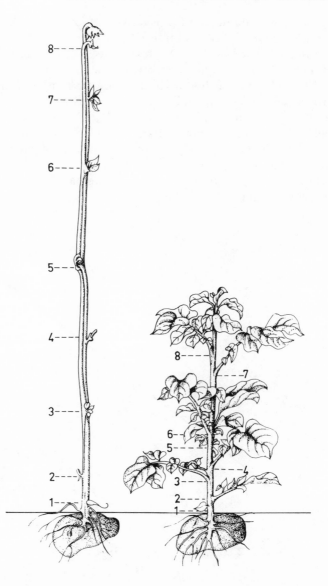

Figure 7.5. Illustrations of alternative developmental strategies. Genetically identical potato plants *(Solanum tuberosum)* were grown in the dark (scotomorphogenesis) or in natural daylight (photomorphogenesis). *Numbers* indicate the position of the corresponding leaves along the main axis to document the constancy of the phyllotactic pattern in light and dark: pattern specification was found to be independent of light even in those cases where pattern realization, i.e., actual growth of leaves, depends on light. (From Mohr, Drumm-Herrel, and Oelmüller 1984.)

As already discussed, blue/UV absorbing pigments (e.g., anthocyanin and/or other flavonoid compounds) have been considered an adaptive mechanism to protect a plant against high levels of short wavelength sunlight. In angiosperms, chlorophyll formation is stimulated by light. Dark-grown plants are etiolated. Etiolation (scotomorphogenesis) is an appropriate survival strategy under conditions where light is lacking or limiting, whereas photomorphogenesis is the appropriate strategy of development under conditions of light affluence (figure 7.5) (Mohr, Drumm-Herrel, and Oelmüller 1984). Upon exposure to the appropriate light stimulus, etiolated plants initiate mass synthesis of anthocyanin. This response has been interpreted as a protective adaptation against short wavelength sunlight. Mohr and Drumm-Herrel (1983) experimentally investigated the nature of light-stimulated anthocyanin synthesis in a number of angiosperm species. They found that although in all cases the synthesis of anthocyanin was under phytochrome control, interactions with pigments in the UV-B and UV-C portions of the spectrum could greatly influence the response. For example, anthocyanin synthesis in *Sinapis alba* was exclusively under the control of phytochrome, *Sorghum vulgare* and *Lycopersicon esculentum* had two nonphotosynthetic photoreceptors (phytochrome and a blue/UV light photoreceptor, "Cryptochrome") involved (Drumm-Herrel and Mohr 1981), and *Triticum aestivum* had phytochrome and a UV-B photoreceptor interacting. If all of these responses represent adaptations which shield the plant from short wavelength sunlight, it is clear that the physiological basis of each response must represent the solution of an evolutionary equation that includes ecological, anatomical, and ontogenetic variables.

CHAPTER EIGHT

Mutation Buffering

The immediate effects of somatic mutations may be neutralized either by mechanisms that enhance the loss or isolation of mutant cells from dividing cell populations or by mechanisms that reduce the phenotypic effects of mutations. In chapters 4, 5, and 6 the various characteristics of plant meristems that may influence diplontic selection were discussed; in this chapter we will consider molecular and developmental mechanisms that may buffer organisms against the phenotypic expression of somatic mutants. Mutations with little or no phenotypic effects are neutral with regard to the viability and vitality of the organism. Many aspects of an organism's biochemistry and development serve to quench the phenotypic effects of mutant alleles. Mutation buffering may occur at two levels: genes and gene products may have characteristics that reduce the phenotypic effects of mutant alleles at the cellular level, and developmental programs may reduce the phenotypic effects of mutant alleles at the organ and organism levels.

MOLECULAR MUTATION BUFFERING

The phenotypic consequences of mutation may be mitigated by the degenerate property of the biological code. Since 61 different codons are available to code for 20 different amino acids, one might have predicted each amino acid to be coded by approximately 3 codons. In actuality amino acids vary greatly in number of codons. Tryptophan and methionine have 1 codon each, whereas leucine, serine, and arginine have 6 codons each. Thus some amino acid sequences are more mutation proof than others. For example, the amino acid sequence Arg-Ser-Leu-Pro-Ala has 3,456 ($6 \times 6 \times 6 \times 4 \times 4$) ways of being coded in DNA whereas Trp-Met-Lys-Asp-Glu has only 8 ($1 \times 1 \times 2 \times 2 \times 2$), thus the former sequence is both more probable and less mutable (Dufton 1986). There is a general correlation between the number of codons for an amino acid and its relative occurrence in proteins (i.e., amino acids occur in proteins roughly proportional to their codon number) (King and Jukes 1969; Doolittle 1981; Dufton 1983) (figure 8.1). Dufton (1986) pointed out that protein interiors are biased in favor of those amino acids with a larger number of codons and, therefore, that the average interior residue is more probable and less mutable than an exterior residue. The significance of codon synonymy as a mutation buffer is a function of how often amino acids with multiple codons occur in proteins.

Assuming that two codons differing in only one base can mutate into each other more often than two codons differing by more than

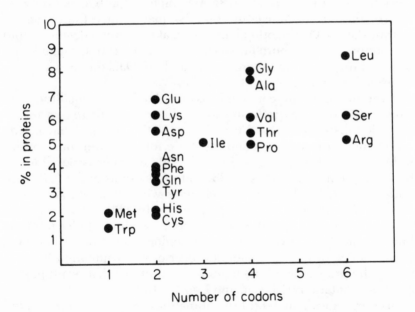

Figure 8.1. The percentages of the 20 amino acids in proteins (Doolittle 1981) compared to the number of synonymous codons that each possesses in the genetic code. Abbreviations: Ala = Alanine; Arg = Arginine; Asn = Asparagine; Asp = Aspartate; Cys = Cysteine; Glu = Glutamate; Gln = Glutamine; Gly = Glycine; His = Histidine; Ile = Isoleucine; Leu = Leucine; Lys = Lysine; Met = Methionine; Phe = Phenylalanine; Pro = Proline; Ser = Serine; Thr = Threonine; Trp = Tryptophan; Tyr = Tyrosine; Val = Valine. (From Dufton 1983.)

one base, mutation buffering is enhanced if chemically similar amino acids are specified by codons differing in only one base. Probabilistic analysis has shown that except for the amino acid serine, the biological code is organized in this fashion (Cullmann and Labouygues 1984; Figureau and Pouzet 1984; Labouygues and Figureau 1984). These authors contend that the code appears to be patterned to confer optimal resistance to the effects of mutations. The usage of synonymous codons is not random, although one might expect otherwise if all synonymous codons for a given amino acid are equivalent informationally. The nonrandom patterns of synonymous codon usage are more conspicuous in highly expressed genes (e.g., genes coding for ribosomal proteins) than weakly expressed genes. In their recent review of this literature, Li, Luo, and Wu (1985) suggest that highly expressed genes require rapid and frequent translation; thus, the mRNAs of these genes are biased for synonymous codons of amino acids which are recognized by the most abundant isoacceptors (tRNAs). Such codon bias not only increases the rate of translation but also may reduce the chance of mistranslation. On the other hand, in weakly expressed genes, rapid translation is not important; thus, synonymous codon usage is not so constrained and, therefore, more random. Data on plant synonymous codon usage is not yet available.

Point mutations are commonly assumed to be random. Studies of animal pseudogenes indicate nonrandomness with regard to nucleotide substitutions (Li, Wu, and Luo 1984). Since each nucleotide can change in three ways and there are four different nucleotides in DNA, a total of $4 \times 3 = 12$ nucleotide changes is possible. Of these 12 possible changes, 4 (33%) are transitions and 8 (67%) are transversions. If all nucleotide changes were equally likely, then transversions should be twice as likely as transitions. The reverse is actually so; the proportion of transitions is almost twice (59%) as high as the value expected under random mutation (33%). One consequence of this nonrandom aspect of point mutation is to increase the proportion of nucleotide substitutions that result in synonymous mutations (Li, Wu, and Luo 1984).

Of course not all nucleotide combinations are equally represented in genomes (if this were the case then the $(A + T)/(G + C)$ ratio would always equal unity). In the human genome, the dinucleotide 5′UA3′ (RNA) or 3′AT5′ (DNA) is discriminated against. The deficiency in this dinucleotide is generally explained by noting that two of the three nonsense codons begin with UA (i.e., UAA, UAG); thus the paucity of 3′AT5′ pairs in the DNA has the effect of lessening the probability that a frameshift mutation will create a nonsense codon (Alff-Steinberger 1987). This argument applies, of course, only to the portion of the genome that codes for proteins. In many plants this portion of the genome represents less than 1% of the total nuclear DNA.

The dinucleotide 5′CG3′ is under represented in many genomes. The cytosine of this dinucleotide is often methylated to give 5-methylcytosine; this results in a DNA modification known as DNA methylation. The significant point to note is that when DNA methylation occurs, the predominant methylated sequence is the CG dinucleotide. Since DNA replication is semiconservative, the replication of methylated CG-containing DNA results in DNA molecules in which only one strand is methylated. Such half methylated DNA appears to be restored to a fully methylated state by an enzyme that only methylates progeny CGs that are opposite methylated CGs on the original parental strand. Thus, DNA methylation patterns may be clonally inherited through many cycles of DNA replication and cell division (see Bird 1983 for review). DNA methylation is a signal on the DNA that is both heritable clonally and still potentially reversible, while not affecting the integrity of the genetic code. The potential importance of such a signal in development is obvious (Razin and Riggs 1980).

DNA methylation has a possible negative aspect as well, 5-methylcytosines may function as mutational "hot spots." The deamination of cytosine gives uracil, which is recognized as abnormal in DNA and removed, whereas the deamination of 5-methylcytosine gives thymine. Since the latter is a normal base of DNA, the original deamination error is not corrected (Coulondre et al. 1978). Because of this increased mutability, genomes in which cytosine methylation occurs generally have the dinucleotide CG under represented. For example, vertebrates have approximately 5% of their DNAs consisting of 5-methylcytosines and have about 20% of the expected frequency of CG dinucleotides; e.g., in human DNA the fraction of $(G + C)$ is 0.4; thus, one would expect CG to occur with a frequency of $(0.2)(0.2) = 0.04$, the observed frequency is 0.008 (Bird 1980). This reduction is attributed to the mutational decay of the 5-methylcytosine to thymine (Salser 1978). In nonsequence-dependent portions of the genome, CGs would be eliminated with time. In sequence-dependent portions, the occurrence of the CG dinucleotides would impose an additional mutational burden in organisms in which methylation occurs. In insects, which have poorly methylated genomes, no CG deficiency occurs (Bird 1980). In plants the proportion of 5-methylcytosines in nuclear genomes varies considerably (Shapiro 1968). In animal cells, 5-methylcytosine accounts for 2 to 7% of the total cytosines, whereas over 25% of the cytosines are methylated in higher plants. Gruenbaum et al. (1981) have shown that in higher plants this difference is due to two factors: the CG dinucleotide sequence is much more common in plant DNA than in animal DNA, and in plants an additional sequence is methylated—the trinucleotide CXG. The extent (if any) to which the elevated levels of 5-methylcytosine impose a mutational burden in higher plants is unknown.

Ohno (1984a, 1985) has shown that it is possible to design genes that are more mutation-proof. If the gene consists of a repeating oligomeric unit of base pairs that is not a multiple of 3 ($N = 3n + 1$ or 2), the resulting gene is more mutation proof than genes without such periodicity. The repeating oligomeric unit usually generates a polypeptide with a repeating sequence of amino acids consisting of more amino acids than the number of codons in the gene oligomer unit. For example, the repeating oligomer CAGCAGCCTGCGA has (4 codons + 1) worth of nucleotides ($13 = (3 \times 4) + 1$) and generates a polypeptide with a repeat pattern of 13 amino acids. In figure 8.2

$N = 3n + 1$

$7 = 3(2) + 1$

ABCDEFGABCDEFGABCDEFGABCDEFG   A

$\underline{ABC}\,\underline{DEF}\,\underline{GAB}\,\underline{CDE}\,\underline{FGA}\,\underline{BCD}\,\underline{EFG}\,\underline{ABC}\,\underline{DEF}G$ B
 1   2   3   4   5   6   7   1   2

$\underline{ABC}\,\underline{DFG}\,\underline{ABC}\,\underline{DEF}\,\underline{GAB}\,\underline{CDE}\,\underline{FGA}\,\underline{BCD}\,\underline{EFG}$ C
 1   ?   1   2   3   4   5   6   7

$\underline{ABC}\,\underline{ZDE}\,\underline{FGA}\,\underline{BCD}\,\underline{EFG}\,\underline{ABC}\,\underline{DEF}\,\underline{GAB}\,\underline{CDE}FG$ D
 1   ?   5   6   7   1   2   3   4

Figure 8.2. A gene "design" which is buffered against deletions and additions. This "design" is based upon a repeating oligomeric unit that codes for 7 triplet codons (A and B). The deletion (C) or addition (D) of a base may disrupt some reading frames, but ultimately the 7 codon repeat pattern is restored.

the mutation-proofing capacity of a simple oligomer of 7 units in a repeat pattern is shown. Such a pattern stores 7 codons worth of information and thus could theoretically code for a polypeptide repeating a 7 amino acid peptidic periodicity. The deletion or addition of a nucleotide early in the gene will result in only one new codon, all subsequent codons are repeats in the same sequence as found in the original sequence. It should be noted that the usual consequence of a base deletion or addition early in a gene without such a $3n + 1$ or 2 organization typically will generate considerable missence and, most probably, nonsense mutations.

Genes that are present in thousands of copies in a plant genome pose special problems regarding mutation. Such genes have many more targets for mutagenesis, and the selective discrimination between cell genotypes with different numbers of mutant and non-mutant copies is difficult (e.g., the metabolic differences between a cell with 1,000 copies of the nonmutant gene vs. 950 copies of nonmutant and 50 copies of mutant are probably very slight). The genes specifying the 18S, 5.8S, and 25S (rRNAs) of ribosomes exist as tandem arrays of thousands at one or a few chromosomal loci in plant genomes (Flavell 1986). Many characteristics of the rRNA genes may be interpreted as conferring mutation buffering qualities to this gene set. The phenomenon of concerted evolution (i.e., the amplification or loss of individual mutations so that the entire tandemly repeated array is relatively homogenous regarding the base sequence of the repeat) has already been discussed (chapter 2). Coen, Thoday, and Dover (1982) have suggested that the phenomenon of concerted evolution of such a multigene family may be a mechanism to promote selective discrimination between genotypes. To quote these authors: "A mechanism which amplifies or reduces the copy number of a variant to different degrees on different chromosomes will increase its contribution to phenotypic variance and hence the effectiveness with which natural selection may act on it" (p. 567).

Thus, the phenomenon of concerted evolution of tandem repeats may promote diplontic selection between cells in an organism and act as a mutation buffer for genes so organized. Perhaps the high frequencies of nontranscribed tandem repeats so characteristic of many plant genomes (Flavell 1985) represent "errors" in this buffering mechanism.

In a recent review Flavell (1986) has summarized the major characteristics of rRNA genes in plants. Many of these characteristics may be related to mutation buffering. The individual sets of 18S,

5.8S, and 25S genes are separated by what is known as intergenic DNA, which may vary in size from 7,000 bp to over 12,000 bp depending upon the species. About half of the intergenic DNA itself consists of tandem repeats (ca. 100 to 300 bp long). Individual species may show considerable polymorphisms for variation in intergenic DNA length (e.g., Saghai-Maroof et al. 1984; Appels and Dvořák 1982a); yet at individual loci there is considerable homogeneity for length of intergenic DNA within the tandem arrays of rRNA genes. In general the rRNA genes are highly conserved between species whereas intergenic DNA sequences vary greatly between species (Dover and Flavell 1984).

The difference in the rates of divergence between rRNA genes and different parts of the intergenic DNA sequences at the same locus has prompted a number of interesting conjectures. Appels and Dvořák (1982b) have suggested the possible involvement of the RNA transcribed from the rRNA genes functioning as a kind of cofactor in a biased gene conversion. Thus, the more frequently transcribed rRNA genes would bias the gene conversion toward those DNA sequences that produce RNA sequences in highest abundance. A difficulty with this hypothesis is that the intergenic DNA sequences contain multiple copies of promoters/enhancers, and the probability of transcription of rRNA gene sequences is a function of the copy number of these intergenic sequences (Reeder 1984; Dover and Flavell 1984). Thus, transcription frequency seems to be a property of the intergenic DNA sequences rather than of the base sequences of the transcribed rRNA genes.

Dover and Flavell (1984) suggested that intergenic sequences evolve more rapidly because they are less selectively constrained than the rRNA genes. The intergenic sequences are more variable because the putative proteins which interact with promoters/enhancers in these sequences could themselves evolve to accommodate and recognize slightly different sequences. Thus, a molecular coevolution is possible for the intergenic sequences, whereas the rRNAs are much more selectively constrained and highly conserved. The number of rRNA genes is considerably in excess of the amount necessary to sustain ribosome synthesis in most plant genomes. This large pool may allow the selection (for transcription) of rRNA genes with the most effective intergenic sequences and may also allow time for the molecular coevolution of proteins to recognize new sequences. The entire system may be viewed as in a state of continual evolutionary flux (Dover and Flavell 1984).

Another aspect of rRNA genes that can function as a mutation buffer is nucleolar dominance. The primary transcript from each rRNA gene (the genes specifying the 18S, 5.8S, and 25S RNAs of ribosomes) becomes associated with proteins to form a ribosome precursor and is processed to form the three stable RNAs. All this occurs in a nucleolus which is associated with a specific rDNA locus or nucleolar organizer (Flavell, 1986, for review). Different nucleolar organizers have different strengths; thus, in species hybrids very often only one of the parental nucleolar organizers will form a nucleolus (Navashin 1934; McClintock 1934; Wallace and Langridge 1971). That the inactive nucleolar organizer is still potentially functional is shown by breeding experiments that remove the chromosome with the active nucleolar organizer. In such genotypes the previously inactive nucleolar organizer forms a nucleolus. Thus, in genotypes with multiple nucleolar organizers with different degrees of dominance, the deletion or inactivation of the active nucleolar organizer may be compensated for by the activation of the previously inactive rDNA locus.

The molecular basis of nucleolar dominance is related to specific sequences in the intergenic DNA. Reeder and Roan (1984) injected cloned amphibian ribosomal genes from *Xenopus laevis* and *X. borealis* into oocytes and embryos from either species. The expression of only the *X. laevis* ribosomal genes in the injected cells was related to the greater number of enhancer sequences between the rRNA genes of this species (Reeder 1984). In *Drosophila melanogaster*, sequence analysis of the rDNA nontranscribed spacer has documented the presence of multiple RNA polymerase I transcription promoters (Coen and Dover 1982). It has been hypothesized that the presence of multiple copies of DNA sequences within the intergenic regions which transiently bind proteins involved in transcription increases their local concentration and consequently promotes the transcription of adjacent rRNA genes (Reeder 1984; Coen and Dover 1982; Flavell, 1986). Nucleolar dominance may be a consequence of the number of copies of these sequences within the intergenic regions of a rRNA gene locus.

Seed storage proteins are coded by multigene families. All seed storage proteins are high in glutamine, asparagine, and proline and are important sources of nitrogen for the germinating seedling. In *Zea mays*, 60% of the total endosperm protein is zein. In a recent review Heidecker and Messing (1986) have pointed out that many characteristics of the organization of the zein gene families may relate to mutation buffering. Based upon protein solubility criteria, two multigene families are recognized, Z1 and Z2. Within each family, subfamilies of genes are recognized. For example, Z1 is subdivided into four subfamilies, Z1A with approximately 25 gene copies, Z1B with 20 gene copies, Z1C with 15 gene copies, and Z1D with 5 gene copies. (Z2 is subdivided into three subfamilies, Z2A, Z2B, Z2C.) Within each subfamily, cohorts of sequences are recognized that are more similar to each other than to other cohorts. Heidecker and Messing estimate that only half of the gene copies are active genes. The zein protein is glutamine rich; for example, the Z1 protein has a central region that contains a tandem repetition of a block of 20 amino acids starting with a row of glutamines. Z1 proteins from different members of the gene subfamilies may vary in the number of repeats of this block. Because Zein protein is rich in glutamine, zein genes have a high proportion of codons (32%) that can mutate to a nonsense codon in one step. The glutamine codons (CAA, CAG) readily mutate to the nonsense codons UAA and UAG since the G → → A transition (at the mRNA level, C → → U) is the most frequently occurring mutation. Consequently, zein genes should be very subject to mutational inactivation. Heidecker and Messing (1986) hypothesize that zein gene organization is an evolutionary response to reduce the impact of this high incidence of nonsense mutations. Zein genes are in that rare class of eukaryotic genes that lack introns (histone genes also lack introns). Gene clustering (tandem repeats) coupled with the lack of introns may facilitate the reconstitution of genes without internal nonsense mutations from mutant genes via unequal crossing over and/or gene conversion. In support of such recombination or conversion repair is the observation that genes with nonsense mutations still retain their informational integrity. Such inactivated genes have not accumulated other forms of mutational damage (deletions, insertions, etc.) typical of inactivated pseudogenes. The zein inactivated genes appear still to belong to the same gene pool as the active genes and are under the same selective constraints. The zein gene family may represent a situation where the evolutionary choice of glutamine as a means of enriching seedling nutrition for nitrogen resulted in a mutation prone genetic information system. Such an error-prone system could be compensated for by either almost never-ending cycles of gene amplification to ensure some level of functional genes or a repair system to form active genes from inactive ones and, thus, maintain a relatively constant ratio of active to inactive genes. The latter seems to have been selected (see chapter 2 for a discussion of the Red Queen effect).

Not all seed storage proteins are enriched with glutamine. In wheat and barley the overall content of proline has been increased at the expense of glutamine. According to Heidecker and Messing (1986), this change should lower the incidence of gene activation by mutation of a glutamine codon to a nonsense codon, and they predict a lower ratio of inactive to active seed storage protein genes in wheat and barley than in corn.

Proteins themselves have characteristics that may allow amino acid replacements to be more or less selectively neutral with regard to a protein's function in the cell's metabolism (Zuckerkandl and Pauling 1965). Such characteristics, in effect, buffer the protein from the effects of mutation. Kimura (1983) has reviewed these protein characteristics from the viewpoint of the neutral theory of molecular evolution. Using data on the amount of intraspecific variability or polymorphism found in proteins with different characteristics, Kimura was able to show that the tolerance for amino acid changes was correlated with structural and functional characteristics of the proteins. In our present discussion, if a protein or portions of a protein exhibit a wide latitude in amino acid content it is assumed that the majority of these changes are neutral with respect to protein function. A critical aspect of mutation buffering at this level is molecular constraint. To quote Kimura: "The probability of a mutational change being neutral depends strongly on molecular constraint. If a molecule or a part of a molecule is functionally less important, then the probability of a mutational change in it being selectively neutral (i.e., selectively equivalent) is higher" (1983:157).

Monomeric proteins show significantly higher levels of heterozygosity and polymorphism than multimeric proteins. Among multimeric proteins, the incidence of polymorphism is especially low among multimeric enzymes in which interlocus molecular hybrids occur (Zouros 1976; Harris, Hopkinson, and Edwards 1977). Kimura (1983) concluded from these correlations that the structural constraints due to intersubunit interactions must be greater for multimers than monomers, and that all other things being equal, the proportion of mutations that are neutral is smaller for multimeric enzyme loci than monomeric ones. Substrate specificity is also correlated to mutation tolerance, functional constraint is stronger for substrate-specific enzymes than substrate-nonspecific enzymes; therefore the latter are probably more mutation buffered (Kimura 1983). Soluble enzymes can tolerate mutational changes much more readily than structural proteins and membrane-bound enzymes, whereas enzymes involved in protein-nucleic acid interactions are the least tolerant to mutation (Langridge 1974). Based upon studies of amino acid replacements on enzyme function in bacteria, Langridge concluded that, excluding nonsense mutations, most mutations in genes coding for enzymes are neutral with regard to enzyme function. The degree to which this generalization applies to eukaryotes is unclear, but it does provoke one to question the generality of the "mutant gene, mutant phenotype" thesis.

The genomic tolerance to mutation has been studied in a very clever way in the yeast *Saccharomyces cerevisiae*. Goebl and Petes (1986) constructed recombinant plasmids with fragments of yeast DNA into which a selectable marker was inserted. Diploid homozygous yeast cells were transformed with a linear segment of these plasmids. This linear plasmid DNA contained the selectable marker inserted into a yeast DNA fragment. When yeast cells are transformed with linear DNA, the fragment integrates by homologous recombination into the genome and displaces the resident chromosomal segment (a "transplacement"). If random DNA fragments are selected for incorporation into the original plasmids, transplacements can be targeted randomly in the genome.

Diploid cells with single transplacements were induced to undergo meiosis and the viability of haploid cells (which are haploid for the transplacement and therefore lack a normal DNA segment) was measured. Of approximately 200 independent transplacements studied, 70% were without phenotypic effect, 12% were haploid lethal, 14% resulted in slow growth, and 4% were associated with a new phenotype (e.g., auxotrophy). These results are not what one would have predicted. As Goebl and Petes (1986) pointed out, the yeast genome is a "lean" eukaryotic genome consisting of only 14,000 kb, 90% of which is single-copy DNA, a large fraction of which (50% to 60%) is transcribed, and few yeast genes have introns. One would have predicted that most of the DNA in this genome would be essential. One can interpret the failure of this prediction in at least two ways, either only a small fraction of the yeast genome is essential (and, therefore, "fat" eukaryotic genomes have even less essential DNA than previous estimates [chapter 2]), or the genome can metabolically compensate for lost genetic material. Perhaps the analogy of an orchestra is applicable. An orchestra can still play many melodies satisfactorily even though some players are sick or absent. This is not to say that all players are equally expendable or that the players who are more expendable are not important. Perhaps the same compensations in cellular metabolism allow for fairly normal phenotypes to develop in spite of the absence of "somewhat

essential" DNA segments. This interpretation leads to the prediction that cells that have lost a number of different "inessential"
genomic segments should exhibit a synergistic loss in viability.

DEVELOPMENTAL MUTATION BUFFERING

The majority of somatic mutations probably have very little immediate consequence to the plant. The expression of such mutants is
inhibited primarily by their recessive nature and because they occur
in diploid cells. Jones (1935) noted that subsequent chromosome
aberrations (e.g., loss of chromosome segments or whole chromosomes) will allow the expression of recessive alleles on the remaining homologue. Mitotic recombination (e.g., mitotic crossing over)
may generate homozygous genotypes from heterozygous genotypes;
such homozygous genotypes will express phenotypes previously
masked in heterozygotes. Given the expression of somatic mutants
by one of the above mechanisms, what are the immediate as well as
the long-term consequences? Will such mutant cells express mutant
phenotypes? will the mutant cells increase in frequency or be lost?
and to what extent will plants (ramets) that are a chimera for such
cells have altered phenotypes? As will be shown subsequently, the
answers to these questions are in large degree determined by the
anatomical and developmental characteristics of the species. It
should be noted that the lessening of the phenotypic impact of
mutant cells may allow higher frequencies of such cells to accumulate and reduce the proportion of mutant-free meiotic products
that the plant produces.

Plasmodesmata

Although plant cells are separated from each other by cell walls,
they have evolved means by which materials may pass from cell to
cell without leaving the cytoplasm. Plasmodesmata interconnect the
majority of cells in higher plants and serve as the cytoplasmic channels between adjacent cells (Gunning and Overall 1983; Gunning
and Robards 1976). Movement of materials from cell to cell through
these channels is documented (Tucker 1982). Of importance to the
present discussion is the possibility of plasmodesmata allowing
cross-feeding of mutant cells. Thus nonmutant cells may compensate for defects in the metabolism of mutant cells. Such cross-feeding

will reduce diplontic selection against mutant cells by reducing the immediate phenotypic consequences of the mutant cell genotypes (Langridge 1958). If such cross-feeding is possible, one could imagine a kind of density-dependent selection occurring within tissues (i.e., the magnitude of the selection coefficient against mutant cells being directly proportional to the fraction of the tissue or organ that consists of mutant cells). It is also possible that different mutant cells could complement each other (i.e., mutant A^-B^+ could cross-feed A^+B^- and vice versa) so that cell mixtures of a number of mutant types would experience less negative diplontic selection in spite of the fact that any given mutant type might be strongly handicapped.

Structured vs. Random Ontogeny

Although ontogenetic patterns usually give rise to organs that are complex internally and externally, plant taxa vary greatly with regard to the patterns of cell divisions and cell lineages that may give rise to these organs. Some of these patterns appear to be more random with regard to cell lineages whereas other patterns are very specific (structured) and development is almost clonal. As an untested hypothesis, one could argue that whether an ontogenetic pattern is structured or random will be a factor influencing the phenotypic impact of mutant somatic cells.

Random ontogenetic patterns could be viewed as being more mutation-buffered since during organogenesis compensatory divisions between wild type and mutant cells may be more likely. On the other hand, structured ontogenetic patterns may be more mutation buffered if pattern is so rigidly determined that it is almost autocatalitic and agenic, thus organogenesis is not as dependent upon the individual cell genotypes. With current ignorance concerning the molecular mechanisms controlling form, little more can be said about mutation buffering and ontogenetic patterns.

Random and structured ontogenetic patterns are not uncommon in vascular plants. The most structured apical meristems are single tetrahedral apical cell patterns which occur in many pteridophytes. Such apical cell-based meristems result in highly predictable patterns based upon a regular pattern of merophytes (cell packets) in which the number and position of cell divisions in the merophyte may be very regular (see figure 4.5). In contrast to this regularity, the cell divisions in the apical meristem of the gymnosperm *Cycas* almost seem chaotic (figure 8.3). The majority of plants have primary apical ontogenies somewhere between these two extremes of the structured-random continuum. The shoot apex of the gymnosperm *Araucaria* has such an intermediate pattern. As Griffith (1952) documented, cell files and division patterns appear much more restrained in comparison to *Cycas* (figure 8.4).

In angiosperms, apical stratification not only promotes the retention of mutants within apical meristems but will strongly influence the spread of mutants in other organs. In some species the tunica-corpus organization results in a very structured pattern of development. In other species with stratified apical meristems, stem growth and other aspects of organogenesis appear more random or chaotic with regard to cell lineages. The experimental induction of periclinal cytochimeras using colchicine (Satina, Blakeslee, and Avery 1940; Blakeslee et al. 1939) has been used to study the ontogeny of various organs in different angiosperm species.

To review: in stratified apical meristems (tunica-corpus organization), the individual component meristems are designated LI, LII, LIII for apices with two tunica layers and an internal corpus. Using

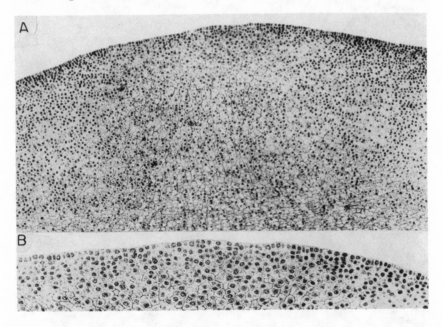

Figure 8.3. Longisection of the shoot apex of a large individual of *Cycas revoluta*. A: low magnification; B: higher magnification of the summit of the apical meristem. Although the demarcation of different zones is possible, one is struck by the lack of clear meristematic patterns and cell lineages. (From Foster 1940.)

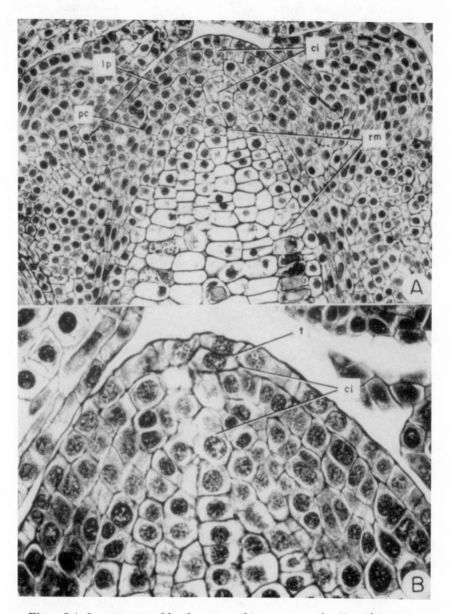

Figure 8.4. Longisection of the shoot apex of *Araucaria excelsa*. Note clear patterns of cell lineages: ci = zone of corpus initials; t = tunica; lp = initiation of a foliar primordium; pc = procambium; rm = rib meristem. (From Griffith 1952.)

colchicine, various cytochimeras can be formed. If diploid cells are designated by the numeral 2 and tetraploid cells by the numeral 4, apices may be 2-2-2 type, where all the component meristems are diploid, or 2-2-4, or 2-4-2, or 2-4-4, etc., for a total of 8 different patterns including 4-4-4. Not all ploidy combinations are stable in most species. Using stable cytochimeras, the ontogeny of various organs in different plant species has been studied to determine the fate of cell lineages derived from the various component meristems of the apical meristem. In *Datura stramonium* the three component meristems (LI, LII, LIII) persist as independent units within the apical meristem, and each gives rise to certain tissues within the various plant organs (see Satina 1945 for review). The ontogenetic pattern in this species is highly structured in comparison to other species. Similar studies on cranberry (Dermen 1947) revealed that although generally the LI gave rise to the epidermis, LII to a single layered hypodermis, and LIII to the rest of the tissue, this was not always the case. Stems were found in which the epidermis was derived from the LI, but the LII might give rise to large portions of the internal tissue of the stem. In peach *Prunus persica* cytochimeras, ontogeny was even more variable (Dermen 1953). Again the LI almost always gave rise to the epidermis of the stem and leaf; thus, this component meristem is ontogenetically related to a specific histogenic layer. With regard to the LII and LIII, no specific tissue was associated exclusively with one or the other of these component meristems (a similar situation was documented in apple cytochimeras, Dermen 1951). Variable degrees of apical stratification were also documented in peach. Leaf development in peach is also interesting since the mesophyll is almost entirely derived from the LII, and the LIII derived tissues are limited to the mid rib. Thus if chlorophyll deficiency due to a mutation affecting the LII occurred, the leaf should exhibit a homogenous pattern (e.g., all white) rather than a pattern of green and white tissues as is often the case in other plants.

The histogenic relationship between a component meristem in a stratified apical meristem and developmental rigidity is also a function of the genotype of the cells in the meristem. In a classic study, Stewart, Semeniuk, and Dermen (1974) compared development of geraniums *(Pelargonium × hortorum)* that were chimeras for different albino mutations, W_1 and W_2. Periclinal chimeras which were GW_1G (LI, LII, LIII) and GW_2G were compared with regard to leaf development. In all leaves the GW_1G chimeras had larger areas of white tissue than the GW_2G chimeras although both leaf types were

similar in final size (figure 8.5). Similar patterns were documented
in other organs for various combinations of green and white component meristems within shoot apices. These results indicate that
the degree of histogenic specificity in cells derived from the component meristems of a stratified apical meristem may be strongly
influenced by the nature of the mutations fixed in these meristems.
Cells expressing the W_2 phenotype were smaller and apparently
less viable than W_1 cells (or G cells which appeared equivalent to
the W_1). The W_2 cells were displaced more readily by G cells from
adjacent component meristems within the apical meristem, thus
diplontic selection was against the W_2 cell type both from the apical
meristem as well as during the development of the mature organs.
As noted by Stewart, Semeniuk, and Dermen: "the competitive
relationships between apical layers and their derivatives did not
upset the genetic control of the morphology of the various organs.
. . . There was a compensation in growth, space not occupied by
derivatives of LIII was filled by derivatives of LII" (1974:64).

In the majority of angiosperms, the cells of the LI divide primarily
in the anticlinal plane. In leaves (and other organs as well), cells
derived from the LI are typically restricted to the epidermis and also
are characterized by anticlinal division patterns during growth. The
leaf mesophyll usually represents cell derivatives from the LII or
LIII. Pohlheim (1983) has shown that the orientation of cell division
planes in the LI-derived epidermal cells is responsive to the relative
vitality of the adjacent LII-derived cells. Periclinal chimeras which

Figure 8.5. Leaves of geranium chimeras. A: leaf from GW$_1$G branch; B: leaf from
GW$_2$G branch. Note the large difference in the degree of LII contribution between
the LII genotypes W_1 and W_2. (From Stewart, Semeniuk, and Dermen 1974.)

have fixed a mutation in the L$_{II}$ that inhibits the growth of cells derived from this component meristem will very often have compensatory divisions in the L$_I$-derived cells during leaf development. In figure 8.6 such compensatory divisions are illustrated for periclinal chimeras in *Pelargonium zonale*. In this case the L$_{II}$ mesophyll is marked with a chlorophyll defect whereas the L$_I$ has the wild-type or normal genotype. As the amount of white tissue (or L$_{II}$-derived cells) decreases, the frequency of periclinal divisions in the epidermis increases. Since the L$_I$ has a wild-type genotype, L$_I$-derived mesophyll occurs as islands of green tissue at the margins of the leaf (Pohlheim 1973).

In the variety *Hessei* of *Prunus pissardi*, a similar L$_I$ proliferation was documented (Pohlheim 1983). In this periclinal chimera, the L$_I$ has a chlorophyll defect and the L$_{II}$ and L$_{III}$ are green. Although green, the L$_{II}$ and L$_{III}$-derived cells appeared to show different growth vitalities. Thus, three cell genotypes and phenotypes were documented:

  1. L$_I$—normal growth during leaf development (X)
  2. L$_{II}$ or L$_{III}$—green, reduced growth during leaf development (Y)
  3. L$_{II}$ or L$_{III}$—green, normal growth during leaf development (Z).

Trichimeras that had the constitution XYZ had isolated islands of white mesophyll at the leaf margins due to the proliferation of the

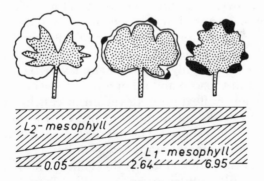

Figure 8.6. *Pelargonium zonale* "Kleiner Liebling." Variegation patterns of the leaves of several periclinal chimeras with a mutated L$_2$ (L$_{II}$) (i.e., with chlorophyll defect). The stronger the inhibition of growth in the L$_2$ (L$_{II}$) components, the more mesophyll areas stemming from L$_1$ (L$_I$) will occur (figures below give the average number of L$_I$ areas per leaf). (From Pohlheim 1981.)

LI. Such periclinal divisions of the LI were nearly absent in XZZ chimeras. (See also examples in Pohlheim and Kaufhold 1985.)

In *Acer platanoides*, the variety *Drummondii* is a chimera with a chlorophyll defect in the LII. The development of the LII-derived leaf mesophyll was more reduced in leaves from long shoots than in those from short shoots. Leaves from long shoots also had greater amounts of LI-derived mesophyll due to periclinal divisions in this layer (Pohlheim 1983).

Although the above examples of growth compensation are limited to leaves, one suspects that similar patterns probably occur in the development of other organs. The development of relatively normal organs from "unusual" contributions of LI-, LII- and LIII-derived cells results in a kind of developmental mutation buffering that allows a functional (and possibly competitively successful) ramet to develop from a mosaic of genetically defective cells. The presence of stratified meristems insures that cells derived from different meristematic cell pools are juxtaposed in a layered sandwich in most organs.

Studies of leaf development using somatic mutations and clonal analysis have documented a much more random (or stochastic) pattern of development than previously envisioned. The classical view has been that leaves arise from a small group of initial cells situated at the tip of the leaf axis and that the lamina arises from a marginal meristem. Differential growth of the marginal meristem was thought to control leaf shape. Recently, Poethig and Sussex (1985a, b) have demonstrated that in tobacco, the leaf axis arises from approximately 100 cells in four layers of the shoot apex and the leaf lamina arises from several rows in each of three layers of the leaf axis. Rather than leaf development based upon specialized meristems with defined cell lineages arising from the division of initial cells, leaf development is largely dependent on intercalary growth and marginal cell lineages which are highly variable in behavior. The patterns of leaf development in tobacco and other angiosperms (Poethig 1984a, b) appear sufficiently flexible with regard to cell lineages to accommodate differential contributions of the LI, LII and LIII and still generate similar leaf shapes. The degree to which other plant organs have such developmental properties is unknown.

It is almost axiomatic in biology that genes control developmental processes in multicellular organisms. What is not so often considered is the degree to which developmental processes or events occur independently of the genes, that is, the extent to which these processes are autocatalytic. The greater the degree of autocatalytic activity in a developmental pathway, the more mutation buffered the pathway becomes. Plants have characteristics that may allow the formation of functional developmental pathways that are more autocatalytic and, consequently, more mutation buffered than those of higher animals.

Trewavas (1982) has summarized the developmental properties of plants and higher animals and has concluded that the epigenetic/transcription paradigms may differ considerably between these organisms. Plants differ from animals in many fundamental ways regarding their growth and organization, and the developmental pathways coordinating these differences have evolved in isolation from animals for a very long time.

In vascular plants there are only about 15–20 cell types, whereas in vertebrates there are more than 100. (Wolpert 1978 estimates 200 different cell types in man.) In plants these cell types are organized into comparatively few different tissue types; e.g., Esau (1976) lists epidermis, periderm, parenchyma, collenchyma, sclerenchyma, xylem, phloem, and various secretory structures as the major tissue categories in plants. These tissues are, in turn, organized in various patterns to generate a relatively limited spectrum of organ types (as well as their modifications, e.g., leaf, stem, root, reproductive structures, etc.).

Cellular differentiation also differs considerably between plants and higher animals. The cell types in animals generally have individually recognizable cytoplasmic characteristics and synthesize numerous cell-specific proteins, whereas plant cell types are distinguishable primarily by their unusual cell wall structure rather than specific cytoplasmic differentiation. Secretion appears to be critical in plant cell differentiation (Trewavas 1982). Plant and animal cells differ greatly in regenerative capacity. Although perhaps not all plant cells are totipotent, many mature tissues contain cells from which whole plants can be regenerated. This developmental plasticity is very different from the regenerative possibilities of higher animal cells. Animal cells follow developmental pathways that are not readily reversible (Blau et al. 1985).

In spite of sometimes remarkable developmental plasticity, the organization of plant cells into a meristem may result in a surprising rigidity of pattern formation. In many woody plants, different growth forms develop on various parts of the same individual. The resulting different shoots acquire a certain "individuality," are highly persistent, and often may be clonally propagated. Brink (1962) and Schaffalitzky de Muckadell (1954) have collected various instances of this interesting aspect of plant development. Some examples of the somatic stability of such phenotypic phase changes include:

1. *Hedera helix.* The juvenile form of this species is a vine with dorsiventral symmetry, palmately lobed leaves, and 2/2 phyllotaxy. The adult form is a semierect or erect radially symmetrical shrub with entire ovate leaves and 2/5 phyllotaxy. Plants grown from rooted cuttings of either retain their respective phenotypes for years (although the juvenile form may give rise to adult forms that are again stable).

2. *Araucaria excelsa.* This conifer has an excurrent growth form. If the main terminal shoot is damaged, it cannot be replaced because the side branches lack the capacity to grow erect. Rooted cuttings from side branches will grow horizontally indefinitely. It is reported by Wareing (1959) that a side branch of this species, rooted by Goebel over fifty years earlier, was still growing horizontally in the Munich Botanic Garden. Similar somatic stability has been noted in *Pinus pinea* and *P. canariensis* (Schaffalitzky de Muckadell 1954).

3. *Fagus sylvatica.* Old beech trees shed their leaves in autumn whereas young trees retain them through the winter. Older trees typically have branches on the lower portions of the trunk that retain their leaves and branches higher on the tree that lose their leaves in winter. Similar patterns of leaf retention and loss are also found in temperate *Acer* and *Quercus* species. Grafts of either leaf-retaining or leaf-shedding branches of *F. sylvatica* retain their phenotypes regardless of the stock to which they were grafted (see table 8.1).

These and other examples (see Brink 1962) suggest that although many plant cells in a mature individual may be totipotent, meristems (which are composed of such cells) have a kind of developmental inertia such that once a developmental pattern is achieved, it is highly persistent and in some sense "autocatylitic."[1] Of course, it is not known to what degree such somatically stable phase changes are gene-regulated or represent alterations in apical meristem topology that become self-perpetuating once established. In other

---

1. Parallel behavior at the cellular level has also been documented in plants. Meins and Binns (1977, 1979) have shown that cultures derived from tobacco pith tissue normally require an exogenous source of the cytokinin kinetin for cell division, but that sometimes cells lose this requirement and can be grown on culture media without added cytokinin. Such "habituated" cells retain this phenotype through many cell generations. The rates of habituation and reversion are higher than normally observed mutation rates. Meins and Binns believe habituation is an epigenetic rather than a genetic phenomenon. Meins (1985) has documented habituation within intact plants.

words, the induction of the phase change may be a gene-regulated response to environment or age, but the persistence of the phase change may be, to some degree, autocatylitic.

Developmental buffering, with regard to the expression of deviant organ phenotypes due to somatic mutation or other internal stresses, is suggested by the responses of plants to chronic doses of ionizing radiation. Previously, we have discussed radiation sensitivity with regard to survival; the question here is not survival but rather phenotypic stability. Dugle and Hawkins (1985) studied *Fraxinus* species experiencing long-term chronic gamma radiation. The leaves of these irradiated plants exhibited a broad spectrum of phenotypic changes, whereas irradiated *Acer negundo* and *Populus alba* specimens showed much fewer phenotypic changes. The balsam fir *(Abies balsamea)* is one of the most radio-sensitive species in the mixed boreal forest, yet very few morphological changes were observed on the leaves of irradiated firs (Dugle pers. comm.). These findings suggest that these species vary with regard to how readily aberrant phenotypes are expressed.

The expression of teratological phenotypes may also be an aspect of developmental buffering. In *Plantago lanceolata* teratologies are not randomly distributed among meristems (van Groenendael 1985). The normal vegetative rosette of this species consists of a primary

Table 8.1. Grafting experiments showing the stability of the juvenile stage in *Fagus sylvatica*. Scions were collected in 1952. (From Schaffalitzky de Muckadell 1954.)

SOURCE OF SCIONS	ROOTSTOCKS	GRAFT PHENOTYPES IN THE SUBSEQUENT YEAR (1953/54)	
		Leaf-retaining	Leaf-shedding
Top of mature tree	Seedlings	0	47
Leaf-shedding epicormics	Seedlings	0	33
Leaf-retaining epicormics	Seedlings	27	0
Leaf-shedding graft from	Seedlings	0	30
1939[a]	Old tree	0	13
Leaf-retaining graft from	Seedlings	58	0
1939[a]	Old tree	7	0

[a]Scions were taken from grafts established in 1939.

meristem (apical) and a population of axillary meristems (or second-order meristems). These second-order meristems may, in turn, form side rosettes with axillary meristems (third-order meristems). No teratological phenotypes involving the primary meristem are known, but teratologies are common in second- and third-order meristems. It appears as though developmental buffering against the expression of deviant phenotypes is strongest in primary meristems. (See Sachs 1982 for further discussion of other stage specific aspects of developmental buffering.)

A particularly interesting case of early developmental buffering has recently been documented in the heterosporous fern *Marsilea vestita*. Fertilized eggs of this species give rise to well-organized embryos within a few hours at 24 C. Early embryogenesis is characterized by highly predictable cleavage divisions and synchronous developmental patterns. The sequence and orientation of the first three planes of cell division of the zygote are almost invariable (figure 8.7). The first division (C_1) is vertical and parallel to the axis of the archegonial canal; the second (C_2) is in a plane perpendicular to C_1 and, again, parallel to the neck of the archegonium. The zygote is thus divided into four cells. The third division (C_3) is horizontal and perpendicular to the first two and results in an eight-celled embryo. Since such highly regular patterns are so characteristic of embryogeny in this species, one might have predicted that normal embryo development is highly dependent upon these patterns. Experimental studies have shown this not to be so. Cold treatment, 16 C, results in highly altered early segmentation patterns in the embryo (figure 8.7 C and D), yet when such embryos are returned to 24 C, most recover normal topographic organization and continue growth (Chenou, Kuligowski, and Ferrand 1986). Similar results were documented when proembryos were treated with colchicine and later grown on colchicine-free medium (Kuligowski et al. 1985). The degree to which such developmental buffering extends to mutations that may alter early embryonic segmentation patterns is unclear. Mutations that act as embryo lethals are, of course, well known in vascular plants (Meinke 1985 and chapter 10). What is unclear is the susceptibility of the genomes of different species for mutations with embryonic effects.

Plant phenotypic responses to ionizing radiation often appear paradoxical. Dose response curves, for example plotting seedling height against dose, show the expected diminution of height with increasing dose until a specific dose threshold. Doses in excess of this threshold result in an increase in seedling height. Such a dose response curve for maize seedling height growth in response to increasing gamma dose is shown in figure 8.8. Schwartz and Bay (1956) explained such responses by noting that with increased dose the average number of chromosome breaks per cell increases and the probability of mitosis decreases. Since chromosome breakage occurs at lower doses than mitotic inhibition, the low dose ranges allow mitosis to occur. Acentric fragments as well as dicentric chromosomes are lost during these divisions, and the consequent loss of this genetic material results in declines of cell and, ultimately, organ viability. When doses are sufficiently high to inhibit mitosis, cell viability is increased because although there are many chromosome breaks, there is little loss of genetic material since the cells do not divide.

Mitotic inhibition and high cell viability at high doses of ionizing radiation have been utilized to decouple cell division from growth and development in plants. Since many developmental events are

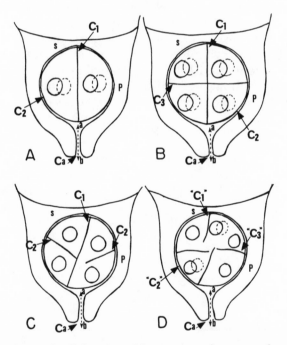

Figure 8.7. Planes of cell division of the zygote and young embryo of the fern *Marsilea vestita*. The normal cleavage pattern is shown in A and B. Abnormal temperature-induced patterns are shown in C and D. See text for discussion. (From Chenou, Kuligowski, and Ferrand 1986.)

associated intimately with cell divisions, it has often been hypothesized that the primary cause of these events was the localization of
cell division. Haber and Foard, in a series of significant papers, were
able to show that very often these critical events of morphogenesis
occurred independently of cell division. For example, the protrusion of a leaf primordium on a shoot apex in wheat was shown to be
independent of the periclinal cell divisions normally associated
with the origin of such primordia (Foard 1971). Using the mitotic
inhibitor colchicine, it also was shown that the initiation of lateral
root primordia in the pericycle was not dependent upon the completion of mitosis (Foard, Haber, and Fishman 1965). The nonessentiality of cell division in the growth and the control of form in
leaf growth was documented in heavily irradiated wheat embryos.
Plants developed from heavily irradiated seeds (800kr of Co^{60}
gamma-radiation) were termed gamma-plantlets. Haber (1962) reported that the first foliage leaf of gamma-plantlets achieved the
same size and had the same allometric constant as the unirradiated
controls. As is shown in figure 8.9, polarized growth occurred without the necessity of concurrent cell divisions.

Botanists have used the concept of the allometric constant as
developed by Huxley (1932) as a measure of polarized growth of a
plant organ (Stephens 1944a, b, c; Haber 1962; Haber and Foard
1963). Huxley's allometric constant was originally developed as a
measure of the relationship of the growth rates of different organs

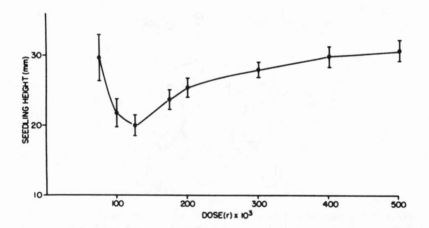

Figure 8.8. Relation between seedling height and gamma dose from Co^{60} source.
Confidence limits are 95%. (From Schwartz and Bay 1956.)

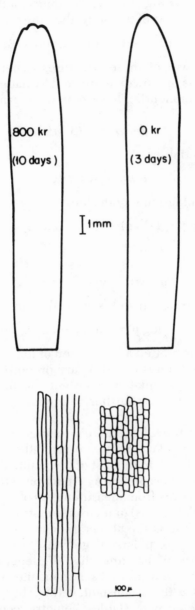

Figure 8.9. Different cell sizes and shapes in gamma-plantlet and unirradiated control leaves of the same size and shape. UPPER: leaves; LOWER: cells of chlorenchymatous mesophyll immediately beneath abaxial epidermis. At left is gamma-plantlet; at right is unirradiated control. (From Haber and Foard 1964.)

and the body. Following Huxley, if y stands for the size of the organ and x for that of the rest of the body, then the differential equations,

$$dx/dt = \alpha x G \quad \text{and} \quad dy/dt = \beta y G$$

describe the growth of these structures (where G measures the general conditions of growth as affected by nature and nurture, i.e., genes and environment). These equations can be rearranged as follows:

$$dy/dx = (\alpha/\beta)(y/x),$$

and separating variables

$$dy/y = (\alpha/\beta)(dx/x)$$

if $\alpha/\beta = k$, then integrating both sides

$$\ln y = k(\ln x) + C, \qquad (x > 0, y > 0)$$

$$y = e^{k(\ln x)+C} = e^C (e^{\ln x})^k$$

or

$$y = cx^k \quad \text{where} \quad c = e^C,$$

which is a power function and is equivalent to

$$\log y = k \log x + \log c.$$

Thus if the log y is plotted as a function of log x, a straight line will result if the growth rates α and β vary proportionately so that the fraction α/β is always equal to a constant, the allometric constant k. (See Batschelet 1971 for further references on the use of k in biology.)

In plant studies the allometric constant generally is determined for the development of a particular organ rather than the relation of the growth of the organ to the rest of the plant. Typically, the log of the length of the organ is plotted as a function of the log of the width. If a straight line results from measurements of successive organs on a stem (for example, leaves) or measurements in a time course, then the slope of that line is the allometric constant, k, and is a useful index of the degree of polarized growth of the organ (Haber and Foard 1963). If k is unchanging, then the length and width of the organ are changing in a proportionate manner so that the ratio of their growth rates, α/β, is a constant, k. It has been shown that many plant organs develop with stable allometric constants throughout the course of their development. These stable allometric patterns are followed despite the occurrence of considerable chronological and histological variability in cell division rates within these organs.

Haber and Foard (1963) documented such an uncoupling of "growth by cell division" and "growth by cell enlargement" during the allometric growth of a number of determinate organs (leaves and fruits) of different taxa. Based upon their allometric analyses, Haber and Foard proposed three "biological conclusions" with regard to plant growth and development: "(1) cell divisions do not directly contribute to or cause growth; (2) cell division plays an essential and immediate role in influencing cell forms, but plays a secondary and much less important role in influencing organ form; (3) there is a fallacy in the usual and accepted manner of interpreting changes in organ size as being due to changes in cell size and changes in cell number" (p. 937). These authors make the important point that just because increase in organ size (growth) can always be *described* in terms of cell numbers and cell sizes, this does not necessarily mean that growth is *determined* by changes in cell number and cell sizes. (See also Haber, Long, and Foard 1964 for further discussion of the difficulty of determining the causes of growth and form.)

The genetic control of plant morphogenesis may follow two broad strategies. Strategy I: plant organ shape may be the sum of the cell division and enlargement patterns of its constituent cells; Strategy II: organ shape may be independently determined and the cells and tissues develop to fill a mold so to speak (figure 8.10).

Of course, these strategies are not mutually exclusive; one can certainly imagine a range of intermediate states. From the viewpoint of morphological stability (or stasis) with regard to somatic mutation (as well as other biotic and abiotic stresses), Strategy II is probably

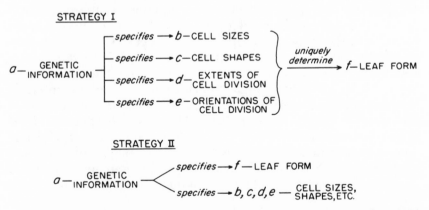

Figure 8.10. Two possible strategies of plant morphogenesis. (From Haber 1962.)

more stable than Strategy I. Strategy I requires more information content per cell, as each cell must function within a cybernetic matrix of sensations and coordinate its development at the tissue and organ level. The final organ shape and organization is thus determined by the sum of the developmental activities of its constituent cells. Strategy II invokes a shape-determining process that is independent of the constituent cells and tissues and their activities. Thus, presumably viable and functional organs could develop from cells that have various genetic impairments. The development of gamma-plantlets, the previously discussed cases of normal leaf development in various *Pelargonium, Prunus,* and *Acer* periclinal chimeras and embryogenesis in *Marsilea* fit within the Strategy II scenario (but do not prove it). The problem with the view of plant development advocated by Haber and Foard (Strategy II) is that the manner in which genes may determine form without invoking control at the level of cell division and expansion was unstated. Thus, although it was clear that Strategy I seemed too simplistic and probably impossible in complex forms, Strategy II had almost a mystical quality since mechanisms were not specified.

Before discussing possible mechanisms for the genetic control of form, it may be best to review the kinds of models proposed for the generation of form through the control of cell division and expansion. These models are based upon some form of communication between cells generating positional information and cellular responses to such positional information. Thus positional information elicits cell responses in the form of cell division expansion and specialization that generates pattern and form. Based upon studies of animal development, Wolpert (1969, 1971) formulated the positional information hypothesis for the generation of pattern and form in multicellular organisms. Recently there has been interest in applying Wolpert's positional information concept to plant development (see the compendium of papers edited by Barlow and Carr 1984). The plant developmental literature is replete with various conceptual models (often in mathematical form) involving intercellular communication to account for pattern and form in plants (see references in Barlow and Carr 1984). These models often are based upon "an interaction of (still hypothetical) molecules" (Meinhardt 1984) and seldom do they deal with the conceptual problems raised by Haber and Foard. Such models are important in a heuristic sense,[2] but it is significant to note that the generation of pattern and

---

2. Heuristic (serving to discover or reveal) applied to arguments and methods of demonstration that are persuasive rather than logically compelling (*Webster's New International Dictionary* 1949).

form through computer simulation is not in itself evidence that plant development is specified in the manner of the models.[3]

Lintilhac (1984) and Poethig (1984a, b) have questioned the applicability of the positional information concept to plants. Lintilhac notes that positionally controlled development relies on a complex reference library (presumably genetic) which must be available to each cell in the developmental field. This library provides the developmental information required to define the final distribution of cell types in the plant. Cells responding to such positional information must be "information-rich" with regard to the genetic information required for the correct interpretation of positional cues. Lintilhac suggests that since plant cells are confined within relatively rigid cell walls and are not motile like animal cells, plant development may be much more "information-poor" than animal development. Because of cell walls, morphogenetic behavior in plants may be responses to stress-mechanical antecedents. To quote Lintilhac:

> Genetic control comes into the picture only insofar as it provides the cells with the appropriate biochemical machinery to sense and respond to mechanical forces. Individual structural events at the cellular level need not be coded for directly. Routine morphogenetic behavior can thus be dealt with directly by the cytoplasmic machinery of growth; no gene transcription or translation is necessary except when needed to maintain the cellular titre of biochemical equipment and to maintain critical tolerance and rates. (1984:94)

The above theory of plant development is called "structural epigenesis." Structural epigenesis, since it is almost "agenic," is a developmental model that confers a very high level of buffering from the phenotypic effects of somatic mutation.

Structural epigenesis as well as related theories of plant development that emphasize physical forces as important determinants in plant morphogenesis (Green 1980, 1985) have recently attracted renewed interest with the finding that ionic channels are activated by membrane stretch (Edwards and Pickard, in Press). Guharay and Sachs (1984, 1985) documented a stretch activated ion channel (mechanotransductive channel) in cultured embryonic chick skeletal muscle. Mechanotransductive channel activity in the plasmalemma of tobacco protoplasts has recently been reported (Falke et al. 1986;

---

3. A characteristic often overlooked in these models is developmental buffering (G. Steucek, pers. comm.).

Edwards and Pickard, in Press). Mechanotransductive channel activity may be the bridge between the stimulus of physical forces
and the differential chemistry necessary for morphogenetic responses to those stimuli. Edwards and Pickard speculate that differential opening of ion channels could exert effects through
changes in transmembrane potential, salt concentration, or levels of
messenger calcium ion (Hepler and Wayne 1985).

As stated previously, the dominant paradigm in developmental
biology is the role of differential gene expression in cellular differentiation. The primary factor in the genesis of different cellular (or
organ) phenotypes is assumed to be differential gene expression.
The structural epigenesis hypothesis is in opposition to this paradigm. Evidence for such an "agenic" developmental program has
recently been presented for cellular differentiation in the dimorphic
yeast *Candida albicans*. This yeast may exist in two alternate growth
forms, the budding growth form and the hyphal growth form. The
single parameter determining alternative growth forms (or phenotypes) is pH (figure 8.11). In a recent review, Soll (1986) presents
evidence that the major distinction between these phenotypes is
not due to differential gene expression. A search for phenotype-
specific polypeptides revealed a preponderance of similar polypeptides in both growth forms. Of 347 individual polypeptides reprod-
ucibly observed on repeat gels, only one hypha-associated polypeptide and only one bud-associated polypeptide were clearly
phenotype-specific rather than pH specific. Extensive chemical
comparisons of both growth forms suggest that phenotypic change
is not due to a change in gene products or qualitative differences in
architectural building blocks. Rather phenotypic change appears
due to a system in which the difference in pH alters the general
expansion mechanism and the temporal and spatial dynamics of the
apical expansion zone resulting in differences in shape and growth
dynamics. Soll relates these pH-induced changes to differences in
actin distribution and cytoskeletal organization but points out that
such biochemical differences need not be the result of differential
gene expression.

Poethig (1984a), although not taking such an extreme "agenic"
view, also stresses the autocatylitic aspects of plant organ formation.
He notes that cell determinaton in plants is a result of the overall
chemical and physical organization of an organ rather than a cell
autonomous property. Plant cells do not seem to receive positional
cues at one time in development and act on these cues at some later
time.

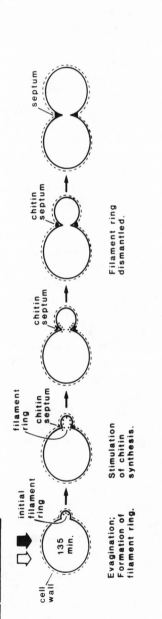

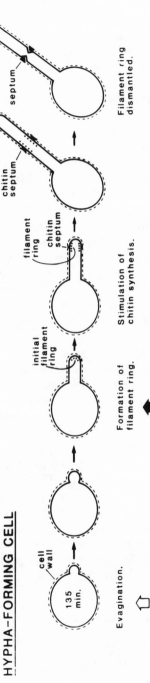

Figure 8.11. A model for the temporal regulation of filament ring and septum position under the regime of pH-regulated dimorphism. Although evagination occurs at the same time in budding and hypha-forming cells, the filament ring begins to form 30 minutes later in the hypha-forming cell. In both cases, the filament ring forms at the apical growth zone and therefore the difference in timing alone dictates the difference in position. The position of the filament ring in turn dictates the position of the chitin containing septum. (From Soll 1986.)

We should now consider the concepts of positional information[4] and structural epigenesis and try to relate these ideas to the paradox posed by Haber and Foard (i.e., that plant organ form is determined directly by genes and is not a consequence of gene-controlled cell division, elongation, and specialization). Two features of plant development may be very critical points, where genes may directly influence organ form: (1) the topology of the apical (or other) meristems; (2) the surface cell layer(s) (epidermis, tunica layers if present).

In our previous discussions of plant fasciations (due to either genetics or other causes), alterations in apical meristem shape resulted in dramatic changes in organ phenotype. Green (1980) has reviewed the literature on plant organogenesis and has concluded that strong theoretical considerations favor a major role for the surface cell layer(s) in morphogenesis, especially the epidermal tangential walls (see Hejnowicz, Nakielski, and Hejnowicz 1984). The importance of the epidermis in determining flower form was demonstrated by Marcotrigiano (1986), using interspecific graft chimeras between *Nicotiana glauca* and *N. tabacum*. These species have apical meristems consisting of two tunica layers and a corpus (LI, LII, and LIII). Periclinal chimeras in which the LI was populated by cells of a different species from the cells in the LII and LIII were studied for floral characteristics. With the exception of style length, all other quantitative traits were consistently altered by the epidermis in the direction of the epidermal species. In other graft chimeras (*Laburnocytisus adamii, Crataegomespilus asniersii*), the central tissue seems to play a greater role in determining form (see Grant 1975 for review).

Of course, plant organogenesis is not simply the "filling in" of forms or shapes determined by the surface or epidermis. Plant organs are not usually internally homogenous but rather consist of complex (and often repetitive) patterns of tissues (e.g., vascular tissue, cortical tissue, etc.). Internal as well as internal-external interactions and communication must occur to coordinate internal patterns of differentiation. An insight on the nature of these interactions is given by the expression of the knotted locus (*Kn1*) in maize. This locus is defined by five dominant mutants that specify

---

4. It is interesting to note that Korn (1982) has extended the positional information paradigm to within plant cells. This model raises the possibility of developmental processes based upon intracellular positional information regarding cell plate and cell wall characteristics and structural epigenesis at the intercellular level governing the positions and orientations of variously specialized cells. Following the example of *Candida albicans*, intracellular positional information need not be based upon differential gene expression.

unexpected cell divisions in cells near lateral veins of the leaf blade. Such divisions result in hollow protrusions from the leaf plane that are known as knots. Knot formation involves cell divisions in all leaf layers. (See Freeling and Hake 1985 for descriptions of other manifestations of the knotted phenotype.) Using x-rays to create somatic sectors lacking *Kn*1, Hake and Freeling (1986) could show that the genotype of the mesophyll and not the genotype of the epidermis was critical for *Kn*1 expression. Mesophyll carrying the *Kn*1 locus induced epidermis lacking this locus to divide. The function of the normal allele of this locus is unknown.

Evidence that genes control plant form is available in the horticultural and agronomic literature on plant mutations. This literature is replete with examples of single gene mutations that have very drastic developmental consequences (Marx 1983). Plants appear to tolerate single gene mutations that cause major differences in morphological structure more readily than animals. In animals, such mutations typically exhibit deleterious pleiotropic effects involving reduced fertility and often other developmental abnormalities (see Gottlieb 1984; Thomas and Greirson 1987 for review). In *Antirrhinum majus*, Harte and Maek (1976) have shown that some mutants which give rise to abnormal leaves also changed the rate of root growth, whereas other leaf mutants did not alter the growth of the root meristem. Also in some leaf mutants with altered root growth, normal root growth occurred when the roots were grown isolated from the plants. Such transorgan pleiotropic effects indicate that the biochemical paths from mutant allele to mutant growth form are complex and diverse. These results also prompt caution in assuming that the wild-type alleles of mutants that alter form are genes involved in the control of form. It is possible that many such "developmental-mutant" genes are involved in more general housekeeping functions and that mutant alleles result in broadscale phenotypic effects, some of which result in altered organ form secondarily.

Even in those situations where a mutation seems to be restricted in its expression to one organ system, the relationship between altered morphology and gene activity is not clear. For example, the dominant gene hooded in barley results in a very dramatic and profound change in the barley spikelet. Stebbins and Yagil (1966) found that the earliest histological difference between mutant and wild type was a change in the tempo of mitotic rhythm and orientation of cell divisions on the adaxial epidermis and subepidermal layers in the distal region of the lemma primordium. Relating these changes to the complex hooded phenotype was not possible; all that could be concluded was that the entirely new epigenetic sequence of development (leading to the hooded phenotype) was in some way initiated by altered division patterns in the primordium. Environmental treatments that suppressed these division patterns reduced to some extent the expression of hooded (Yagil and Stebbins 1969). Of course, the nature and controls of the epigenetic sequence are the central and still unresolved problem in plant morphogenesis.

Although much of plant development and organogenesis is characterized by high levels of homeostasis (i.e., once a developmental path is chosen it has a certain momentum so to speak), plants also have extreme sensitivities to weak physiological and environmental signals. Such strong homeostasis coupled to sensitivity to specific weak signals was documented by Mauseth (1979) in the apical meristem of a cactus species. Seedling apical meristems were able to function normally in the presence of numerous hormones but responded quickly to a specific treatment with cytokinin which induced the formation of a pith-rib meristem. Responses to such weak signals often result in a total restructuring of the developmental program into another mode. Examples of weak signal-induced responses include flowering and its relationship to phytochrome and photoperiodism, the various tropisms (e.g., phototropism, the bending response to light, geotropism, the response to gravity, thigmotropism, the curvative toward or away from a point of mechanical stimulation), hormonal responses (e.g., ethylene and fruit ripening, gibberellin and the hydrolysis of starch in seeds), wounding responses, and aging effects (e.g., changes in leaf morphology between juvenile and adult forms, organ or ramet abscission, organ senescence). Thus, any overall theory of plant development must consider both high levels of developmental buffering against environmental and genetic noise (the latter including mutations) coupled with very high sensitivities to specific weak signals (including environmental, physiological, and genetic cues). Because of ease of experimental access, more is known about the genetic and molecular responses to these signals than the control and determinants of long-term and repetitive developmental patterns.[5]

The emphasis that different researchers have placed upon the

---

5. The use of gene transfer systems to introduce foreign genes into a plant genome is a new approach that is rapidly increasing our understanding of genetic responses to so-called weak stimuli. Such foreign genes can be introduced with all or portions of associated *cis* regulatory sequences, thus allowing fuller documentation of the path from signal reception to transcriptional response (see Kuhlemeier, Green, and Chua 1987 for review).

autocatalytic vs. gene-controlled aspects of plant development often is prejudiced by whether the homeostatic patterns of organ development or the developmental responses to selected weak stimuli are being investigated. It is interesting to note that although it may be argued that autocatalytic systems may be genetically information poor and, conversely, gene-controlled systems are information rich, such a relationship is not obligatory. At least theoretically, a set of cybernetically interacting and correlated subsystems (a system of sets of weakly interacting elements) can have the properties of homeostasis as well as high sensitivity to some weak signals (Fukshansky and Wagner 1985; Fukshansky pers. comm.). Such a developmental system may be relatively information-rich in terms of gene activity. This kind of genetic and physiological organization could account for the large responses to weak or small signals such as the environmental stimuli already discussed, the relatively small set of gene mutations that have large phenotypic effects, as well as the general homeostatic qualities of development. The homeostasis exhibited by much of plant development may be one of the reasons that biochemical mutations (primarily auxotrophs) are so difficult to induce in plants in contrast to lower eukaryotes and animals. Perhaps plant metabolic pathways are so interconnected cybernetically that genetic lesions in one pathway are metabolically compensated to the extent that mutant phenotypes are not expressed. Schmalhausen in a classic book on the consequences of stabilizing selection and development came to similar conclusions. To quote,

> Stability of organization is evidence of the existence of a more or less complex regulating mechanism that protects the normal structure from possible disturbances due to accidental changes in external environment. Since morphogenetic processes are unique and are determined primarily by internal developmental factors, these regulatory mechanisms also protect the normal structure from any chance variations in the internal factors; that is, from mutations. Accordingly, many slight mutations do not manifest themselves in the stable organism. (1949:43)

Regardless of the molecular mechanisms involved in plant development, the observations summarized in this chapter support the thesis that plants can develop functional organs and organisms from mixtures of variously mutant and wild-type cells. Also because of developmental homeostasis, cross-feeding in mutant cells in meristems, and diploidy, diplontic selection within the plant is probably not very efficient in purging mutant cells from meristems. One suspects that more mutants are lost through stochastic events during plant growth than through the various forms of diplontic selection.

As the following chapter will document, the most effective mutant purging systems are expressed during the plant's reproductive development. In fact, many of the developmental manifestations associated with sexual reproduction in plants appear to have evolved as adaptations in response to mutation buffering. Thus, mutation persistence and transmission during somatic development is compensated to some degree by mutation loss during reproductive development.

CHAPTER NINE

Soft Selection and Life Cycles

DEVELOPMENTAL SELECTION

As discussed already probably the most significant aspect of somatic mutations is the progressive decline of the genotype due to the accumulation of disadvantageous (deleterious, lethal, etc.) mutations with continued growth. The frequency of such mutations may be lessened somewhat depending on apical meristem organization and growth or through ramet competition, but these mechanisms are not foolproof and often not very efficient or effective. Other aspects of plant biology, especially those dealing with sexual reproduction, are probably the most significant parameters for preserving the genetic integrity of the organism (in the broad sense) against the continuous onslaught of the forces of disorder (mutation pressure).

Continued clonal growth in conjunction with mutation pressure may result in the formation of ramets that are heterozygous for disadvantageous mutations. Such mutations may reduce embryonic viabilities (V) in homozygotes ($V = 1 - s$, where $0 < s < 1$) or have effects in heterozygotes ($V = 1 - hs$, where h is degree of dominance). Table 9.1 shows the progressive self-sterilization of a ramet with increasing heterozygosity for mutations. For simplicity, only fully recessive embryonic lethals ($s = 1$, $h = 0$) are considered. Although at first sight it may seem that many plants produce sufficient numbers of propagules (spores or seeds) to accommodate large numbers of lethals, it is important to remember that not all of these propagules actually ever have a chance to audition their genes, so to speak. Among the vast numbers of seeds and spores present in the soil and arriving through dispersal, only a tiny fraction germinates to give rise to seedlings. Harper (1977) has suggested that the numbers of seedlings appearing can be thought of as a function of the number of "safe sites" offered by the environment.[1] One would

1. To quote Harper (1977:112), "A 'safe site' is envisaged as that zone in which a seed may find itself which provides (a) the stimuli required for breakage of seed dormancy, (b) the condi-

suspect that, in general, seeds or spores chance upon safe sites independent of their genotypes; thus, the probability that a viable seedling emerges in a safe site is the product of the probability of a viable zygote times the probability a seed is dispersed to a safe site. Thus if 1 percent of a plant's seeds are dispersed to safe sites and the seeds are the product of a ramet heterozygous for 10 different recessive lethals selfing, then the ramet must produce 3,570 seeds to insure that just two seeds containing viable embryos land in a safe site. Of course, all seedlings growing in safe sites do not necessarily reach sexual maturity and reproduce. The true seed number necessary for reproduction is a function of a series of multiplicative probabilities (i.e., the probability of forming genetically viable embryos times the probability of a seed landing in a safe site times the probability of life cycle completion). Thus a common stratagem for effective reproduction often is a series of escalations of seed or spore output. Since the capacity to form such propagules is ultimately finite, very often the effectiveness of reproduction is enhanced by a series of secondary adaptations. These include adaptations that allow seeds to be more specifically targeted for safe sites (seed and fruit characteristics related to dispersal) and, more importantly to the present discussion, adaptations that increase the frequency of genetically vigorous zygotes.

Table 9.1. Consequences of increasing the number of heterozygous recessive lethals (n) in the genotype. Probability of a lethal-free haploid product (spore) is $(1/2)^n$; probability of a lethal-free zygote resulting from selfing is $(3/4)^n$.

	n		
	5	10	15
Lethal-free spores	0.031	0.001	3.052×10^{-5}
Number which must be formed to yield on the average two lethal-free spores	64.5	2,000	65,536
Lethal-free zygotes	0.237	0.056	0.013
Number which must be formed to yield on the average two lethal-free zygotes	8.4	35.7	153.8

tions required for the germination processes to proceed, and (c) the resources (water and oxygen) which are consumed in the course of germination. In addition a 'safe site' is one from which specific hazards are absent—such as predators, competitors, etc."

Wallace (1981) in contrasting hard and soft selection brought out the important concept of "unit biological spaces," spaces in which only one individual can survive and develop normally. Unit biological spaces exist in the external environment and if occupied by more than one individual, only the most adapted genotype will be the surviving individual. Selection in this way is soft since, generally, one individual always survives. The viability of a genotype is scaled against its immediate competitors and not against some arbitrary ideal. In figure 9.1 hard and soft selection schemes in plant

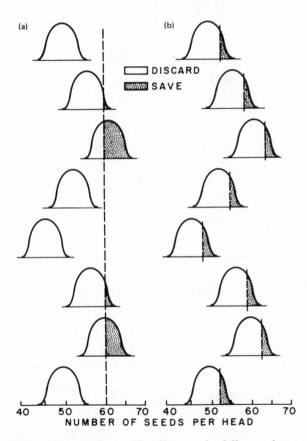

Figure 9.1. Two selection schemes that illustrate the difference between hard and soft selection based seed set per head or inflorescence. A: Hard selection. Plants are saved (or selected) that form 60 or more seeds per head. Thus in some populations no plants will be selected for the next generation. B: Soft selection. In each population the plants with the highest numbers of seeds per head are selected. Consequently, some individuals in each population will contribute seeds for the next generation. (From Wallace 1981.)

breeding are illustrated. Wallace envisioned soft selection in terms of the genetic load associated with genotypes with viability differentials when in competition within unit biological spaces. Genotypes that are lethal in all environments are always under hard selection.

The concept of unit biological space can be extended beyond the external environment to include the internal environment of the organism. Many plants have life cycle characteristics that generate internal unit biological spaces and, thus, result in the operation of soft selection through various and sequential phases of the reproductive cycle. This internal soft selection may eliminate lethals as well as mutations with less harmful consequences without seriously reducing the reproductive capacity of the plant. Buchholz (1922) was the first to note that this type of selection could be operative in plants. He coined the term developmental selection to describe these selective processes within the reproductive cycle of the plant.

Perhaps the best way to visualize these modes of hard and soft selection is with a hypothetical example. Consider two conifers, each heterozygous for 10 unlinked recessive embryonic lethals. Conifer B forms 1 embryo per ovule, initiates and matures 2 ovules per cone, and initiates and matures 50 cones per plant for a total maximum reproductive capacity of 100 seeds. Since conifer B is heterozygous for 10 different nonallelic lethals, its reproductive capacity is 5.6% (i.e., it forms 5.6 viable seeds rather than 100, table 9.1).

In contrast, the reproductive cycle of conifer A has a number of levels of unit biological spaces allowing soft selection to operate. Conifer A forms 3 embryos per ovule but matures only one, initiates 4 ovules per cone but matures only 2, and initiates 100 cones but matures only 50. In figure 9.2 the operation of soft selection (or, more specifically, developmental selection à la Buchholz) is illustrated. Again the individual plant is heterozygous for 10 recessive lethals. Because of the opportunities for selection within the various phases of the reproductive cycle, the reproductive capacity is 62.0% rather than 5.6% as it is in conifer B. Soft selection, therefore, results in an eleven-fold gain in reproductive capacity.

One might argue that this gain is not without its costs and, of course, this is so, but it is instructive to compare the costs for conifers A and B. For conifer B to have a reproductive capacity similar to A (assuming again heterozygosity for 10 lethals), it would need to form 1,175 embryos, similar numbers of ovules, and 587.5 cones. In contrast, conifer A formed 1,200 embryos, 400 ovules, and only 100 cones for an identical reproductive capacity. Since, in general, ovules and cones represent a greater biological investment than young embryos, the advantage is with the life cycle represented in conifer A.

Applying the unit biological space concept to plant reproductive development may, perhaps, show other characteristics of this process. In figure 9.3 soft and hard selection are compared. Sequential junctures or developmental stages are designated a unit biological space. If more than one unit enters a space (or initiates a stage in development) and only one unit survives (the best genetically) to enter the next developmental stage (or unit biological space), then soft selection may operate (e.g., the formation of three embryos in an ovule in which only one embryo survives). Also the more competitive units for each unit biological space and the more sequential unit biological spaces, the more effective soft selection becomes (figure 9.3A). In figure 9.3B a developmental pattern of sequential unit biological spaces with only one unit entering and leaving is illustrated, selection is limited to the final developmental stage. In terms of plant reproductive development, such an organization allows only hard selection to operate and is a biologically costly

---

3 zygotes per ovule $\longrightarrow$ 1 or more viable embryos per ovule

$$1 - (1 - 0.056)^3 \qquad = \qquad 0.1588 = A$$

- -

4 ovules per young cone $\longrightarrow$ 2 ovules per mature cone

$$(A + B)^4 \quad = \quad \underbrace{A^4 + 4A^3B + 6A^2B^2}_{} \quad + \quad \underbrace{4AB^3}_{} \quad + \quad \underbrace{B^4}_{}$$

Cones with two ovules with viable embryos = 0.1212	Comes with one ovule with a viable embryo = 0.3781	Cones with all ovules aborted = 0.5007

- -

100 young cones $\longrightarrow$ 50 mature cones

50 cones without viable seeds
37.8 cones with a single viable seed $\Big\}$ 50 mature cones
12.1 cones with two viable seeds

- -

50 mature cones $\longrightarrow$ 62 viable seeds

Figure 9.2. Developmental selection in a hypothetical conifer. The results of selfing an individual heterozygous for 10 unlinked nonallelic recessive lethals. The frequency of ovules with one or more viable embryos is A, B = 1 − A.

method of maintaining reproductive capacity. It should be noted, however, that the biological cost (in the currency of plant parts, ovules, cones, etc.) is significant only if plants are heterozygous for disadvantageous mutations. When this is not the case the reverse is true, the developmental pattern represented in figure 9.3A is less efficient biologically than the pattern in figure 9.3B. To illustrate this let us reconsider the two conifers, A and B, discussed previously (figure 9.2). If neither plant is heterozygous for lethals, then conifer

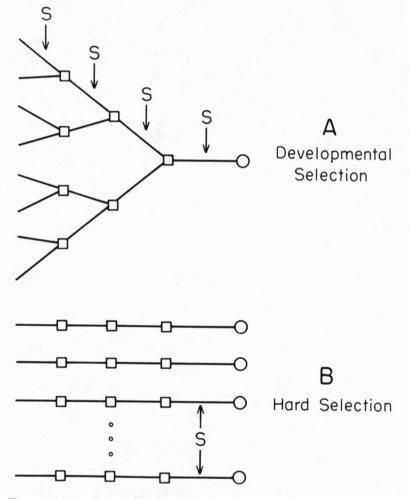

Figure 9.3. Developmental selection and hard selection. S = timing of selection; □ = unit biological space; ○ = end product or propagule.

B requires 100 embryos, 100 ovules, and 50 cones to mature 100 seeds whereas conifer A requires 1,200 embryos, 400 ovules, and 100 cones to mature the same number of seeds (100)! In the following discussion, plant reproductive cycles will be analyzed to determine if they conform to the patterns in figure 9.3. Aspects of the reproductive cycles that generate unit biological spaces with more than one unit competing allow the operation of soft selection, these developmental stages are termed "soft selection sieves." The degree to which Buchholz's developmental selection operates is a function of the number of competing units at a soft selection sieve and the number of such sieves in the reproductive cycle.

Recently Willson and Burley (1983) have reinterpreted Buchholz's ideas in terms of the concepts of sexual selection and sociobiology. Thus, various aspects of sexual reproduction are viewed as being the result of conflict between either males and females, parents and offspring or kin-selected altruistic self sacrifice. Willson and Burley take the position that such intraspecific competitive interactions should tend to accelerate evolution rather than stop it. Thus, the conflicts between different genders and generations result in a kind of evolutionary cold war where each side must match the new advantages of the other. In contrast with the sexual selection models, developmental selection is primarily an aspect of stabilizing or normalizing selection. In this book, it is interpreted as an aspect of mutation buffering whereby genetically defective gametes and zygotes can be lost without proportionate losses of reproductive capacity. Of course, aspects of sexual selection can have similar consequences but, as Willson and Burley point out, the unique significance of the sexual selection concept is its emphasis on gender and generation conflict as a significant evolutionary force. Thus, the emphasis on evolutionary change (directional selection) rather than stasis (normalizing or stabilizing selection). In the following discussion of plant reproductive characteristics, Buchholz's ideas of developmental selection are used since their emphasis is on preserving genetic integrity through time rather than change. Where appropriate, differences of interpretation between developmental selection and sexual selection will be noted.

REPRODUCTIVE CYCLES
AND SOFT SELECTION SIEVES

According to the hypothesis of Buchholz:

Developmental selection expresses itself in some form or other in the sexual reproductive cycle of practically all vascular plants. It is probably also involved in the life cycle of the cryptogamic forms, and is a factor to be reckoned with among animals as well. (1922:253)

Buchholz considered this form of selection to be different from natural selection; current usage would recognize developmental selection as a form of natural selection. This distinction between the two forms of selection is shown in figure 9.4. In the following text, the various life cycle characteristics promoting developmental selection will be discussed and brought up to date with new observations. Included will be life cycle characteristics and implications that Buchholz omitted (but certainly would have thought of given our current vantage point).

GAMETOPHYTE SELECTION (VASCULAR PLANTS)

Seed Plant Megagametophyte

The megagametophyte in seed plants (or any heterosporous vascular plant) originates through the mitotic division of one or more megaspores. The quartet of megaspores is the product of the meiotic division of the megaspore mother cell. The latter differentiates in the nucellar tissue of the ovule. From the viewpoint of soft selection, ovule ontogeny appears to have a number of levels of "unit biological spaces" where competition and selection may occur. Regarding the megagametophyte, two features may allow such competition: 1) the megagametophyte may develop from more than one megaspore; and 2) more than a single megagametophyte may form per ovule. Since only one cell in the megagametophyte ultimately functions as the egg and one embryo generally matures per ovule, soft selection is possible at either or both of the junctures.

In gymnosperms megagametophytes are generally monosporic (i.e., develop from a single meiotic product, megaspore) (Chamberlain 1935). In angiosperms megagametophyte (embryo sac) development is more variable. Although monosporic embryo sacs are typical, bi- or tetrasporic embryo sacs are not uncommon (Maheshwari 1950; Rutishauser 1969). These three general patterns, with regard to how many meiotic products are included within a single megagametophyte, occur in many variations in different taxonomic groups (see figure 9.5). In monosporic embryo sacs, soft selection is possible if the selection of the single megaspore giving rise to the embryo sac is not rigidly determined spatially within the ovule. For example in *Oenothera*, the chalazal or the micropylar megaspore may develop into the embryo sac. This choice may reflect megaspore viability. In the majority of angiosperms with monosporic embryo

NATURAL SELECTION Environmental process occurring in external physical and biological environment of organism, where conditions of struggle for existence are very complex	Struggle against unfavorable environment of physical surroundings. Struggle against other species; extraspecific competition. Struggle against fellows; intraspecific competition. ——— Selection between vegetatively branching parts of either the gametophyte or sporophyte; buds and branches of trees, which later give rise to reproductive parts.
DEVELOPMENTAL SELECTION Occurring during early embryonic or gametophytic stages within tissues of parent plant, under conditions uniform for competing individuals	Interovular selection, between ovules within same ovary: (1) after fertilization, largely due to activities of contained embryos; (2) before fertilization, due in part to activities of contained female gametophytes, megaspores, or archesporial cells. Embryonic selection, between embryos within the same ovule, or within tissues of parent gametophyte. Gametophytic selection: (1) between male gametophytes, such as pollen tubes within carpellary and nucellar tissues; (2) between female gametophytes within the same ovule. Gametic selection: (1) between male gametes or sperms; (2) between female gametes or eggs.

Figure 9.4. The concept of developmental selection. (As defined by Buchholz 1922.)

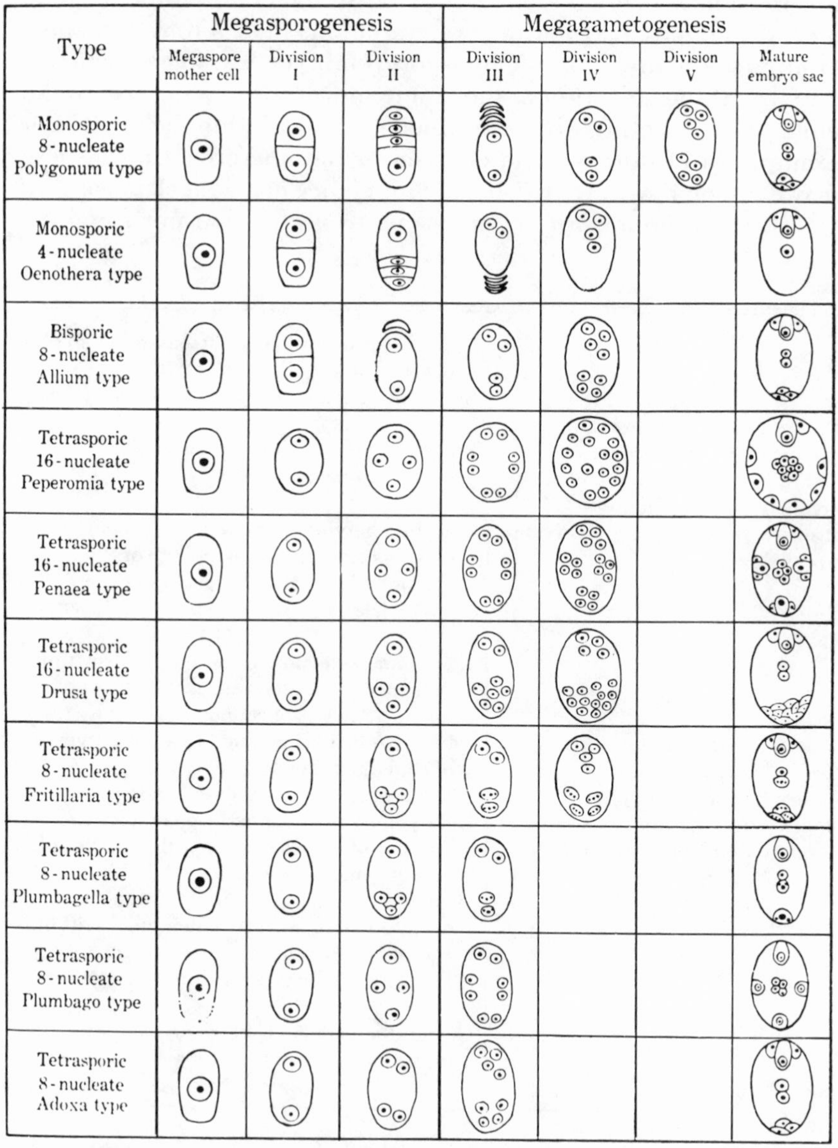

Type	Megasporogenesis			Megagametogenesis			
	Megaspore mother cell	Division I	Division II	Division III	Division IV	Division V	Mature embryo sac
Monosporic 8-nucleate Polygonum type							
Monosporic 4-nucleate Oenothera type							
Bisporic 8-nucleate Allium type							
Tetrasporic 16-nucleate Peperomia type							
Tetrasporic 16-nucleate Penaea type							
Tetrasporic 16-nucleate Drusa type							
Tetrasporic 8-nucleate Fritillaria type							
Tetrasporic 8-nucleate Plumbagella type							
Tetrasporic 8-nucleate Plumbago type							
Tetrasporic 8-nucleate Adoxa type							

Figure 9.5. The major patterns of embryo sac development found in the angiosperms. (From Maheshwari 1950.)

sacs, the chalazal megaspore of the linear meiotic tetrad functions to give rise to the embryo sac. In these cases, the megaspore appears to be selected on the basis of its spatial position in the ovule rather than its viability.

Because megaspores are the immediate products of meiosis, sib megaspores are very likely to be dissimilar genetically. Thus embryo sacs developing from two or four megaspores are probably genetic chimeras. The presence of different genotypes of haploid nuclei in the same embryo sac undergoing various numbers of mitotic divisions may generate a situation in which some form of competition is possible. Presumably such competition increases the probability of the most viable haploid genotype functioning as the egg. The degree to which such competition is possible among the haploid nuclei (or cells) of the embryo sac is inversely related to the rigidity of development within this generation. If the spatial placement of the megaspore nuclei during meiosis is what obligatorily determines the egg genotype, then developmental selection is impossible.

Multiple megagametophytes per ovule are not uncommon in seed plants. These result from different megaspores within the ovules and, thus, competition between megagametophytes as well as competition between embryos (polyembryony, the occurrence of more than one embryo per ovule) is possible. In the gymnosperm *Taxus canadensis*, it is common to find more than a single megaspore mother cell per ovule. All four megaspores may divide, as well as one or two of a second tetrad. Multiple megagametophytes have been observed in *Sequoia, Cunninghamia, Sciadopitys, Taxodium, Cryptomeria* and, rarely, in *Pinus* (see Chamberlain 1935 for review). In sexual angiosperms there may be two to several megagametophytes in the nucellus of the same ovule. Such multiple megagametophytes giving rise to multiple embryos per ovule has been termed false polyembryony (Johansen 1950) in contrast to polyembryony in its strict sense, which refers to more than one embryo per embryo sac. Davis (1966) notes that seeds with multiple embryos are common in the angiosperms, but very often the published data are not sufficient to distinguish between true and false polyembryony (nor between sexual and asexual embryos). Thus the overall importance of false polyembryony as a soft selection sieve in the angiosperms is unknown (see Coulter and Chamberlain 1912 for reports of multiple embryo sacs in monocotyledonous and dicotyledonous genera, e.g., *Lilium, Agraphis, Uvularia, Loranthus, Casuarina, Salix, Fagus, Corylus, Delphinium, Ranunculus, Rosa, Astilbe, Pyrethrum* and *Senecio*).

Seed Plant Microgametophyte

In seed plants the microgametophyte is the pollen grain and the pollen tube that it forms. The potential significance of pollen competition in seed plants was considered initially by Buchholz (1922) in his original formulation of developmental selection. In later papers Buchholz documented the effectiveness of pollen selection in *Datura* in collaboration with A. F. Blakeslee (see Jones and Tippo 1952 and Jones 1928 for a bibliography of this literature). Buchholz recognized that in the angiosperms pollen competition may be especially important because of the presence of styles and associated insect pollination. To quote:

> It is very evident that a good selective process . . . should meet at least the following four requirements. It should (1) start the competition simultaneously, (2) take place under uniform conditions, (3) measure comparable merit, and (4) rigidly eliminate the great majority that fall below standard. . . . Pollination, especially when the pollen is transferred in masses or clumps by insects, is another more or less sudden event, which launches the competition of pollen tubes. (Buchholz 1922:277)

Perhaps one of the clearest examples of the influence of pollen competition on the transmission of traits through the male parent is the inheritance of aneuploidy in *Datura* (Buchholz and Blakeslee 1930).[2] Such plants produce pollen grains with n and $(n + 1)$ chromosomes that generate a bimodal distribution of pollen tube lengths after 16 hours of growth through the stigma. It was determined that pollen grains with $(n + 1)$ chromosomes were slower germinating and growing than those with n chromosomes (comparative growth rates were 1.9mm and 2.6mm per h). The influence of these differentials on pollen transmission was demonstrated in two ways (see table 9.2). Plants were sparingly pollinated with n and $(n + 1)$ pollen and, after a period of time, the styles were excised. The prediction was that the faster growing pollen tubes would enter the ovary prior to the slower growing tubes which would be lost with the style excision and, consequently, the gamete pool should be enriched for pollen with (n) chromosomes. The data in table 9.2 show that this

---

2. The length of the style in *Datura* is very long, making it an ideal organism to study pollen tube interaction.

procedure completely eliminated $(n + 1)$ gamete genotypes. The second demonstration of pollen competition was based upon the observation that the ovules in the upper half of the capsule were fertilized prior to those in the lower half. Thus if plants are sparingly pollinated with normal and aneuploid pollen, one would expect seeds from the upper half of the capsule to have fewer aneuploid embryos than those from the lower half. As shown in table 9.2, this is exactly what was observed. More recently Meinke (1985) performed a similar analysis on the fruits (siliques) of *Arabidopsis thaliana* plants heterozygous for embryo-lethal mutants. Twenty-five percent of the seeds from selfing such genotypes usually were aborted. Assuming no pollen competition, these aborted seeds should be distributed randomly in the siliques. Of 32 different embryo mutants so screened, 24 showed random seed abortion in the siliques. Eight mutants gave nonrandom patterns of seed abortion, in 7 cases the mutants were more frequent in the upper portion and

---

Table 9.2. Pollen competition and the transmission of aneuploids in *Datura stramonium*, $2n$ females were crossed with $2n + 1$ males. The males formed two classes of pollen, n and $n + 1$. Progeny that are $2n + 1$ result from the union of an egg with n chromosomes and a sperm nucleus with $n + 1$ chromosomes. Thus, the frequency of $2n + 1$ progeny is a measure of the transmission of the aneuploid male gamete. (From Buchholz and Blakeslee 1930.)

	Capsule	Progeny number	Percentage $2n + 1$
SPARINGLY POLLINATED, STYLES NOT EXCISED			
24a	Upper ½	188	0
	Lower ½	205	65.4
25a	Upper ½	147	4.1
	Lower ½	158	52.5
26a	Upper ½	167	0.6
	Lower ½	167	49.1
SPARINGLY POLLINATED, STYLES EXCISED			
21a		34	0
12a		19	0
22a		87	0

in 1 case, in the lower portion. Since fertilization order was not reported in these siliques, these 8 mutants formed either more or less competitive pollen tubes in contrast to the nonmutants.

Buchholz and Blakeslee (1930) also performed an experiment documenting the importance of pollination order that has relevance to natural patterns of pollination in *Datura*. The flowers of white plants were pollinated with pollen (20–60 grains) from purple plants and repollinated after 24h by selfing. Thus pollen carrying the dominant purple allele had a day's start over pollen carrying the recessive white allele. The seeds obtained were scored as to their position in the capsules and the seedlings were grown and scored for phenotype. In the total, 92.2% of the purple plants were found in the upper portion of the seed capsule, while only 6.7% were found in the middle section and only a single purple plant was obtained from seeds of the lower portion. These experiments, together with those discussed previously, clearly show the interactions of pollination order, timing of pollen germination and pollen tube growth rate in affecting the transmission of genes from the male parent and the distribution of these genes among the embryos within the ovary.

Pollen competition was well known and often discussed in the early days of plant genetics and evolution but, as with so many topics in science, eventually interest waned and the area was neglected both by experimentalists and theorists. Recently there has been a renaissance of interest in this area inspired by the work of Mulcahy and his colleagues (Mulcahy 1979; Ottaviano, Sari Gorla, and Mulcahy 1980; Mulcahy, Bergamini Mulcahy, and Ottaviano 1986). The importance of pollen competition is a function of the degree of genetic overlap between genes expressed in the pollen and those in the sporophyte. Tanksley, Zamir, and Rick (1981) studied the degree of this overlap in *Lycopersicon esculentum* using isozymes. For dimeric isozymes, zymograms of sporophyte tissue and pollen allow one to distinguish whether genes are expressed pre or postmeiotically. Genes expressed premeiotically will form heterodimers whereas those expressed postmeiotically will form only homodimers. Of the 30 isozymes present in one or more sporophyte stages, 18 were also found in the pollen and only 1 was found to be unique to the pollen. These data suggest that the majority of genes expressed in the pollen are also expressed at some stage during sporophyte development and are indicative of the potential of pollen tube competition as a soft selection sieve. Snow (1986) studied pollination in *Epilobium canum,* a long-styled hummingbird pollinated species that disperses pollen tetrads. She demonstrated that the potential for pollen competition certainly exists in this species under conditions of natural pollination.

Polyembryony

The formation of multiple embryos within a single ovule was noted by Buchholz as being a significant aspect of developmental selection in vascular plants.[3] Generally three modes of polyembryony are recognized depending upon the origin of the embryos:

　　1. *False Polyembryony:* multiple megagametophytes in an ovule, each of which may form an embryo,
　　2. *True Polyembryony:* multiple embryos formed by a single megagametophyte;
　　　　(a) simple polyembryony—embryos arise through the fertilization of different egg cells on the same gametophye;
　　　　(b) cleavage polyembryony—multiple embryos originate through the division or cloning of a single embryo into a group of embryos.

Both false and true polyembryony generate the possibility of soft selection since generally (but not always) only one embryo per ovule survives to give rise to the young sporophyte.

The three kinds of polyembryony differ in significant ways with regard to the genotypes of the competing embryos. False polyembryony results in genetically dissimilar embryo genotypes because the individual megagametophytes arise from different megaspores, thus they are likely to be genetically dissimilar. The eggs that these megagametophytes form are, therefore, genetically distinct and these eggs are fertilized by sperm of yet different genotypes. In the case of simple polyembryony a single megagametophyte (or gametophyte in a homosporous plant) forms a number of eggs of identical genotypes. These eggs are fertilized by sperm of potentially different genotypes, thus generating genetically variable embryos competing on the same gametophyte. Finally, cleavage polyembryony occurs when a single embryo divides to make a number of identical embryos, barring mutation this situation generates a group of competing embryos that are genetically identical. In figure 9.6 these

---

3. In humans (and perhaps in other mammals) soft selection based embryonic competition may also occur. Ayme and Lippmand-Hand (1982) presented evidence supporting the hypothesis that the stringency of embryonic selection is inversely related to maternal age. Thus, the higher rate of live births with autosomal and X trisomes in older females is, at least in part, related to the decreasing probability of the spontaneous elimination of such genotypes before term in women over age 29.

three forms of polyembryony are compared genetically. Assuming the simplest case, heterozygosity of the parental sporophyte at a single locus and selfing, it is clear the three forms of polyembryony differ greatly regarding the genetic variability of the embryos within a single ovule (conovulate embryos). False polyembryony generates the greatest genetic variability, simple polyembryony generates comparatively less genetic variation, and cleavage polyembryony results in no genetic variability among conovulate embryos. In addition, false polyembryony really represents two soft selection sieves since it allows for megagametophyte competition as well as embryo competition. The occurrence of false polyembryony in the seed plants has already been discussed in the section on megagametophyte competition.

Simple polyembryony results when a single gametophyte gives rise to a multiplicity of egg-containing archegonia and more than one egg is fertilized. Thus the gametophyte must form a multiplicity of archegonia, allow multiple fertilizations and sustain the development of multiple embryos for at least a short period of time. Where these three features occur, there is the possibility of competition between embryos with different genotypes and consequent soft selection. These characteristics are met in the majority of the pteridophytes (both homosporous and heterosporous) with the possible exception of *Marsilea*. Most pteridophyte gametophytes form a mul-

Egg genotypes per ovule	Possible embryo genotypes per ovule
False Polyembryony	
A,a	AA,Aa,aa
True Polyembryony	
Simple	
A	AA,Aa
or	
a	Aa,aa
Cleavage	
A	AA,AA,AA or
	Aa,Aa,Aa
or	
a	aa,,aa,aa or
	Aa,Aa,Aa

Figure 9.6. Genetic consequences of the various forms of polyembryony. The parental sporophyte has the monohybrid genotype, Aa.

tiplicity of archegonia and very often more than one embryo forms during the life of the gametophyte (see chapter 10). The earliest record of simple polyembryony was discovered by Smoot and Taylor (1986) in Permian seeds from Antarctica. In the gymnosperms, simple polyembryony also appears to be common. Multiple archegonia occur in many genera (Chamberlain 1935). For example in the Coniferales, *Thuja* has 6 archegonia per megagametophyte, *Juniperus communis* has 4 to 10, *Libocedrus decurrens* has 10 to 15, *Pinus strobus* and *P. resinosa* have 3.

The significance of simple polyembryony as a factor in sheltering lethals from selection can be readily appreciated by the following calculations (see also Lindgren 1975). If p is the frequency of the wild-type allele A and q the frequency of the recessive allele A′, then the array of genotypes after random mating and simple polyembryony is

Genotype	AA	AA′	A′A′
Frequency	p^2	$2pq + q^2$	—

The frequency of AA′ is zero because gametophytes with A′ genotypes generally participate in the breeding pool long enough for the viable embryo AA′ to be formed not once but many times. The selection coefficient (s) is unity for a recessive lethal and the viability of heterozygotes is $V = 1 - hs$ where h is the degree of dominance. In *Drosophila* the viability of such heterozygotes is reduced from 1 to 0.98 indicating $h = 0.02$ (Crow 1958). If a gametophyte forms a number of AA′ zygotes, the probability that at least one forms a viable embryo is $1 - (hs)^n$ where n is the number of archegonia containing fertilized eggs. It is obvious that n does not have to be very large before the occurrence of a viable heterozygous embryo in an ovule becomes unity. Since recessive lethals behave as if fully recessive in heterozygotes, h can be taken as zero in the following equations. Therefore, average fitness $(\overline{W})$ is unity and, consequently, genetic load (L) is zero. The frequency of A′ after one generation is

$$q_1 = (2pq + q^2)/2\overline{W}.$$

The change in allele frequency is

$$\Delta q = q_0 - q_1,$$

$$\Delta q = q/2.$$

If u is the forward mutation rate of A $\rightarrow \rightarrow$ A′ and v is the reverse mutation rate A $\leftarrow \leftarrow$ A′, then the equilibrium allele frequency occurs when $\Delta q = \Delta p$. The equilibrium frequency of the recessive lethal $(\hat{q})$ may be calculated as follows:

$$\Delta p = up - vq,$$

Since $\Delta q = \Delta p$, then

$$q^2/2 = u(1 - q) - vq;$$

then following Klekowski (1982):

$$\hat{q} \cong \sqrt{2u}.$$

In the absence of simple polyembryony and developmental selection,

$$\hat{q} \cong u/(hs),$$

where $h > 0$, see Crow and Kimura (1970). For example if $u = 1 \times 10^{-5}$, $h = 0.02$ and $s = 1$, then

$$\hat{q} = 5 \times 10^{-4}$$

without simple polyembryony, and

$$\hat{q} = 4.5 \times 10^{-3}$$

with simple polyembryony.

Simple polyembryony allows lethal and other disadvantageous alleles to become more frequent in the gene pool without seriously reducing the reproductive capacity of individuals (Lindgren 1975). This is an important point applying to all aspects of plant life cycles functioning as soft selection sieves at the zygotic or postzygotic levels. Although such sieves preserve reproductive capacity despite the presence of disadvantageous mutations, they often are not effective in reducing the frequency of disadvantageous mutations in the gene pool. The latter is primarily a consequence of micro- or megagametophyte competition, soft selection at the prezygotic level.

Cleavage polyembryony, the subdivision of a single embryo into a group of competing embryos, appears to have no significance in terms of selection since all embryos have the same genotype. If one considers genotypes whose viability is less than one but greater than zero $(0 < V < 1)$, it is possible to construct a model to demonstrate soft selection. For example, consider a genotype with such an intermediate viability. If $V = 0.4$ and if moribundity occurs early but is not restricted to the very earliest stages of embryonic development, then subdividing the embryo increases the proportion of ovules that eventually will have a viable embryo. The probability of at least one viable embryo (E) is

$$E = 1 - (1 - V)^n$$

where n is the number of embryos derived from the cleavage of a single zygote. Thus in our example, if an embryo cleaves into eight embryos as is typical in *Pinus*, the probability that an ovule has at least a single surviving embryo is

$$1 - (1 - 0.4)^8 = .9832,$$

thus a genotype may greatly increase its viability through cleavage polyembryony.[4] Again it should be noted that although reproductive capacities greatly increased, there is a reduced selection against the disadvantageous allele. As noted by Chamberlain (1935), there may be as many as 24 embryos starting in the same ovule in *Pinus* (3 eggs fertilized, simple polyembryony, each of which cleaves to form 8 embryos—cleavage polyembryony).

In the angiosperms, although polyembryony occurs (very often associated with apomixis) it is not common. True cleavage polyembryony has been occasionally reported in sexual species (Maheshwari 1950).

Ovule-Ovary Competition

The formation of seeds and fruits (or strobili) is a process that very often represents a compromise between a variety of conflicting adaptive needs. Thus the problems of pollination, resource allocation, dispersal, and seed predation are superimposed upon any selective forces that may increase offspring quality (Janzen 1977; Bertin 1982; see Stephenson 1981 for a recent review). It is not the goal of the present discussion to reanalyze the total biology of seed and fruit formation but rather to consider to what extent soft selection may be possible during these two developmental stages. It also should be noted that although seed and fruit maturation may have properties promoting the loss of genetically defective offspring without seriously reducing reproductive capacity, this does not mean that this is the primary "adaptive significance" of these aspects of plant reproductive biology. These phases of the reproductive cycle of

4. Discussions with D. L. Mulcahy contributed in the development of this interpretation of cleavage polyembryony. Under the sexual selection model, cleavage polyembryony is interpreted as kin-selected altruistic self sacrifice of multiple identical embryos to increase the nutrition and competitive ability of the remaining embryo (Willson and Burley 1983).

seed plants are too critical in too many biotic and environmental interactions to presuppose simple one-dimensional adaptive explanations.

The majority of seed plants produce more ovules than seeds. Although seed to ovule (S/O) ratios vary, Wiens (1984) documented an interesting pattern relating to life span. Annuals have an S/O ratio of 85%, herbaceous perennials have lower S/O ratios (57.2%), and woody perennials have the lowest ratio (32.7%). The ratios generally are independent of breeding system (table 9.3). It is interesting to note that it appears, therefore, that S/O ratios are inversely related to the life span of the individual genet. Of course, one must make the usual caveat that some herbaceous perennials live longer than some woody forms, but as a statistical generalization the reverse is more common. Cumulated mutational load is also a function of genet age (as well as meristem and ramet growth characteristics). It is interesting, therefore, that the life forms with the highest predicted mutational load also require more ovules to form a single viable seed than those life forms with lower mutational load.

Low S/O ratios allow for developmental selection at two levels, gametophytic and sporophytic. Competition between unfertilized ovules within an ovary may allow for megagametophyte competition, i.e., those megagametophytes developing from megaspores with defective genotypes may cause the ovule to abort. Competition between fertilized ovules within an ovary may allow discrimination of embryo viabilities. Thus, excess ovules allow the operation of two soft selection sieves.

The majority of seed plants also produce more flowers (specifically ovaries) than mature fruits (Stephenson 1981). Darwin (1883) was the first to hypothesize that the production of excess fruit enables plants to regulate the quality of their offspring through selective fruit abortion (see Janzen 1977; Charnov 1979; Lee 1984; Bawa and Webb 1984 for more recent variations of this hypothesis). It is known that at least some plant species abort fruits with low numbers of seeds (e.g., Lee and Bazzaz 1982) and, more importantly, that the seeds in the retained fruits contain embryos of higher quality than those within the aborted fruits.

Stephenson, Winsor, and Davis (1986) pollinated *Cucurbita pepo* with low, medium, and high pollen loads and measured fruit set and seedling characteristics. In table 9.4 is shown a summary of their data. Low pollen loads per stigma were calibrated at ca. 240 pollen grains, medium loads were twice that value, and the stigma was saturated with pollen grains for the high pollen loads. It is clear

Table 9.3. Seed/ovule ratios and brood size or the mean number of seeds matured per ovary in plants with differing life histories and breeding systems. (From Wiens 1984.)

| | SPECIES PAIRS | | | | | | TOTAL SAMPLE | | | | | |
| | S/O RATIO | | | BROOD SIZE | | | S/O RATIO | | | BROOD SIZE | | |
	$\bar{X}$	S.D.	N	$\bar{X}$	S.D.	N	$\bar{X}$	S.D.	N	$\bar{X}$	S.D.	N
Annuals												
Inbreeders	92.2	5.4	28	24.9	40.1	26	85.2	18.6	61	20.6	33.5	59
Outcrossers	91.5	10.6	11	6.4	6.0	11	81.4	16.9	27	24.2	49.9	27
Significance	$P = 0.42$ (N.S.)			$P = 0.008$			$P = 0.07$			$P = 0.31$ (N.S.)		
Perennials												
Inbreeders	63.9	—	2	1.1	—	1	62.1	16.7	6	2.3	1.5	5
Outcrossers	52.3	18.9	31	10.2	15.0	31	49.1	20.3	102	9.9	14.8	100
Significance	—			—			—			—		

NOTES: S/O ratios of all annuals compared with all perennials ($P \ll 0.003$).
Brood sizes of all annuals compared with all perennials ($P \ll 0.001$).

that in *C. pepo* fruit abortion is inversely related to seed number per fruit and that pollen load determines seed number. It also could be argued that increased numbers of pollen grains per stigma maximizes the opportunities for pollen competition and this, in turn, insures higher seed sets. The following will show that pollen competition is probably *not* a significant factor in seed set in this example.

At this point it may pay to digress with a human parable illustrating the difference between competition and inviability. Consider a sample of runners in a marathon; those who finish the 26.2 mile ordeal may be scaled according to their finishing times and can be assigned relative positions, 1st, 2d, 3d . . . *n*th. The runners who fail to finish cannot be so scaled, such runners presumably would fail to finish regardless of the efforts of their fellow runners. Thus in our original sample of runners, the relative position of the finishers is a function of the fellow runners who also finish; those runners who fail, do so independently of the success or failure of the other runners. Let us term the finishers viable entrants and the nonfinishers, inviable entrants. Competition only truly occurs among viable entrants, inviable entrants do not compete with other entrants (only with themselves).

Now to return to our example in *C. pepo*. If a mature fruit can form approximately 300 seeds, then at least 300 ovules are available for fertilization in an ovary. If such a pistil is pollinated with approximately 500 pollen grains (medium load) and, on the average, forms only 77 seeds per fruit, perhaps only 77 pollen grains could affect fertilization while the remaining grains were inviable in this regard and their failure had nothing to do with competition. This point is even clearer under the conditions of low pollen load, where approximately 1 pollen grain was available for each fertilizable ovule and yet only 13 seeds were formed!

Embryo-endosperm interactions also may be involved in the *C.*

Table 9.4. Relationship between pollen load, seeds per fruit, and fruit abortion in *Cucurbita pepo*. (From Stephenson, Winsor, and Davis 1986.)

Pollen Load	Seeds per Fruit	% Fruits Aborted
low	13 ± 7.8	82
med	77 ± 30.7	72
high	290 ± 26.5	29

pepo example. If the ovary forms ovules in excess of the 300 it may mature as seeds, increasing pollen load may result in a parallel increase in fertilized ovules. If many ovules abort because of variously defective embryo/endosperm genotypes, then the actual *number* of viable embryo/endosperm genotypes per ovary will be a function of pollen load. In this case again, it is not pollen competition that is the cause of high seed sets per fruit. Stephenson, Winsor, and Davis (1986) also showed that various measures of offspring vigor were inversely related to the degree of fruit abortion on the maternal plant.

In an ingenious experiment based upon *Lotus corniculatus*, Stephenson and Winsor (1986) demonstrated that offspring quality and fruit abortion are related. Each inflorescence on *L. corniculatus* commonly aborts half of its immature fruits. These authors reasoned that if this species regulates seed quality via selective fruit abortion, the natural patterns of fruit abortion (self-thinned) should produce offspring that are more vigorous than those produced by random patterns of fruit abortion (hand-thinned). In table 9.5 are shown measures of the vigor of seedlings produced by self- and hand-thinned inflorescences. In both cases the flowers were pollinated naturally by bees. Although the offspring from self- and hand-thinned inflorescences failed to show differences in germination percentage, they were different in a number of vegetative and floral

---

Table 9.5. The vigor of seedlings produced by self- and hand-thinned inflorescences of *Lotus corniculatus*. Mean ± S.D. (From Stephenson and Winsor 1986 and pers. comm.)

Year	Treatment	Number of Seedlings	Stem Length 60 Days (mm)	Leaf Number 60 Days
1982/83	Self-Thinned Inflorescences	23	774.8 ± 282.8[a]	90.0 ± 30.8[b]
	Hand-Thinned Inflorescences	25	606.2 ± 308.7	64.5 ± 31.4
1983/84	Self-Thinned Inflorescences	133	988.9 ± 281.9[c]	65.5 ± 16.2[d]
	Hand-Thinned Inflorescences	67	902.4 ± 315.1	58.6 ± 17.6

[a] t = 1.98; df = 46; p < 0.05
[b] t = 2.84; df = 46; p < 0.005
[c] t = 1.97; df = 198; p < 0.025
[d] t = 2.76; df = 198; p < 0.005

characteristics. These results support the hypothesis that seed quality is inversely related to fruit abortion.

Studies documenting the phenotypic and competitive vigor of seedlings grown from seeds matured under conditions of maximum fruit competition are of critical importance. Also, the extent to which these differences in vigor may reflect maternal effects rather than seedling genotypes must be determined before the importance of fruit abortion as a mechanism of soft selection can be assessed. Thus, second generation effects, i.e., the offspring of vigorous seedlings should be more vigorous, must be measured also.

As to the morphogenetic signal or signals that determine which fruits are to be aborted and which matured, many factors are important. The number of fruits initiated and matured per inflorescence is influenced by the environment (resource availability, seed predation, etc.) (Lee and Bazzaz 1982). Fruit maturation is often related to when fruit initiation and pollination occur during the growing season. In many species the fruits are matured in the order initiated, with the probability of abortion highest for late-initiated fruits. Such "coarse" morphogenetic signals may be modulated and refined through subtle hormonal signals between reproductive organs. The most important hormonal stimuli acting on fruit growth are the pollen and pollen tubes and ovules (Nitsch 1952) (figure 9.7). The deposition of pollen on the stigma and the growth of pollen tubes in the style promote fruit development (this promotion is independent of fertilization). Pollen stimulation of fruit growth appears fairly nonspecific since it can be demonstrated with interspecific pollen and inviable pollen (as well as pollen extracts) (see Nitsch 1952 for review). The development of fruits in response to hormonal stimuli from ovules maturing into seeds is well known in many species (Nitsch 1963).

A consequence of the pollen/ovule control of fruit development is the generation of a number of soft selection sieves in the typical angiosperm life cycle. High pollen loads which enhance the probability of fruit maturation by functioning as a hormonal stimulus may allow for pollen competition (Lee 1984) and, since a fruit usually forms more ovules than can be matured as seeds, also maximize seed competition within the developing fruit. Whether such pollen/ovule/fruit interactions were selected because of their potential to enhance offspring quality is unknown. The same morphogenetic signals maximize the probability that the plant commits resources only to those fruits with a full complement of seeds, regardless of their genetic quality.

Delay of Fertilization or Embryo Development

Developmental selection is based upon competition between spores, gametophytes, embryos, ovules, etc., for a limited number of biological spaces. Such competition is a more meaningful measure of differential vitality (or embryonic viability) if the competing elements initiate competition simultaneously. Buchholz (1922) stressed the need for simultaneous pollination by many pollen grains to maximize the effects of pollen competition. In many plants (especially wind pollinated species) pollination is not effected by

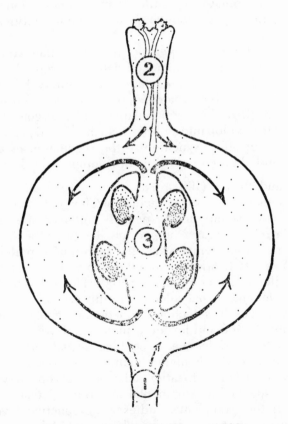

Figure 9.7. Diagrammatic representation of three possible sources of hormonal stimuli acting on fruit growth. (1) the vegetative parts; (2) the pollen and pollen tubes; (3) the ovules. (From Nitsch 1952.)

clumps or masses of pollen. In such species pollination occurs over a long time interval with each pollination event characterized by one or a few grains and many individual pollination events making up a successful total pollination of the reproductive structure. Under such conditions, pollen competition is all but negated by the random association of the time of pollen deposition and pollen vitality.

Delay of fertilization or zygote to embryo maturation may be a means of "synchronizing" postpollination events (fertilization, embryonic development) to maximize competitive interactions from the postpollination event onward. Delays may be of two kinds:

1. The postpollination event may occur after a set time interval following pollination (e.g., fertilization occurs two days after pollination). Such delays have little effect on competition since they do not synchronize the population of developing entities (zygotes, embryos).

2. The postpollination event occurs after a defined time interval that is keyed to the developmental state of the reproductive organ (ovary or ovule). Thus, after a defined time interval during which pollination is possible, whatever pollen grains are present simultaneously fertilize the egg population (or the zygotes simultaneously start to develop into embryos). Such delayed postpollination events will in effect synchronize the population of zygotes or embryos and, therefore, promote developmental selection.

Willson and Burley (1983), using a sexual selection argument, come to a somewhat similar conclusion. They note that when prezygotic selection is imperfect (especially pollen competition) fertilization delay can increase competition. To quote the authors, "By delaying fertilization, females may give high-quality pollen the maximum opportunity to arrive before they have to "decide" to accept or reject pollen of lesser quality" (p. 100). The difference between developmental and sexual selection in this case is as follows. The former emphasizes the selection of functional zygotes resulting from the syngamy of male and female gametes, both or either of which may be genetically defective. In contrast, sexual selection emphasizes the selection of only higher quality male gametes.

The significance of fertilization delay or embryonic development delay as an aspect of developmental selection is trivial in the pteridophytes. In the seed plants, studies of reproductive development are often insufficient to determine whether the delays are keyed to the time of pollination or the reproductive stage of the ovule or ovary (#1 or #2) (see Willson and Burley 1983 for review). This distinction is crucial before one can assess the relative importance of this aspect of plant development as a mutation buffer.

Flower Position

The mechanisms of developmental selection already described either enhance the loss of deleterious mutations without proportionate losses of reproductive capacity (prezygotic mechanisms) or reduce the immediate consequences of mutation expression by allowing homozygous zygotes or embryos to be easily displaced by heterozygotes (postzygotic mechanisms). Such mechanisms function for mutants inherited from the previous generation as well as those that have arisen since the parental zygote was formed. When plants are a chimera for postzygotic mutations, flower position and gender can also reduce the immediate consequences of mutation expression in the offspring. For example, the ontogenetic distance separating male and female flowers on a monoecious plant can strongly influence the expression of mutant genotypes in offspring. The closer male and female flowers are with regard to being derived from the same meristematic cell pools, the more likely that the meiocytes will share identical postzygotic mutant alleles. Thus male and female flowers occurring on the same branch (e.g., *Fagus sylvatica*) will form a higher frequency of zygotes homozygous for postzygotic mutations than when male and female flowers occur on different branch systems. An interesting example is the conifer *Pinus silvestris;* in this tree male and female strobili occur on the same branches. Eiche (1955) documented considerable tree to tree variation for the segregation of albino seedlings. These segregation patterns are compatible with the hypothesis that the parental trees were chimeras for postzygotic chlorophyll mutations and that the extent of the mutant sectors varied from tree to tree. If this pine species formed male and female strobili on branch systems that were ontogenetically very distant, the chlorophyll mutations would probably never have been detected.

When stems are mericlinal or sectorial chimeras for postzygotic mutations, flower position and origin may reduce the frequency of homozygous mutant genotypes in the offspring. Consider the case of axillary vs. terminal flowers formed by such a sectorial branch. If Aa is the mutant genotype (A → a), then the mutant sector area is a function of the number of apical initials (i.e., with two apical initials half the tissue is AA and half Aa, with three apical initials two thirds is AA and one third Aa, etc.). In table 9.6 the theoretical results from selfing axillary flowers on such a stem are compared with the selfing of terminal flowers. Since terminal flowers sample from the total pool of cells in the apical meristem (mutant and nonmutant) and axillary flowers sample primarily a single genotype from a sector (of course, some flowers will originate at a sector margin), the frequency of homozygotes is greater in axillary flowers when the number of apical initials exceeds one. Axillary flowers will segregate α times more aa homozygotes (where α is the number of apical initials). If axillary flowers are randomly mating then the frequency of homozygotes is identical in both flower types. It is clear from this simple example that the demography of flower types formed by a single plant may strongly influence the expression of postzygotic mutations in the following generation.

Developmental Selection Caveat

From the previous discussions the reader may have the impression that the multiplicity of reproductive stages that can function as

Table 9.6. Mutation expression from different flower positions.

NUMBER OF APICAL INITIALS	GENOTYPES OF INITIALS	FREQUENCY OF aa AFTER SELFING[a]	
		Terminal flowers	*Axillary flowers*
1	Aa	1/4	1/4
2	Aa AA	1/16	1/8
3	Aa AA AA	1/36	1/12
.	.	.	.
.	.	.	.
.	.	.	.
	Aa = 1 AA = $\alpha - 1$	$1/4\alpha^2$	$1/4\alpha$

[a]Consider the case of two apical initials. Terminal flowers will randomly combine the following gamete pool; 1/4a, 3/4A. Thus the probability of the aa homozygote is $(1/4)^2$. Axillary flowers will form flowers that yield either 3:1 ratios or 4:0 ratios. Since both ratios are equally probable, the sectorial branch will give a 7:1 ratio for AA to aa (i.e., the aa genotype occurs with a 1/8 probability). (Li and Rédei 1969.)

soft selection sieves in plants (especially angiosperms) coupled with diplontic selection within meristems would all but preclude the transmission of harmful mutations. In reality nothing could be further from the truth; the botanical and genetic literature is replete with examples of the frequent transmission of such genotypes to the next generation (see chapter 10). Thus, although many characteristics of vascular plant life cycles can be interpreted as functioning as soft selection sieves, the overall effectiveness of these sieves is unclear. Perhaps the best that can be said is that various vegetative and reproductive characteristics can certainly modify the frequency of transmission of presumably harmful alleles and genotypes in some contexts.

As an illustrative example, consider the case in *Pinus resinosa* (red pine) documented by Fowler (1964). In the course of a study of the effects of inbreeding in this species, a tree was found that consistently segregated for a distinctive abnormal seedling phenotype. Assuming the genetics to be monohybrid and the mutant allele to be recessive, one would have anticipated a 1:3 ratio for mutant to wild type upon selfing this tree. Ths was not what was found. Selfing this tree consistently gave 1:5 ratios. When the pollen of this tree was diluted with dead pollen of another pine species and this mixture used in the self-pollination, 1:3 ratios were documented in the offspring (table 9.7). Fowler interpreted these results as follows: selfing with undiluted pollen resulted in a number of different embryo genotypes competing within an ovule because of simple polyembryony. Not infrequently embryos homozygous for the mutant allele were in competition with embryos having at least one

Table 9.7. Controlled self-pollination on a red pine tree heterozygous for a marker gene. (From Fowler 1964.)

Pollen	Strobili Pollinated	Cones Matured	Normal Seedlings	Mutant Seedlings	Ratio mutant:normal
Self	10	5	114	24	1:4.8
1/2 Self 1/2 Carrier	10	8	130	40	1:3.3
1/8 Self 7/8 Carrier	11	1	—	—	

NOTE: Carrier = dead pollen of another pine species.

dominant wild-type allele. In such ovules, the embryos with the wild-type allele were often more competitive than those homozygous for the mutant allele. Thus the frequency of mutant embryos declines from the expected 25% to approximately 16.7%. The reason mixed pollination with dead pollen reduced selection against the mutant is the pine ovule has a micropyle that can accommodate only a restricted number of pollen grains. Introducing a proportion of dead pollen increased the probability that the micropyle contained only one viable pollen grain and, consequently, prevented simple polyembryony. Elimination of simple polyembryony and consequent embryonic competition resulted in the Mendelian ratio of 1:3 (see also Matheson 1980 and Moran and Griffin 1985 for other examples of biased gamete contributions in conifers).

The numerous reproductive stages which may function as soft selection sieves in vascular plants should have an important genetic consequence, deviations from Mendelian expectations should be common. Many of the mutants that have been subjected to Mendelian analysis are deleterious in one aspect or another, thus one would predict skewed Mendelian ratios in their inheritance. Mulcahy and Kaplan (1979) have noted that in spite of strong evidence for the possibility of pollen competition, such skewed or deviant ratios are relatively uncommon. These authors utilized a statistical analysis developed by Sokal and Rohlf (1969) to determine the sample sizes necessary to satisfactorily discriminate between true and deviant Mendelian ratios.

In table 9.8 the sample sizes necessary to detect deviations from a 3:1 or 1:1 with an 80% certainty at the 5% level of significance have been calculated. With regard to the monohybrid 3:1 ratio, detecting differences between 0.75:0.25 and 0.80:0.20 or 0.70:0.30 requires in excess of 1,000 offspring. On the other hand, detecting differences between 0.75:0.25 and 0.65:0.35 or 0.85:0.15 requires less than a few hundred offspring. The 1:1 ratio follows a similar pattern, deviant ratios in excess of 0.40:0.60 should be readily detected with approximately 400 offspring.

If soft selection is occurring during the various reproductive stages and if such soft selection is a significant factor in reducing the frequency of deleterious alleles, one would anticipate deviations from Mendelian ratios to be of sufficient magnitude to be detectable with sample sizes of a few hundred. Zamir and Tadmor (1986) surveyed segregation ratios for nuclear genes in the genera *Lens*, *Capsicum* and *Lycopersicon*. Of the 52 isozyme markers segregating in intraspecific crosses, 7 (13%) showed unequal segregations (interestingly, unequal segregations were documented in 54% of the genes in interspecific crosses). Whether unequal segregations were due to the gene markers or due to linkage to deleterious mutants is unknown (see also Sorensen 1967 for other causes of deviant Mendelian ratios). Clearly, this is an area of plant genetics in which more experimentation is needed before the full significance of the importance of soft selection and sexual reproduction can be assessed.

Finally, developmental selection may promote mutants that are not in the organism's best interest. Whyte (1965) pointed out that internal selective forces can both restrict and dictate the patterns of evolutionary change since such changes must fit into existing metabolic and ontogenetic pathways. What is often unforeseen is that these same pathways may generate internal forces of selection that can promote disadvantageous mutations (with reference to the individual). The previously discussed study of Ruth, Klekowski, and Stein (1985) provides an example of such selection within the dynamics of apical meristems in *Juniperus davurica* (see chapter 5). In this example internal selective forces favored the mutant albino cells over the nonmutant cells despite the fact that the inability to form chlorophyll is strongly disadvantageous to an autotrophic vascular plant. The juniper situation is not unique. Many horticultural forms of gymnosperms and angiosperms are known in which similar kinds of mutant cells are maintained through very long periods of growth. Some of these species are long-lived trees which maintain

---

Table 9.8. Sample sizes (n) necessary to be 80% certain of detecting a true difference between two proportions p_1 and p_2 at the 5% level of significance. (Following Sokal and Rohlf 1969:609.)

$$n = \frac{12,884.8}{(\arcsin \sqrt{p_1} - \arcsin \sqrt{p_2})^2} .$$

The true Mendelian ratio is p_1. The modified ratio due to the operation of one or more soft selection sieves is p_2.

$p_1 = 0.75$	p_2	n	$p_1 = 0.50$	p_2	n
	0.60	151		0.35	169
	0.65	327		0.40	386
	0.70	1,251		0.45	1,559
	0.80	1,092		0.55	1,559
	0.85	247		0.60	386
	0.90	96		0.65	169

periclinal chimeras in spite of cell displacements between the component meristems. These examples suggest that in vascular plants, hormonal gradients and responses may occasionally and inadvertently promote harmful cell genotypes either within apical meristems or within the soft selection sieves associated with sexual reproduction.

The various aspects of developmental selection (e.g., diplontic selection within meristems, competition between ramets, soft selection during sexual reproduction) should reduce the frequency of deleterious alleles transmitted to a plant's progeny and ultimately influence the frequency of such alleles in the gene pool. Thus, the molecular infidelities associated with gene replication and transmission within an organism ultimately will have important consequences at the population level. These consequences will be manifested as genetic load, the subject of the following chapter.

CHAPTER TEN

Genetic Load

As indicated in chapter 9, the leakiness of the various levels of developmental selection should result in an accumulation of deleterious mutations within the gene pool of a species. The frequency of such mutations is a function of the fidelities of gene replication and transmission, the persistence or loss of mutations as a consequence of the species' meristem and developmental characteristics, the degree of ramet competition, and the strengths of the various soft selection sieves manifested during sexual reproduction. Finally, whether the species predominantly inbreeds or outcrosses is also a very important factor governing the frequency of deleterious mutations which can accumulate in the gene pool. Thus, practically the "total biology" of an organism contributes factors into the equation that governs the levels of deleterious mutations or genetic load.

The first nine chapters of this book have dealt with mutation and its consequences at the cellular, meristem, organ, and individual organism levels. The subsequent chapters will consider the consequences of mutation at the population or gene pool level. Genetic load is best understood as a population characteristic since it reflects not only the mutations that have accumulated in the course of an individual's lifetime but also includes mutations inherited from previous generations.

GENERAL REMARKS ON LOAD

In practically all species of sexually reproducing, outcrossing organisms, a low frequency of extreme deviant phenotypes segregate out each generation. This observation, and the finding that inbreeding such species usually results in a higher frequency of deviant phenotypes in the offspring, has led to the view that most individuals are heterozygous for deleterious or lethal factors (genetic load). Numerous hypotheses have been proposed to explain the frequency and occurrence of load levels in populations (Crow and Kimura 1970; Simmons and Crow 1977; Wallace 1970, 1981). Only two of these hypotheses or kinds of load will be discussed with reference to plants: mutation load and balanced or heterotic load.

Balanced load was defined initially by Dobzhansky (1955, 1970) as the case where the load components are deleterious in the double dose (homozygotes) but give rise to heterosis in heterozygotes (see also discussion of heterotic load in Wright 1977). Deleterious alleles are maintained in the population at equilibrium frequencies by the selective advantage of the corresponding heterozygotes. This view of population structure is in contrast to that which results from mutational load. In the latter, "load consists of dominant, semidominant, and completely recessive genetic variants which are deleterious to their carriers in most environments in which the population lives" (Dobzhansky 1955:5). The mutational load concept is attributable to H. J. Muller. In a paper entitled "Our Load of Mutations," Muller (1950) stressed the negative consequences to human welfare of increased mutation pressure (due to ionizing radiation) and the relaxation of natural selection. Muller emphasized the need to consider mutation as a significant cause of impairment of human functions. Further studies of human consanguineous marriages led to the hypothesis that the majority of genetic damage expressed as a result of inbreeding is mutational in origin (Morton, Crow, and Muller 1956). The balanced and mutational load concepts are based upon different philosophical premises. Implicit in the former is the view that the majority (if not all) of an organism's traits (including recessive lethals) ultimately have some function or utility. The mutational load concept embodies the premise that organisms are less than perfect biological machines which can and do accumulate errors. As pointed out by Wallace (1981), Muller's treatment was fundamentally qualitative and it was Crow (1958) who defined genetic load more formalistically as the "proportional amount by which the average fitness of the population is depressed for genetic reasons below that of the genotype with maximum fitness." Thus

$$\text{Genetic Load} = (W_{max} - \overline{W})/\overline{W}$$

where W_{max} is the fitness of the optimum genotype (generally $W_{max} = 1$) and $\overline{W}$ is the average fitness of the population. This formalistic definition has several important consequences, for not only does it allow the incorporation of Muller's and Dobzhansky's ideas into a population genetics framework, it also considerably broadens the load concept beyond the original observations of inviability and sterility (Brues 1969). Genetic load concept now includes lethals, deleterious mutants, and morphological mutants as well as mutants that decrease fitness because of ecological interactions, e.g., mutants contributing to decreasing viability because of predation (the mutants taste better) or competition (the mutants grow slower). In figure 10.1 this broadened concept of genetic load is shown. Two extremes of load are noted: "physiological load" (mutations that cause inviability or decreased fitness regardless of the environment, e.g., albinism in green plants); and "ecological load," in which the reduction in fitness of a genotype is dependent upon environmental parameters (e.g., mutations that alter flower morphology, and thus reduce the frequency of effective pollination, mutations for heavy metal tolerance, etc.).[1] In nature a continuum exists from cases where genotypes are disadvantaged due to environmental causes to genotypes that are disadvantaged in all environments because of intrinsic physiological disturbances. The majority of genetic load estimates in *Drosophila, Homo sapiens*, ferns, conifers, and flowering plants are estimates of physiological load (i.e., lethals, deleterious mutants, morphological disorders, and sterility). The majority of such muta-

$$L = \frac{W_{MAX} - \overline{W}}{W_{MAX}}$$

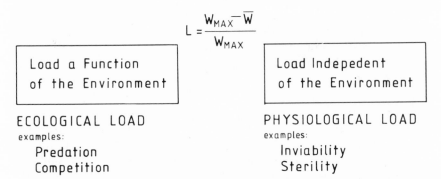

Load a Function of the Environment	Load Indepedent of the Environment
ECOLOGICAL LOAD	PHYSIOLOGICAL LOAD
examples:	examples:
Predation	Inviability
Competition	Sterility

Figure 10.1. Two aspects of genetic load.

1. The author apologizes to the reader for introducing yet another load concept since, to the uninitiated, the literature is already sufficiently burdened with the mutational loads, segregational loads, heterotic loads, substitutional loads, etc. Implicit in these load concepts is causality, why the loads are present in a population. The loads in figure 10.1 are concerned with how a genotype is less fit in comparison to other genotypes in the population and not *why* a given genotype is present in the population.

tions are a genetic burden since they reduce the viability of homo-zygous (and, perhaps, heterozygous) carriers. (The reader is referred to Crow and Kimura 1970, Wallace 1970, Solbrig and Solbrig 1979, for the algebraic relationship of mutation, selection and genetic load.)

Genetic loads are seldom measured on a per locus basis but rather per genome or per chromosome. If the genetic load at the ith locus is L_i and if the effects on viability are multiplicative, then, following Jacquard (1974), the frequency of load-free individuals is

$$1 - Ł = (1 - L_1)(1 - L_2)(1 - L_3) \ldots (1 - L_i) \ldots$$

where $Ł$ is the total genetic load. If n is the number of loci contributing to total load and L_i the average load contribution per locus, then

$$1 - Ł = (1 - L_i)^n$$

$$1 - Ł = 1 - \left(\frac{\sum L_i}{n}\right)^n.$$

The limit of this equation as n approaches infinity is

$$1 - Ł = e^{-\sum_i L_i}.$$

This relationship is the basis of an important genetic load metric, the lethal equivalent. As formulated by Morton, Crow, and Muller (1956), a lethal equivalent is defined as: "a group of mutant genes of such number that, if dispersed in different individuals, would cause on the average one death." Following these authors, the total load expressed by a population may also be written as

$$1 - Ł = e^{-(A + BF)}$$

where A is the amount of expressed load or damage in a randomly mating population $(F = 0)$ and B is a measure of the hidden genetic damage that would be expressed fully only in a complete homozygote $(F = 1)$.[2] B is the number of lethal equivalents per gamete. Sorensen (1969) was the first to apply this concept to plants, specifically the conifer Douglas-fir. The results of two levels of mating, random $(F = 0)$ and selfing $(F = 0.5)$, were analyzed in the following manner.

2. Wright's inbreeding coefficient, F, is the probability that an individual has inherited two alleles of a locus that are identical by descent, i.e., share a common pedigree (see Crow and Kimura 1970 for discussion).

$$R = \frac{\text{Proportion of good seeds at } F = 0.5}{\text{Proportion of good seeds at } F = 0.0}$$

$$R = \frac{e^{-A-0.5B}}{e^{-A}}$$

$$R = e^{-0.5B}$$

$$B = -2 \ln R$$

where B is the number of lethal equivalents per gamete. The number of lethal equivalents per zygote is $2B = -4 \ln R$.

Lindgren (1975) has pointed out an important distinction between recessive embryonic lethals and embryonic lethal equivalents. The former can best be viewed as lethal alleles that in homozygotes result in embryonic death with a probability of almost 100%. Lethal equivalents are lethal alleles which are less lethal so to speak, i.e., homozygous embryos have a probability of death that is much less than 100%. The greater the number of homozygous lethal equivalents in the genotype the more likely the embryo will die. This distinction is best illustrated by the genetic load expressed during sequential generations of selfing. If P is the parental generation, F_1 the first generation resulting from P, and F_2 the second generation resulting from selfing the individuals in the F_1, with reference to the average homozygosity for recessives, $P < F_1 < F_2$. The average homozygosity for recessives is $(0.25)n$ for the F_1 and $(0.375)n$ for the F_2 when n is the number of loci heterozygous for recessives in P. The loads expressed in these generations under the *lethal equivalent* model are

$$F_1 \text{ load} = 1 - (1 - s)^{0.25n}$$

$$F_2 \text{ load} = 1 - (1 - s)^{0.375n}$$

where s is the average selective disadvantage of individual recessive homozygous pairs of alleles and is much less than one. Under this model the F_2 load normally will exceed the F_1 load for many values of n. The two loads are similar at very high n values (and load levels). The selective loss of alleles in the F_1 has not been included since it only makes the calculation more complex without changing the overall trends.

In contrast, under the *recessive lethal* model considerable amounts of selection occur at each selfing generation since $s = 1$. Thus, in the F_1 resulting from selfing a dihybrid ($n = 2$) parent ($P = AaBb$) will consist of the following familiar array:

A-B-   0.563

$$\left.\begin{array}{l} \text{aaB-} \\ \text{A-bb} \\ \text{aabb} \end{array}\right\}\quad 0.437.$$

Thus, the load exhibited by the F_1 is 0.437. Selfing the viable F_1 (A-B-) gives the following F_2:

A-B-   0.694

$$\left.\begin{array}{l} \text{aaB-} \\ \text{A-bb} \\ \text{aabb} \end{array}\right\}\quad 0.306.$$

The load, therefore, exhibited by the F_2 is 0.306. Under the recessive lethal model the load of the F_1 will exceed the load of the F_2 for many values of n. For high n values the loads of both generations become very high and converge. Thus, if the genetic load is caused by recessive embryonic lethals, the F_1 generation should express greater load than the F_2. In contrast, if the genetic load is due to embryonic lethal equivalents, since the F_2 is more homozygous than the F_1, the F_2 should express greater genetic load than the F_1. In the conifer *Picea abies* (Norway Spruce), F_2 generation homozygotes exhibited greater genetic load than the F_1 generation (Andersson, Jansson, and Lindgren 1974). Thus, at least in this example, embryonic lethal equivalents are a more biologically meaningful metric of genetic load (Lindgren 1975). (See also Griffin and Lindgren 1985 for an epistatic genetic model to explain genetic load expression in trees.)

GENETIC LOADS IN VARIOUS PLANT GROUPS

Homosporous Ferns

The homosporous ferns are vascular plants with life cycles that are ideal for genetic load studies. These plants form free-living autotrophic gametophytes which, in most instances, are capable of forming both male and female sex organs (antheridia and archegonia) simultaneously. Thus, two levels of mating are possible, intergametophytic (exchange of gametes between different gametophytes) and intragametophytic (the self-fertilization of a gametophyte) (figure 10.2). While intergametophytic mating occurs in all plants, intragametophytic mating is restricted to homosporous plants. Since

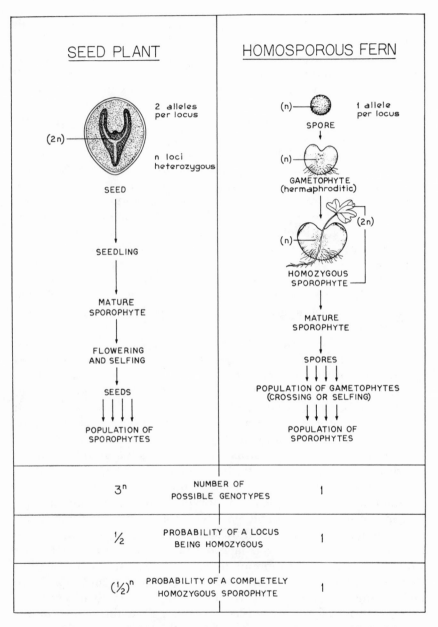

Figure 10.2. Genetic comparison of sporophyte populations established from a single seed and a single spore.

mitotic divisions give rise to the gametes in vascular plants, intra-gametophytic mating results in totally homozygous zygotes ($F = 1$). Whether such extreme inbreeding is common in nature varies with the fern species (Hedrick 1987).

Many homosporous fern species readily form functionally her-maphroditic gametophytes in culture. The self-fertilization of such gametophytes (intragametophytic selfing) results in the formation of completely homozygous zygotes (inbreeding coefficient, $F = 1$). Thus, recessive deleterious mutations for genes that are significant in the development of the zygote and early development of the embryo will be expressed in these homozygous genotypes. Based upon this aspect of the biology of ferns, the frequency of sporophytes heterozygous for deleterious mutations can easily be measured. Such data can form the basis of estimating the genetic load of a population. Estimates can be made of the number of genotypes heterozygous for deleterious mutations as well as the number of lethal equivalents per spore. Implicit in such studies is the assump-tion that the fern genotype contains genes that are restricted in their expression or critical primarily to the sporophyte generation. The recessive lethals usually studied are those genes expressed only in the sporophyte generation and not in the haploid gametophyte generation.

The logic of genetic load studies in ferns is as follows: spore samples are collected from individual sporophytes in nature and populations of sib-gametophytes are established in the laboratory. These cultures are randomly sampled and three different kinds of subpopulations of hermaphroditic gametophytes are established: isolated gametophytes where only intragametophytic selfing is pos-sible ($F = 1$), paired sib-gametophytes where intergametophytic selfing occurs ($F = 0.5$), and pairs of non-sibs where intergameto-phytic crossing occurs ($F = 0$). In the paired gametophyte cultures, the maximum amount of permissible outcrossing occurs because of the phenomenon of simple polyembryony. Thus, gametophytes that self and form a lethal zygote genotype continue to form viable ar-chegonia and to be fertilized by spermatozoids of the other game-tophyte. Evidence that the lack of sporophyte formation by the various gametophyte subpopulations is due to genetic causes rather than the environment is presented in tables 10.1 and 10.2. Two populations of *Osmunda regalis* and a composite study of a number of populations of *Athyrium filix-femina* all show the same pattern; there is an inverse relationship between the level of inbreeding and the frequency of sporophytes. Thus as homozygosity increases, vi-ability decreases.

Table 10.1. Genetic load studies in *Osmunda regalis*. Sporophytes NW40, NW41, NW42, NW43, NW44, and NW45 are individuals from one population and sporophytes CON58, CON61, CON62, CON63, and CON64 are individuals from another population (NW = South Deerfield, Mass.; CON = Conway, Mass.). Numbers of sporophytes formed from populations of isolated gametophytes, sib pairs, and interpopulation crosses are shown. Isolated gametophytes can only form completely homozygous zygotes, in sib pairs the probability of an individual locus being homozygous is 0.5, and high levels of heterozygosity are possible in the crosses. Note that the frequency of sporophytes is inversely proportional to the level of homozygosity. (From Klekowski 1973.)

	ISOLATES		PAIRS		CROSSES	
	Number of gametophytes	*Number of sporophytes*	*Number of gametophytes*	*Number of sporophytes*	*Number of gametophytes*	*Number of sporophytes*
NW40	50	5	100	62		
CON58	50	2	100	54		
NW40 × CON58					100	90
NW41	50	15	100	60		
CON61	50	14	100	70		
NW41 × CON61					100	92
NW42	50	24	100	59		
CON62	50	4	100	33		
NW42 × CON62					100	84
NW43	50	9	100	65		
CON63	50	7	100	48		
NW43 × CON63					100	90
NW44	50	2	100	69		
CON64	50	4	100	41		
NW44 × CON64					100	90
NW45	50	5	100	68		
NW45 × CON64					100	91
Total	550	91	1100	629	600	537
Percent		16.5%		57.2%		89.5%

The data in table 10.1 are restricted to sporophytic and gametic lethals. When ferns are inbred many mutations are expressed in addition to direct-acting lethals. A variety of abnormal embryonic phenotypes has been encountered including the absence of organs (root or leaves) as well as abnormally formed organs (Klekowski 1970). Sporophytes that appear normal initially often exhibit decreased viabilities with continued growth. In table 10.2 are shown the results of continued culture of sporophytes of *A. filix-femina* with differing levels of homozygosity (from Schneller 1979). Again, it is apparent that long-term viability is inversely related to the level of homozygosity. These results parallel similar studies in *Drosophila* where lethals have been shown to block specific stages in development (Poulson 1945; Hadorn 1948). As shown in table 10.2, genetic load estimates based upon the viability of zygotes and embryos may underestimate physiological load. Based upon Schneller's data, only 10.3% of the homozygous zygotes (isolated gametophytes selfed) were without apparent load, yet the proportion of viable embryos was 27%. It should be noted that the interactions of simple polyembryony and developmental selection apply only to early-acting lethals. Components of genetic load that reduce viabilities of postembryonic stages are not subject to soft selection due to simple polyembryony. Load expressed after embryogenesis is subject to hard selection.

Another aspect of fern genetic load is the varying expressivity of

---

Table 10.2. Long term survival and development of *Athyrium filix-femina* sporophytes resulting from various levels of mating. The data are recalculated from Schneller (1979) and are presented on the basis of 100 zygotes. (Adapted from Schneller 1979.)

Level of Mating	Isolated Gametophytes	Pairs of Sib Gametophytes	Pairs of Nonsib Gametophytes[a]
Initial number of zygotes	100	100	100
Viable embryos	27	79	100
Sporophytes alive after 1.5 years	15.6	53.7	93
Normal phenotype	10.3	48.2	93

[a]Pair of gametophytes from different populations.

some of the lethal genotypes. Some isolated hermaphroditic gametophytes will allow homozygous embryos to develop readily
whereas others do so only after many repeated fertilizations. These
latter genotypes are called leaky lethals and they have been found
in *Osmunda regalis* (Klekowski 1970), *Pteridium aquilinum* (Klekowski 1972), *Acrostichum aureum* (Lloyd 1980), and *Onoclea sensibilis* (Saus and Lloyd 1976). Gametophytes exhibiting the phenomenon of leaky lethality produce sporophytes only after a
prolonged culture period. During this time numerous gametangia
are produced and many self-fertilizations have occurred. Since each
fertilization results in a homozygous zygote identical genetically to
the others, why one zygote finally develops into an embryo cannot
be attributed to genetic differences. Figure 10.3 shows a gametophyte of *Pteridium aquilinum* with a leaky lethal genotype; in this
case the gametophyte exhibits simple polyembryony and ten embryos have formed. All these embryos have the same homozygous
genotype, yet only one has formed leaves. Although the biochemical
basis of these genotypes is unclear, it is not without precedent.
Lethals with varying penetrance and expressivity are very common
in genetic load studies in *Drosophila* (Dobzhansky 1970; Wallace
and Madden 1953). Hadorn (1948) has given the name *"Durchbrenner"* (breakthrough) to cases where individuals homozygous for true
lethals occasionally escape from the deleterious action of their genotype. The phenomenon of leaky lethality in ferns parallels the
"Durchbrenner" in *Drosophila*.

Leaky lethality presents a number of problems in genetic load
analysis in ferns. In a genetic load analysis, cultures of gametophytes
are established and grown until the gametophytes are hermaphroditic. At that point in time, a periodic watering schedule is established to allow the fertilizations to occur. Sporophyte production in
such cultures is not simultaneous but rather occurs over a prolonged
culture period. Table 10.3 shows the formation of homozygous sporophytes of *Phegopteris decursive-pinnata* (Masuyama 1979). Two
duplicate experiments were established for each culture and diploid
and tetraploid forms are compared. In this case, the tetraploid forms
were without genetic load whereas considerable load was evi-
denced by the diploid. The timing of sporophyte formation in diploid and tetraploid cultures differs, with sporophyte formation in
the former spread over a broader period of time. Thus, the length of
time of culture is a factor influencing the frequency of sporophytes
formed in genetic load studies. This does not mean that with an
infinite culture period all of the isolated gametophytes will form

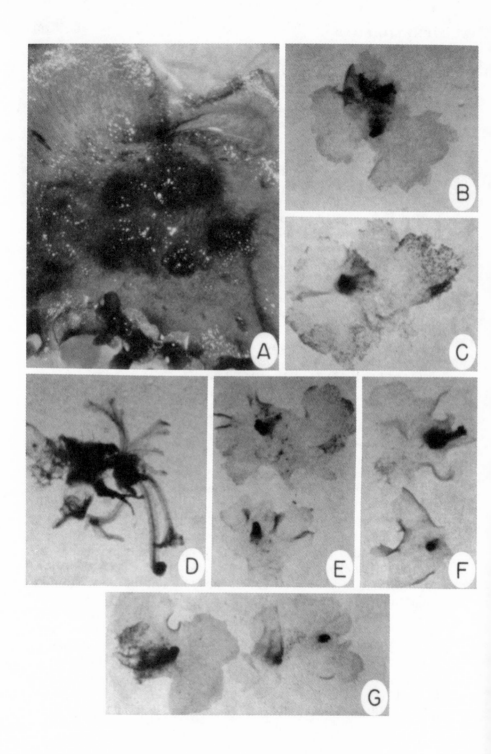

sporophytes. Many lethal genotypes have uniformly high expressivity and never form sporophytes. In the past, culture periods were established by the investigator based upon previous experience with the species. In most instances, the periods of culture exceeded the length of time that the species was reproductive in nature. These long gametophyte culture periods and the fact that in culture gametophytes generally exhibit increased simple polyembryony suggest that load levels are conservatively estimated.

A number of fern species have been studied extensively enough to at least allow tentative generalizations concerning genetic load levels in their populations. As will become clear from the following discussion, load levels vary greatly in fern species, and the causes of this variability are not always clear.

Pteridium aquilinum is a cosmopolitan fern species inhabiting both temperate and tropical environments (Tryon 1941). The fern has an extensive capacity for vegetative reproduction based upon rapidly growing underground rhizomes (Conway 1957; Watt 1976), and populations often consist of thousands of ramets and very few genets (Harper 1977). The spore samples studied represent a single frond sample (ramet) per population. Individuals (genets) are very long lived with life spans of centuries or possibly even millennia (Oinonen 1967). A genetic self-incompatibility system was proposed for this species based upon the breeding behavior of three spore collections from Scotland (Wilkie 1956); these results could not be reproduced with other spore collections (Klekowski 1972). Of the 59 sporophytes sampled for spores,[3] more than 30% essentially lack lethals in their genotypes. Summing all of the data for this species,

Figure 10.3. Lethals in the fern *Pteridium*. A: Six inhibited embryos on a single isolated gametophyte. B–G: Variously inhibited embryos formed on a single gametophyte which has been cloned. All embryos have the same genotype, yet one (D) managed to form rudimentary leaves. (From Klekowski 1972.)

3. Previously unpublished data of *Pteridium aquilinum* will be analyzed; thus, taxonomic collection data and S values follow: ssp. *aquilinum* var. *typicum*, Richmond, England, 0.46; Kew, England, 0.48; Edinburgh, Scotland, 0.275, 0.098, 0.167; var. *latiusculum*, Shutesbury, Massachusetts, 0.72; Cape Cod, Massachusetts, 0.0, 0.12, 0.12; Marshalls Creek, Pennsylvania, 0.76; Montague Plains, Massachusetts, 0.585, 0.563; Port Clyde, Maine, 0.64; var. *Feei*, Puebla, Mexico, 0.875; var. *pseudocaudatum*, 0.54; Raleigh, North Carolina, 0.76; ssp. *caudatum* var. *caudatum*, Miami, Florida, 0.98; the following collections are from southern Mexico, Xaccb, 0.48; Palenque, 0.90, 1.00, 1.00, 1.00, 1.00; La Venta, 1.00, 1.00, 1.00, 1.00, 1.00, 0.98; Coatzacoalcos, 0.92, 0.98; Acayucan, 0.86, 0.50; Xcabacab, 0.98, 1.00; Vera Cruz, 0.92; Catemaco, 0.85, 0.96, 0.96, 0.86; Los Mangas, 0.92; Xbacb, 1.00; San Andreas Tuxtla, 0.96. Each S value is for a single frond spore collection from a population and represents the frequency of isolated gametophytes that formed viable homozygous embryos. On the average, 50 gametophytes were isolated from each spore sample.

Table 10.3. The formation of homozygous sporophytes by isolated gametophytes of *Phegopteris decursive-pinnata* as a function of culture time in replicated experiments. Diploids and tetraploids are compared. (From Masuyama 1979.)

Parent Plant	No. of Isolates Tested		No. of Isolates Yielding Sporophytes								Total No. and % of Isolates Yielding Sporophytes
			Weeks from the first watering								
			1	2	3	4	5	6	7	8–10	
No. 2 (2X)	53	first test	0	0	4	7	11	4	3	4	33 (62.3%)
		second test	0	1	4	6	10	5	5	3	34 (64.2%)
No. 16 (2X)	54	first test	0	0	2	6	6	2	1	2	19 (35.2%)
		second test	0	1	1	2	7	2	4	1	18 (33.3%)
No. 35 (2X)	51	first test	0	0	0	7	10	3	2	2	24 (47.1%)
		second test	0	1	0	3	8	5	4	4	25 (49.0%)
No. 98 (2X)	55	first test	0	1	6	8	2	1	1	2	21 (38.2%)
		second test	0	3	4	2	3	2	2	2	18 (32.7%)
No. 53 (4X)	52	first test	0	0	22	11	6	5	2	2	51 (98.1%)
		second test	0	1	26	15	7	2	5	–	52 (100%)
No. 66 (4X)	53	first test	0	5	37	3	4	4	–	–	53 (100%)
		second test	0	4	24	12	8	3	2	–	53 (100%)
No. 71 (4X)	53	first test	0	2	40	5	3	1	1	1	53 (100%)
		second test			(the second test failed)							
No. 97 (4X)	55	first test	0	9	41	3	1	1	–	–	55 (100%)
		second test	0	3	19	27	4	1	1	–	55 (100%)

the mean lethal equivalents per gamete (or spore) was 0.474, 62.3% of the spores are lethal free, and 37.7% carry one or more lethal equivalents.

Osmunda regalis is a member of the ancient filicalean fern family Osmundaceae whose fossils are found in rocks of Upper Permian age (Taylor 1981). Based upon data from western Massachusetts (Klekowski 1970, 1973; Klekowski and Levin 1979), the average frequency of lethal equivalents per gamete (or spore) is 1.20 and the frequency of lethal-free spores is 0.303. It should be noted that the modal value of lethal equivalents is in the range of 1.05 to 1.25. *O. regalis* is characterized by a lack of vegetative reproduction in the populations studied, slow growing rhizomes are present which dichotomize to form fairy rings 1 to 1.5 meters in diameter with 8 to 16 shoot apices. Individual plants may be centuries old. In terms of sexual reproduction, abundant spores are formed in the late spring, but they are green and very short lived (48h under field conditions) (Lloyd and Klekowski 1970). Thus, unless suitable environments for germination are encountered within 48h, the spores lose their viability. Why this species has such a high genetic load is unclear. In this case clonal growth and high genetic load are unrelated, and the high load is not compensated for by effective and efficient sexual reproduction.

Saus and Lloyd (1976) and Ganders (1972) reported the results of genetic load studies for populations of *Onoclea sensibilis* from the northeastern part of the United States. The mean frequency of lethal equivalents per gamete was 0.292 and the frequency of lethal-free spores, 0.731 (Klekowski 1984). *O. sensibilis* has the capacity for extensive vegetative reproduction. Rapidly growing rhizomes are present which grow at, or close to, the soil surface. Individuals may be large clones composed of hundreds of apices, all of which may form fertile fronds every year. The fertile fronds are woody and remain erect in the winter. Spores are shed in the winter and early spring. Although the spores are green, they have long viabilities (months). This species thus has an extensive capacity for clonal and sexual reproduction.

In the genus *Ceratopteris*, *C. thalictroides* and *C. pteridoides* have been extensively studied for genetic load. Both species are aquatics with short life cycles in terms of gametophyte and sporophyte maturation and have a high capacity for vegetative reproduction through the differentiation of plantlets from the sinuses of the fronds (Kny 1875). In contrast to *Pteridium*, *Osmunda* and *Onoclea*, *Ceratopteris* populations essentially lack a genetic load. The mean frequency of lethal equivalents per gamete was 0.003 in *C. pteri-doides* and 0.01 in *C. thalictroides*. The frequency of lethal-free spores was 0.997 and 0.998, respectively. These two species have the lowest load level yet recorded in any sexual fern species (and probably the lowest load levels ever measured in any sexually re-producing species with a dominant diploid generation) (Lloyd and Warne 1978; Warne and Lloyd 1981).

Low genetic loads have been found in two members of the genus *Acrostichum*, *A. danaeifolium*, and *A. aureum*. The former had 0.119 lethal equivalents per gamete and 0.887 lethal-free spores and the latter, 0.057 lethal equivalents per gamete and 0.946 lethal-free spores (Lloyd and Gregg 1975; Lloyd 1980). These species grow in brackish habitats in tropical lowlands where they may be weedy in habit. New individuals are readily established from spores in dis-turbed sites. Vegetative reproduction is also important (see Tomlin-son 1986 for further discussion of the ecology of this fern genus).

It is instructive to compare these species in light of the various generalizations that have been put forward concerning load levels in organisms. Harper (1977), in contrasting genet- and ramet-based plant reproduction, indicated that the death risk for a genet declines continuously from birth whereas that of a ramet is constant. Harper suggests that the high mortality risk of young genets is due to genetic load expression whereas ramets are portions of genets that have already proven themselves genetically. Thus, young genets would be expected to suffer higher mortalities than young ramets. From this relation, one may conclude that species reproducing primarily by ramets could tolerate higher genetic loads than species whose reproduction is genet based. This generalization is roughly so in the ferns, those essentially without genetic load *(Ceratopteris)* are often considered to be annuals in nature (Lloyd 1974). The low loads found in *Acrostichum* are a bit more anomalous. Lloyd and Gregg (1975) noted that the gametophytes of this genus are characterized by very rapid growth and early attainment of sexual maturity (as are the gametophytes of *Ceratopteris*). On the other hand, *Acrostichum* is a perennial. Whether populations are maintained primarily by sexual or vegetative reproduction is unknown.

Other load correlations have also been reported. Polyploidy has sometimes been associated with low load levels. Masuyama (1986) reported that diploid forms of *Phegopteris decursive-pinnata* have high load levels whereas tetraploid forms are loadless. Schneller (pers. comm.) has studied sympatric populations of *Dryopteris filix-mas* (tetraploid) and *Athyrium filix-femina* (diploid); the former were loadless, whereas the latter exhibited considerable load. Thus, it seems that polyploidy is at least a short-term means of reducing load. Ecological correlations between pioneer (loadless) and non-pioneer (loaded) ferns have also been documented (Lloyd 1974; Holbrook-Walker and Lloyd 1973; see also Cousens 1979; Crist and Farrar 1983 for other ecological studies).

The generalization that plants relying upon ramet-based reproduction can tolerate higher genetic loads than plants reproducing via genets begs the question as to the causes of load. The critical question is whether the genetic load detected in these "more tolerant" life forms is due to mutational load. If mutational load is the most significant component of genetic load in plants, one would also predict higher load levels in species with the longest-lived sporophytes, all other factors being equal.

Circumstantial evidence that the genetic loads measured in fern clones are primarily mutational load is available for two species, *Onoclea sensibilis* and *Matteuccia struthiopteris*. Many clones of these ferns have apices heterozygous for gametophyte mutations. In figure 10.4 the phenotypes of the most typical kinds of gametophytic mutations are shown. The most common mutant phenotype consisted of gametophytes capable of only two or three mitotic divisions resulting in a few prothallial cells and a rhizoid. At this stage of retarded development, the mutant gametophyte terminated growth but remained alive for at least three more weeks (and sometimes even longer). Such mutants remained neuter and, consequently, must represent post-zygotic mutations as the lack of gamete formation precluded their transmission to zygotes. Such mutants were used to estimate the per ramet generation mutation rate for gametophyte mutations in fern clones (Klekowski 1984). To calculate the mutation rate two numbers are necessary: the mean number of shoot apices per clone, and the frequency of clones that are mutant-free. One then assumes that the distribution of mutational events follows a Poisson distribution where the probability of a clone with x mutational events equals

$$\frac{(rU)^X e^{-rU}}{X!}$$

where r is the number of ramet-doubling generations (r = number of apices per clone minus one) and U is the mutation rate per ramet generation. The mutation rate is conveniently calculated from the zero term ($X = 0$) of this distribution.

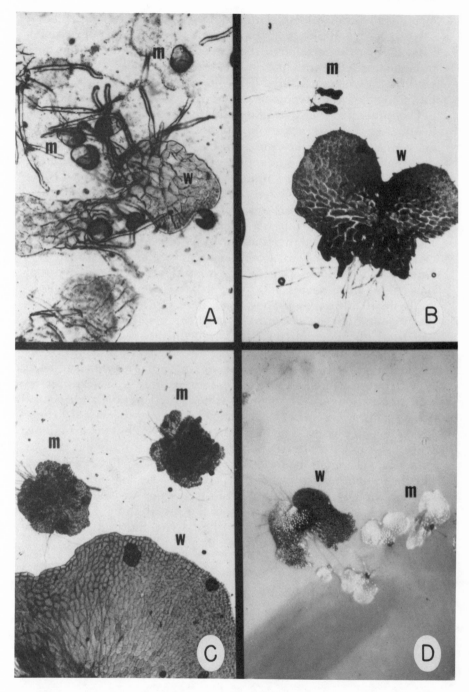

Clone sizes in nature are very difficult to delimit. In these studies, a clone was defined as a discrete patch or population of shoot apices. A sample of shoot apices from a patch was screened for gametophyte mutations. Each shoot apex was individually sampled for spores, the spores sown, and gametophyte populations scored. If an apex was heterozygous, a 1:1 ratio of mutant to wild-type gametophytes was found in the gametophyte culture. Clone limits were usually overestimated, thus what was operationally defined as a single clone was either a single clone or a number of clones grown together. As is shown in figure 10.5, such a sampling procedure will overestimate the number of ramet doubling generations (r). Since the zero term of the Poisson equals e^{-U}, overestimating r reduces the calculated mutation rate. Thus, this sampling protocol conservatively estimates the mutation rate per ramet doubling generation. These mutation rates were given in table 2.5. These rates are, at first sight, remarkably high, 2% to 3%, until one recalls that they are estimates of the cumulative rates for all of the genes that may be mutant and result in abnormal gametophytes.[4]

Figure 10.4. Gametophyte mutants in homosporous ferns. A: Two-week-old gametophytes of *Osmunda regalis*. The sib gametophytes were grown from the spores of a single frond and exhibited a 1:1 ratio for mutant (m) and wild-type (w) phenotypes. The mutants underwent 1 to 3 mitotic divisions before growth was arrested. B: Three-week-old sib gametophytes of *Onoclea sensibilis*. The gametophyte population exhibited a 1:1 ratio for mutant (m) and wild-type (w) phenotypes. The mutants terminated growth after 3 to 4 mitotic divisions. C: Three-week-old sib gametophytes of *Matteuccia struthiopteris*. A 1:1 ratio of callus-like or ameristic (m) and wild-type (w) phenotypes was documented. D: Three-week-old sib gametophytes of *Onoclea sensibilis*. A 1:1 ratio of albino-like (m) and wild-type green (w) phenotypes was documented.

4. The mutation rate U has been used to predict the level of mutational load expected in fern populations (Klekowski 1984). Unfortunately, as the following will demonstrate, such predictions are fraught with problems. If one assumes that the mutation rate U is a meaningful estimator of the mutation rate for sporophytic lethals, then since

$$1 - Ł = \exp\left(-\sum_i L_i\right)$$

and since the mutational load for recessive lethals at a single gene locus expressed under conditions of intragametophytic mating and simple polyembryony is (see Klekowski 1982)

$$L_i = \sqrt{2u}$$

where u is the mutation rate for a single locus, thus

$$u = U/n$$

where n is the number of loci whose mutation rates have been summed to give the value U. Substituting and rearranging terms gives

$$Ł = 1 - \exp(-\sqrt{n2U})$$

since n is totally unknown, $Ł$ cannot be realistically estimated.

Despite the fact that both *O. sensibilis* and *M. struthiopteris* have relatively similar mutation rates (0.034 and 0.018, respectively), the species do differ greatly in their genetic load levels. Based upon previously cited genetic load studies, *O. sensibilis* has 0.292 lethal equivalents per spore and, on the average, 0.731 of the spores are free of sporophytic lethals. In contrast, *M. struthiopteris* populations have very high genetic loads, populations in western Massachusetts average 2.9 lethal equivalents per spore and only 5.5% of these spores are free of sporophytic lethals.[5] These two species also differ in their reproductive patterns. Although *O. sensibilis* has very effective vegetative reproduction, the species commonly reproduces sexually as well and, in some environments, is almost weedy (D. Buckley, pers. comm.). In western Massachusetts, *M. struthiopteris* reproduces almost entirely through vegetative stolons. Clones are large and old, and sexual reproduction is very rare (despite the fact that many fertile fronds are formed every year). The similar mutation rates and yet very different genetic load levels in both species prompt the hypothesis that the key factors are clone longevity and

ONE CLONE

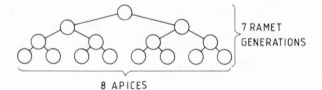

7 RAMET GENERATIONS

8 APICES

TWO CLONES INTERMIXED

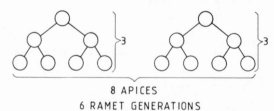

8 APICES
6 RAMET GENERATIONS

Figure 10.5. The relationship between ramet number (apices) and ramet generations.

5. Unpublished studies by the author.

frequency of sexual reproduction. The latter periodically purges the genotype of lethals and the former allows their accumulation through a chemostat-like effect (see chapters 2 and 11).

Gymnosperms

By far the most important of the gymnosperms are members of the order Coniferales or the conifers. The order is comprised of 7 families, 48 genera, and approximately 520 species, with a large number of monotypic and endemic genera (Lawrence 1951). Two families of conifers have been studied genetically, the Pinaceae and Taxodiaceae. Inbreeding depression is very common in both of these families (see Franklin 1970 for review). In a study of coastal Douglas-fir (*Pseudotsuga menziesii* var. *menziesii*), genetic load was measured for 35 trees. The relative self-fertility varied widely between trees, with a median of 10 embryonic lethal equivalents per zygote and a range of 3 to 27 lethal equivalents per zygote in the sample (Sorensen 1969). In another member of the Pinaceae, loblolly pine (*Pinus taeda*), a sample of 116 parent trees revealed a similar range of embryonic lethal equivalents per zygote per tree and a median of 7 embryonic lethal equivalents per zygote was documented (Franklin 1972). Franklin also suggested that this genetic load was mutational load. The expression of similar embryonic genetic loads from inbreeding has been documented in *Larix laricina, Picea mariana, Pinus silvestris, Picea glauca, Metasequoia glyptostroboides,* and *Picea abies* (Park and Fowler 1982, 1984; Wilcox 1983; Johnsson 1976; Mergen, Burley, and Furnival 1965; Kuser 1983; Koski 1973). Although inviability in the offspring resulting from selfing is typically expressed in the embryonic stages, a loss of vitality also is observed in the survivors. Data on the magnitude of such postembryonic inbreeding depression is available for a 61-year-old experimental planting of *Picea abies* in Sweden. Self- and open-pollinated progeny of four trees were established in 1910, plant lethality during the first summer was 30% for the selfed and 12% for the open pollinated. In figure 10.6 the average trunk volumes of these trees at sixty-one years of age is shown (Eriksson, Schelander, and Akebrand 1973). In *Pinus radiata* a similar postembryonic inbreeding depression was observed (Wilcox 1983). In table 10.4 the phenotypic characteristics of self and cross pollinated progeny of this species at seven years of age are compared (see figure 10.7 as well). Postembryonic inbreeding depression also has been documented in *Picea glauca* (Ying 1978). In *Abies procera,* self-fertility studies have documented low levels of embryonic load and relatively high levels of inbreeding depression in subsequent growth (Sorensen, Franklin, and Woollard 1976; Sorensen 1982). In redwood, *Sequoia sempervirens,* selfing produced no additional cone abortion and no consistent effects on number of seeds per cone or seed viability. The apparent lack of embryonic load was associated with considerable inbreeding depression expressed during postembryonic growth (Libby, McCutchan, and Millar 1981). It is interesting to note that redwood is one of the few polyploid conifer species ($2n = 6X = 66$) (Schlarbaum and Tsuchiya 1984).

Since conifer species typically express high levels of load upon inbreeding, the occurrence of essentially loadless species is of considerable interest. Species may lack a genetic load because of a number of reasons; these include: 1) reduced mutation rates; 2) mutant phenotypes are not expressed because of either genomic or developmental constraints; 3) inbreeding and loss of recessive mutants due to selection.

Two species are known in which the progeny resulting from outcrossing and selfing are essentially similar in viability and vitality, *Picea omorika* and red pine, *Pinus resinosa.* In a series of studies,

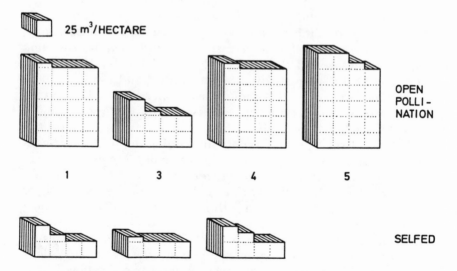

Figure 10.6. The mean trunk volume of four selfed and four open-pollinated progenies of different mother trees of *Picea abies* at age 61 years. It should be noted that the selfed progeny from mother tree number 5 consists of only one tree. (From Eriksson, Schelander, and Akebrand 1973.)

Fowler (1965a, b, c) documented that red pine is essentially load-less, although trees carrying recognizable mutations are not uncommon. The conventional explanation as to why this species is loadless has been to postulate episodes of high levels of inbreeding in nature. This, of course, may not be the entire explanation. In *Picea omorika*, Langner (1959) found that the viability of seeds developed from selfing or crossing was similar. Although little genetic load was documented in the embryonic or juvenile stages of growth, clear inbreeding depression was found in one- and two-year-old seedlings as well as older trees (Geburek 1986). The *P. omorika* populations in Yugoslavia are thought to be postglacial relicts of formerly larger populations that existed in Europe in preglacial times. Thus again the lack of genetic load is attributed to inbreeding and selection.

As discussed in chapter 9, gymnosperms (including conifers) undergo simple polyembryony (i.e., the fertilization of multiple eggs within a single ovule). Although all of the egg genotypes produced by a single ovule are identical, pollen chambers usually contain more than one pollen grain; therefore, different male gamete genotypes fertilize these eggs (Lindgren 1975; Sorensen 1982). Consequently, a number of different embryo genotypes are in competition within the same ovule. Bramlett and Popham (1971) calculated

---

Table 10.4. Means of self- and cross-pollinated *Pinus radiata* progenies. (From Wilcox 1983.)

Trait	Selfs	Crosses	Inbreeding[a] Ratio
Height: in nursery (cm)	45	55	0.82**
Height: 1 yr (cm)	76	83	0.92**
Height: 4 yr (dm)	43	47	0.91**
Diameter: 4 yr (cm)	6.9	8.0	0.87**
Volume: 4 yr (dm^3)	8.6	12.2	0.71**
Straightness: 4 yr (1–9)	5.65	5.88	0.96**
Malformation: 4 yr (1–6)	5.03	5.14	0.98**
Diameter: 7 yr (cm)	16.2	18.4	0.88**
Straightness: 7 yr (1–9)	5.23	5.96	0.88**
Branching quality: 7 yr (1–9)	4.24	5.29	0.80**
Needle retention: 7 yr (%)	53.4	63.8	0.84**
Wood density: 7 yr (kg/m^3)	302	298	1.01**

[a]Ratio of self/cross.
** = ratio is significantly different from 1.0; $P = 0.01$.

the probabilities of unsound seed after self-fertilization, assuming an unsound seed occurs because all of the embryos are homozygous for recessive lethal alleles. In table 10.5 the results of their calculations are shown. It is clear that in trees heterozygous for small numbers of lethal alleles, simple polyembryony (or multiple fertilization) significantly reduces the impact of these alleles on seed viability. What is not so intuitively obvious is that as the number of lethal alleles increases in a tree's genotype, the relative importance of simple polyembryony as a buffering agent decreases.

The effectiveness of simple polyembryony as a soft selection sieve is limited to inbreeding depression expressed during the embryonic phase. In species without significant embryonic load but with considerable postembryonic load (e.g., *Picea omorika, Abies procera,*

Figure 10.7. Four-year-old cross- and self-pollinated progenies (S) of *Pinus radiata.* The pronounced inbreeding depression showed by this parent was already evident in the nursery. (From Wilcox 1983.)

Sequoia sempervirens), simple polyembryony is of little consequence (Sorensen 1982).

The self-pollination of individual trees often results in the segregation of distinct mutant phenotypes in the offspring. For example, Franklin (1969) reported that in *Pinus taeda* the self-pollination of over 100 trees revealed that 25% were heterozygous for mutant markers expressed in the seedlings. Mutant phenotypes included cotyledonary chlorophyll deficiencies, abnormally colored primary and secondary foliage, and hypocotyl color as well as various morphological abnormalities. In 35 families in which 2 seedling phenotypes segregated, 3 (8.5%) families had ratios significantly different from 3:1 (41:5, 109:14, 62:10). Such "disturbed" segregation ratios also have been documented in other taxa for morphological mutants (Franklin 1970) and allozyme segregations in the megagametophytes of single trees (Rudin and Ekberg 1978; Adams and Joly 1980; Stewart and Schoen 1986; Furnier et al. 1986). Sorensen (1967) suggested several possible causes to account for such seedling ratios. These included: 1) deviant seedling phenotypes may be difficult to identify; 2) sampling error, since most selfed families are small; 3) pregermination selection after multiple fertilization, sim-

Table 10.5. Probability of unsound seed after self-fertilization as a result of the homozygosity of embryonic lethal alleles after multiple fertilization per ovule. (From Bramlett and Popham 1971.)

Lethal Alleles	Number of Fertilizations				
	1	2	3	4	5
1	.25	.12	.06	.03	.02
2	.44	.27	.17	.11	.08
3	.58	.40	.29	.22	.16
4	.68	.52	.41	.33	.27
5	.76	.62	.52	.44	.38
6	.82	.70	.61	.54	.48
7	.87	.77	.69	.63	.58
8	.90	.82	.76	.71	.66
9	.92	.86	.81	.77	.73
10	.94	.90	.86	.82	.79
15	.99	.97	.96	.95	.94
20	1.00[a]	.99	.99	.99	.98

[a]Rounded to 1.00; actual value = .9968.

ple polyembryony; 4) more than one gene locus involved (e.g., dihybrid, trihybrid, etc.); 5) linkage of the phenotypic marker to a recessive lethal (figure 10.8).

With regard to distortion in megagametophyte ratios, megaspore competition is also a possibility. Deviations from simple 3:1 and 1:1 ratios also can be caused by postzygotic somatic mutation. Thus, if the individual tree is a chimera for two cell genotypes (Aa and AA) some branches may consist wholly of one genotype and others may be mixtures of both genotypes. Meiosis will give rise to distorted ratios of megagametophyte genotypes per tree. The self-fertilization of such trees will result in disturbed segregation ratios (such ratios are common in plant mutation experiments, Li and Rédei 1969; Weiling 1962).

Angiosperms

Angiosperms are the dominant group of land plants. Despite the well-documented occurrence of lethals and inbreeding depression in many crop plants (especially *Zea mays*) (Crumpacker 1967), investigations of natural populations of angiosperms for load and load-related phenomena are meager).

Recently Levin (1984) reported the results of an extensive genetic load study of natural populations of *Phlox drummondii*, a self-in-

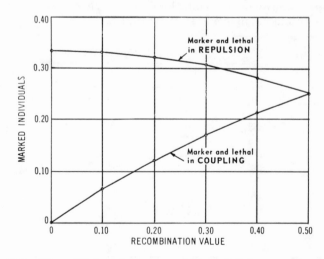

Figure 10.8. Effect of linkage between embryonic lethal gene and marker gene on percent of marked individuals in a selfed progeny. (From Sorensen 1967.)

compatible annual. Self-fertilization was accomplished in this species by removing the corolla of an immature flower and placing
pollen from open flowers on the immature stigma. Such bud pollination reduced the average seed set per capsule. The *P. drummondii*
flower forms 3 ovules; on the average, 2.25 develop into seeds from
the cross-pollination of open flowers and approximately 1 ovule
develops into a seed from bud pollination. Based upon a study of
28 different populations, Levin reported that the number of lethal
equivalents per zygote varied from 0.08 to 2.26 with a mean of 0.79.
These values may be an underestimate if each seed produced
through bud pollination represents 3 fertilized ovules and each seed
resulting from cross-pollination represents 3/2.25 = 1.33 ovules fertilized. If such multiple fertilizations occur, then the seed to ovule
ratio may represent a soft selection sieve. If A_s is the frequency of
seed abortion from selfing and A_c the frequency of seed abortion
from crossing then the relative survivorship (R) with regard to zygote
viabilities is

$$R = \frac{1 - \sqrt[3]{A_s}}{1 - \sqrt[1.33]{A_c}} .$$

The number of lethal equivalents per zygote $(2B)$ is calculated from
the formula $(2B) = -4 \ln R$. In *P. drummondii* the mean seed inviability from selfing was 0.126 and from crossing was 0.275. Using
these values as A_s and A_c, the number of lethal equivalents per
zygote is 3.25. In table 10.6 these two estimates of lethal equivalents
are compared. It is clear that for low load levels soft selection results
in gross underestimates of the number of lethal equivalents. On the
other hand, when loads are high soft selection has less consequence
(see table 10.5 for a similar situation in gymnosperms).

In *Eucalyptus regnans* both an embryonic and postembryonic
genetic load have been documented (Eldridge and Griffin 1983,
Griffin, Moran, and Fripp 1987). In table 10.7 the seed sets from
selfing, crossing, and open-pollination are presented. Based upon
seed germination per capsule, the relative survivorship (R) is 0.53,
thus there are 2.54 lethal equivalents per zygote expressed during
the embryonic phase. This estimate is conservative as the possibilities of soft selection have not been considered. A considerable
postembryonic load was documented in this tree species also.In
figure 10.9 these progeny have been phenotypically classified after
9 and 12.5 years of growth. At 12.5 years the majority of cross-
pollinated progeny were classed as dominant or codominant phen-

otypes whereas the minority of selfed progeny fell into this category. Most of the selfed progeny were divided evenly among two categories, dead and suppressed. Interestingly, the progeny resulting from open-pollination express intermediate embryonic and postembryonic loads, indicating that these offspring probably represent a mixture of selfs and crosses. In almost every respect the load expressed in *E. regnans* fits the general pattern that characterizes many conifers.

Genetic load has been documented in *Kalmia*, a small North American genus of woody shrubs belonging to the Ericaceae. Jaynes (1968) studied three species, *K. latifolia*, *K. angustifolia* and *K. polifolia*. Seed set from self-pollinations was approximately 85 to 90% less than cross-pollinations for the first two species. Seed sets from crosses and selfs were identical in *K. polifolia*. These differences in self-sterility are correlated with chromosome number, *K. latifolia* and *K. angustifolia* are diploids ($2n = 24$) and *K. polifolia* is tetraploid ($2n = 48$) (Jaynes 1969). Although seed germination was similar in all three species, the progeny resulting from selfing exhibited a marked reduction in vigor and survival in the diploids. The genetic load in *Kalmia* in some ways resembles the load in the hexaploid conifer *Sequoia sempervirens*. Polyploidy is correlated with reduced embryonic load yet seems to have little influence on postembryonic load (see also Lyrene 1983 for genetic load in another member of the Ericaceae, *Vacinium*).

In the tree species *Liquidambar styraciflua*, a small study of seven

Table 10.6. Lethal equivalents per zygote in *Phlox* and multiple fertilization. (Modified from Levin 1984.)

	Seed/Ovule Ratio	
Seed abortion	1	<1
AS = 0.05 AC = 0.03	0.08[a]	1.54[b]
AS = 0.275 AC = 0.126	0.79	3.25
AS = 0.50 AC = 0.12	2.26	5.41

[a]Lethal equivalents per zygote calculated without soft selection.
[b]Lethal equivalents per zygote calculated with soft selection.

individuals documented high levels of self-sterility. The relative seed yield of selfs to crosses was approximately 5%. Anatomical study found 4- to 6-celled embryos in selfed flowers (Schmitt and Perry 1964). A relative self-fertility of 5% corresponds to a genetic load of 11 to 12 lethal equivalents per zygote. In conifers, individual trees with relative self-fertilities of 5% or less are not uncommon in many species, e.g., *Pseudotsuga menziesii* (Sorensen 1969); *Pinus taeda* (Franklin 1972); *Larix laricina* (Park and Fowler 1982).

Preliminary studies of other tree species *(Ulmus americana, Fagus sylvatica, Caragana arborescens)* have documented considerable tree-to-tree variation in self-sterility (Lester 1971; Blinkenberg et al. 1958; Cram 1955). Such variability parallels the genetic loads in conifers. All of these investigations were based upon small numbers of trees, and because control pollinations and anatomical studies to determine whether embryos actually formed were often not carried out, little can be concluded about genetic load in these species.

Additional evidence that genetic load is an important aspect of angiosperm biology comes from studies of the proportion of zygotes which develop into viable progeny (seeds) in inbreeding and outcrossing plant species (typically the inbreeders are annuals and the outcrossers, perennials). Wiens et al. (1987) and Wiens (1984) found that in a broad sample of approximately 200 species, spontaneous abortion reduced embryo survivorship about 50% or more in outcrossers, but only approximately 10% among inbreeders. These researchers recognized that some of these abortions are fixed developmentally and are related to dispersal systems (Casper and Wiens

Table 10.7. Seed set from self-, cross- and open-pollination in *Eucalyptus regnans*. (From Eldridge and Griffin 1983.)

Treatment	Selfed Emasculated	Crossed Emasculated	Open-Pollinated
Number of trees	4	3	4
Flowers	402	321	937
Mature capsules per 100 flowers	20	15	13
Seed germinated per mature capsule	2.3	4.3	2.8
Seed germinated per 100 flowers	47	90	71

1981). Adjusting for this limitation, the authors still felt that the majority of the differences in abortion between inbreeders and out-crossers was due to the expression of genetic load.

MUTATIONAL VS. HETEROTIC LOAD IN SEED PLANTS

To what extent genetic load represents mutational or heterotic load is still unclear in seed plants. Based upon the classic example of heterotic load in humans (sickle cell anemia), one would anticipate that the selectively advantaged heterozygote would be relatively common within one or more populations of a species. In contrast, mutational load represents rare detrimental mutations which persist in rare heterozygotes. Thus, the individual mutant alleles are in low frequency within a population and seldom are similar in different populations. Mutational load consists of a diversity of mutant alleles

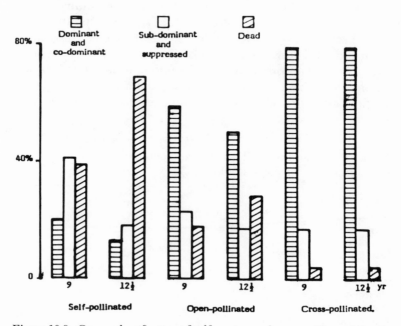

Figure 10.9. Crown classification of self-, open-, and cross-pollinated *Eucalyptus regnans* trees in field trial at 9 and 12.5 years. (From Eldridge and Griffin 1983.)

at different loci, and the population consists, therefore, of a variety of heterozygous genotypes.

A possible case of heterotic load has been documented in natural populations of the orchard grass, *Dactylis glomerata*, in Israel (Apirion and Zohary 1961). This species is diploid ($2n = 14$) and self-incompatible. Many populations were found to have individuals heterozygous for chlorophyll deficiency (4 to 30% heterozygosity depending upon the population). The widespread occurrence and relatively high frequencies of heterozygotes in some populations were taken as strong evidence for heterotic selection.

As already indicated, mutational load consists of a diversity of mutant alleles at different loci. The population, therefore, consists of a variety of heterozygous genotypes. Since the deleterious mutant alleles generally represent unique mutational events and the alleles persist for a few generations in heterozygotes, the distribution of these families of heterozygotes is likely to be clustered in the population (this distribution is, of course, strongly dependent upon seed and pollen dispersal and vegetative growth in clonal forms). Thus, the amount of genetic load expressed should be inversely proportional to the spatial distance separating the parents in the population if the load is mutational load. The results of such a study in *Picea glauca* are presented in table 10.8. Coles and Fowler (1976) studied two populations, a natural upland stand of mixed conifers (AFES) and a pure stand colonizing abandoned farmland (Tay). The average genetic load for both populations combined was 8.73 lethal equivalents per zygote. As is shown in table 10.8, there is evidence for considerable allelism of recessive lethals for trees growing within

Table 10.8. Relationship between percent sound seed and spatial separation of the parents in *Picea glauca*. (From Coles and Fowler 1976.)

Population	Distance Class	Sound Seed (%)
AFES	Self-pollination (0m)	8.1
	1 to 100m	32.7
	101 to 500m	50.7
	32 km	50.7
Tay	Self-pollination (0m)	4.1
	1 to 100m	29.8
	101 to 500m	40.6
	32 km	35.9

100m of each other; beyond that distance seed set is similar to interpopulation crosses. A somewhat similar study was carried out by Levin (1984) in *Phlox drummondii*. A negative correlation between percent seed abortion and distance between parents was also documented. Proximity-dependent loads in both *Picea* and *Phlox* are certainly consistent with the expectations of mutational load but do not, of course, prove the case.[6]

In cultivated buckwheat, *Fagopyrum esculentum*, many aspects of genetic load fit the mutational load model. Ohnishi (1982) studied the frequency of chlorophyll deficient mutants in cultivated populations in Japan. Farmers in Japan cultivate buckwheat every year from the seeds they harvested the previous season. The cultivated population of a village can be considered to be a local Mendelian population of 10^4 to 10^6 individuals. Buckwheat is heterostylous, consisting of pin and thrum forms, thus these large populations are primarily outcrossing. Ohnishi estimated the frequency of chlorophyll-deficient mutant gametes using controlled inbreeding experiments and the frequency of chlorophyll deficient homozygous seedlings occurring spontaneously in cultivated fields through field surveys (the latter arise as chance homozygotes during random mating). If n is the number of loci involved in chlorophyll abnormalities and p_i the frequency of the detrimental allele at the ith locus, then the inbreeding experiment estimates

$$\sum_{i=1}^{n} p_i$$

and the results of the field survey for chlorophyll-deficient homozygotes is an estimate of

$$\sum_{i=1}^{n} p_i^2.$$

Assuming equal allele frequencies at each locus, the number of loci involved in chlorophyll abnormalities is estimated by

$$\hat{n} = \frac{\left(\sum_{i=1}^{n} p_i\right)^2}{\sum_{i=1}^{n} p_i^2}.$$

6. It should be noted that in both of these studies, maternal effects and dominant embryonic lethal genotypes (which may arise from epistatic or additive interactions of different mutant alleles, from dominant embryonic lethal somatic mutations, or due to chromosome aberrations such as inversions or translocations) would tend to make loads independent of parental distances.

The average allele frequency per locus is estimated by

$$\hat{p} = \frac{\sum\limits_{i=1}^{n} p_i^2}{\sum\limits_{i=1}^{n} p_i} .$$

In table 10.9 the data for the Japanese populations are summarized for different categories of chlorophyll abnormalities. Two points are clear from these data, many loci are involved in chlorophyll abnormalities (ca. 100) and the maximum deleterious allele frequency per locus is 0.001 to 0.003. Such low allele frequencies support the hypothesis that the maintenance of these detrimental alleles in the populations is due to mutation-selection balance. Ohnishi and Nagakubo (1982) reported that mutations at five different loci could cause dwarfism in buckwheat. The frequencies of dwarfism alleles at four of these loci in Japanese populations was estimated as 0.00235, 0.00067, 0.00294, and 0.00196. These low allele frequencies imply that recurrent mutation is balanced by selection against the dwarf alleles. Thus the characteristics of both kinds of deleterious phenotypes (chlorophyll abnormalities and dwarfism) fit the mutational load model.[7]

Table 10.9. Estimated number of loci ($\hat{n}$) and the gene frequency per locus ($\hat{p}$) for chlorophyll-deficient mutants in buckwheat populations. (From Ohnishi 1982.)

Type of Abnormality	$A = \sum\limits_{i=1}^{n} p_i$	$B = \sum\limits_{i=1}^{n} p_i^2$	$\hat{n} = A^2/B$	$\hat{p} = B/A$
Cotyledons				
albino	0.0037	0.000004	3.4	0.0011
yellow	0.0356	0.000126	10.1	0.0035
pale yellow	0.0486	0.000090	26.2	0.0019
pale green	0.0740	0.000093	58.9	0.0013
variegated	0.0237	0.000076	7.5	0.0032
Foliage leaves				
albino	0.0025	0.000006	1.0	0.0024
pale yellow	0.0294	0.000091	9.5	0.0031
pale green	0.0350	0.000130	9.4	0.0037
variegated	0.0192	0.000086	4.3	0.0045

---

7. Buckwheat populations have relatively high frequencies of detrimental mutants such as chlorophyll abnormalities, dwarfism and sterility, yet low levels of allozyme variability (Ohnishi 1985).

SELF-INCOMPATIBILITY OR LOAD

"To prove adaptation one must demonstrate a functional design"
(Williams 1966:212).

In the literature on plant reproductive biology, there is some confusion concerning genetic load, self-sterility, and self-incompatibility. These three terms are not synonymous and thus cannot be used interchangeably. A plant may be self-sterile for many reasons; self-sterility is a generic term describing a distinct pnenomenon, the inability to form viable offspring upon self-fertilization. The term does not imply mechanisms or causes; it is neutral in this respect. Self-incompatibility, on the other hand, is generally based upon prezygotic factors that prevent self-fertilization (Lewis 1954). Self-incompatibility implies the presence of genetic factors that "interfere with the sexual differentiation process prior to fertilization" (Van Den Ende 1976). Self-incompatibility acts by interposing a physiological barrier preventing self-fertilization (Bateman 1952); self-incompatibility exists because it is part of the "design for survival." Where self-sterility is caused by genetic load, the lack of viable offspring from self-fertilization is due either to the formation of genetically defective gametophyte, gamete or zygote genotypes that are incapable of normal development or to embryogeny. Genetic load is "faulty-design." The distinction between self-incompatibility and genetic load is clearly shown by the response of these two forms of self-sterility to mutation. Since self-incompatibility is part of a "design," induced mutations of the self-incompatibility mechanism should mainly damage the design. Consequently, the most common mutational response should be the generation of genotypes that have a faulty self-incompatibility mechanism and are, therefore, self-compatible to some degree. After reviewing a number of induced mutation experiments with plants having homomorphic, gametophytic self-incompatibility systems, Lewis (1954) noted that the most typical genetic change was to go from a self-incompatible to a self-compatible genotype. In contrast, when self-sterility is due to genetic load, the most common (if not only) mutational response should be to increase self-sterility.

Embryo abortion has sometimes been considered as a post zygotic incompatibility system (see reviews in de Nettancourt 1977 and Seavey and Bawa 1986). Perhaps the best way to consider the problems with such an interpretation is by applying it to the case of inbreeding in humans. Incest in humans almost invariably leads to a higher frequency of premature deaths and offspring with birth defects (Bodmer and Cavalli-Sforza 1976). Applying the postzygotic incompatibility concept, one could view premature deaths and birth defects as part of the "design for survival" to prevent incest. This kind of explanation also attributes an "adaptive significance" to the causes of premature deaths and birth defects as mechanisms to prevent incest! Of course, such an interpretation is not generally held by geneticists. As will be subsequently discussed, some angiosperm reproductive systems have been interpreted in this way.

Three criteria may be used to distinguish genetic load from self-incompatibility: 1) self-incompatibility prevents fertilizations, selfed plants should lack aborted embryos; 2) since self-incompatibility is an adaptation, one would expect the majority of individuals of a species to exhibit similar levels of self-sterility. A high genetic variance for self-sterility is typical for genetic load; 3) different responses to induced mutations.

In the homosporous ferns, the pioneering study of Wilkie (1956) on *Pteridium aquilinum* documented high levels of self-sterility for three clones from Scotland. Preliminary genetic analysis supported a single locus, multiple allelic self-incompatibility system. Further studies of other populations failed to confirm this interpretation. *P. aquilinum* populations vary greatly in self-sterility (0 to ca. 100%) and self-sterility is associated with high levels of embryo abortion (Klekowski 1972). Studies of other fern species have documented similar patterns of self-sterility (Masuyama 1986) and support the general conclusion that genetic load is the basis of self-sterility in this plant group. In the conifers self-sterility also fits the genetic load model. In most species there is a considerable tree-to-tree variability for self-sterility and high levels of embryo abortion have been documented following selfing (see Orr-Ewing 1957; Mergen, Burley, and Furnival 1965). Whitehouse (1950) concluded that for theoretical reasons self-incompatibility could not evolve in archegoniates. The data from ferns and conifers support this view.

In a provocative paper, Seavey and Bawa (1986) have argued that many examples of embryonic genetic load in angiosperms may be examples of postzygotic self-incompatibility. These authors base their hypothesis upon the observations that clear segregation patterns are seldom encountered for lethals and very often the expression of embryo abortion is confined to a single period of embryogenesis. Considering how few genetic load studies have been published on angiosperms, generalizations about load characteristics are premature. Regarding the lack of clear segregation ratios for lethals, the original formulation of the lethal equivalent concept was based upon the homozygosities of many genes with small detrimental effects. Thus, the genetic load model is compatible with the presence or absence of clear ratios in the progeny resulting from selfing. With regard to timing of embryonic abortion, not enough is known about the interactions of embryo endosperm and maternal tissues to make even reasonable predictions. For example, in the conifer *Picea glauca* selfed and outcrossed embryos develop normally until the two-tiered stage, and then the majority of selfs abort prior to the differentiation of cotyledons (Mergen, Burley, and Furnival 1965). Thus, in this case genetic-load-based embryo abortion occurs within a relatively short developmental interval.

The postzygotic self-incompatibility hypothesis[8] is not to be confused with the rejection or differential abortion of genetically defective embryos due to soft selection (see chapter 9). To prove that postzygotic self-incompatibility exists, one must demonstrate the presence of lethal alleles whose primary (if not only) function is to abort embryos that happen to be homozygous for those alleles. It also must be shown that these lethals exist "by design," that is, are part of an adaptive genetic system and not random mutational load.

Two species of legumes exhibit patterns of self-sterility which do not appear to be straightforward genetic load manifestations, *Vicia faba* and *Medicago sativa*. In both species the original breeding system appears to have promoted outcrossing and in both species, domestication has generally enforced inbreeding. Selfing either species results in considerable embryo abortion.

In *V. faba* 12 chromosomes are present at mitosis, five pairs with a subterminal centromere and a very much larger chromosome pair with a median centromere and a subterminal secondary constriction. This latter pair of chromosomes are the M chromosomes. Rowlands (1958) investigated meiosis in *V. faba* and reported a very high frequency of cytological irregularities. The M chromosomes as well as one of the smaller pairs were frequently involved in bridge formation during anaphase I or II and acentric fragments were common. In some individuals 50% of the meiotic figures showed irregularities whereas others showed 5.7%. Along with this meiotic system is a high degree of embryo abortion associated with selfing,

8. It should be noted that the timing of self-incompatibility reactions may not be limited to the style; Kenrick, Kaul, and Williams (1986) report that in *Acacia retinodes* the site of pollen tube arrest is the nucellus. Thus, the documentation of pollen tube growth in styles is not in itself evidence of zygote formation.

although occasional individuals may be quite self-fertile (Rowlands 1960a, b; 1963). Although it has been suggested that a balanced lethal system has evolved in this species (Rowlands 1963), one is tempted to relate embryo abortion to the high mutability of the karyotype, especially the pair of M chromosomes.

In alfalfa, *Medicago sativa*, sixteen pairs of chromosomes are present at meiosis and there is no evidence of meiotic irregularities. Nevertheless, many alfalfa plants form a high proportion of aborted pollen (Brink and Cooper 1936). The number of ovules per flower is 9.3, and although all form embryo sacs the majority of ovules fail to form viable seeds. Contrasting plants that exhibited high seed set after selfing with those that had low seed set after selfing, an average of 3.1 ovules per flower were fertilized and 1.25 seeds per flower were formed in high seed setters; 2.5 ovules per flower were fertilized and 0.7 seeds developed per flower in low seed setters. Thus, the majority of the self-sterility is due to embryo abortion (Cooper, Brink, and Albrecht 1937). Fertilized ovules are not randomly distributed in ovaries; the probability that an ovule is fertilized is highest for ovules found in the apex of the ovary and lowest for those found at its base. Embryo abortions are more common for ovules in the base of the ovary in spite of the lower frequency of fertilized ovules occurring in this position. Both of these observations suggest pollen quality and competition are important features of self-sterility in this species. The growth of embryos of high seed setting plants is faster than the embryos of low seed setters. Although a greater proportion of embryo abortion takes place during young embryo stages, some abortion occurs later (Cooper, Brink, and Albrecht 1937). Whether the self-sterility in alfalfa represents an adaptive system to promote outcrossing or is a consequence of the chance fixation of a maladaptive mechanism(s) that results in genetically defective pollen (and possibly eggs) is a moot point.

LOAD LEVELS

In *Homo sapiens*, Bodmer and Cavalli-Sforza (1976) summarized a number of studies and concluded that the load of lethal equivalents per gamete varied from 0 to 4 with a median of slightly above 1. Thus, perhaps each of us carries more than two lethal equivalents in our genotype, which, of course, is an underestimate of the actual number of recessive detrimental genes. In vascular plants different species and life forms vary enormously in genetic loads. In the homosporous ferns some species are essentially loadless (e.g., *Ceratopteris*) and others have as many as 2.9 lethal equivalents per gamete. In conifers loads vary from loadless *(Pinus resinosa)* to as many as 13 lethal equivalents per gamete in some trees of *Pseudotsuga*. In angiosperms, based upon very few and often inconclusive studies, 2 to 3 lethal equivalents per gamete have been reported. But these estimates are very conservative and may be much too low. It should be remembered that load is estimated on the projected viability of a gamete genotype converted into a homozygous zygote. The numerous possibilities of soft selection in plant life cycles (especially angiosperms) may result in large underestimates of load because of nonrandom gamete and/or zygote selections in the progeny resulting from selfing.[9]

Another problem with comparisons of human and plant genetic loads relates to the mathematical formulations used in calculating lethal equivalents. As discussed earlier,

$$S = 1 - Ł$$

where S is survivorship and $Ł$ is load, thus

$$S = e^{-(A + BF)}$$

where A is a measure of expressed genetic damage $(F = 0)$ plus environmental damage, B is a measure of hidden genetic damage that could manifest itself only with complete homozygosity $(F = 1)$, and F is the inbreeding coefficient. The load (due to environment and genetics) in a randomly mating population is

$$1 - S_1 = 1 - e^{-A}$$

---

9. A particularly clear example of such selective bias against lethals is the inheritance of Renner complexes in *Oenothera*. *Oenothera parviflora* race Charlottesville is a complex translocation heterozygote in which all the chromosomes associate in a single circle (14 chromosomes) at meiosis. The chromosomes segregate alternately at anaphase I of meiosis and two types of gametes are formed. The α gamete which is regularly transmitted through the egg and the β gamete which is transmitted through the pollen. Epp (1974) hybridized *O. parviflora* (female parent) and *O. hookeri* (male parent); the latter species forms seven bivalents at meiosis and was a lethal-free laboratory strain. Lethals are present on the α Renner complex of *O. parviflora*. The F_1 hybrid formed a ring of 14 chromosomes and segregated two kinds of haploid genotypes at meiosis, α eggs and *hookeri* eggs and sperm. One would have expected 1:1 ratios in the F_2 based upon random Mendelian expectations. In actuality, there was a consistent deficiency of α/*hookeri* heterozygotes, the genotypes with lethals. Epp scored 9,773 F_2 offspring and found 4,221 α/ *hookeri* heterozygotes and 5,552 *hookeri* homozygotes or 42.5% to 57.5%. In the genus *Oenothera* such deviations from typical Mendelian ratios are common. The major causes of these deviations are megaspore competition, pollen competition and the expression of lethals in the eggs, pollen, or in homozygous and heterozygous embryos (Cleland 1972).

lethal equivalents per gamete (B) are calculated from the ratio of the viable progeny resulting from inbreeding and random mating

$$S/S_1 = e^{-(A+BF)}/e^{-A}$$

$$S/S_1 = e^{-BF}$$

Let us apply these formulations to an actual set of breeding data. Gabriel (1967) published the results of an extensive crossing and selfing program in sugar maple, *Acer saccharum*. In table 10.10 the mean survivorships in terms of seed set from cross- and self-pollination are shown (we will not consider soft selection due to the low seed ovule ratio in this species). The important point to note in table 10.10 is that the value of S_1 varies between 0.42 to 0.25. This is in marked contrast to the S_1 values reported for humans which are typically greater than 0.9 (see Vogel and Motulsky 1986). In many plants the survivorship of outcrossed progeny is very often less than 0.9, see table 10.11. In general as life span increases, the value of S_1 generally decreases. In addition, in plants, S_1 values may show considerable intraspecific variation. In table 10.12 this variation is documented. The seed set from cross-pollination (an estimate of S_1) varied from 0.18 to 0.47. The four sets of reciprocal crosses were analyzed with a 2×2 chi-square contingency test, one (M13 × M4; M4 × M13) was significant at the less than 1% level. Thus, plant S_1 values have three characteristics that differ from animal S_1 values: 1) plant S_1 values are often much lower (especially long-lived plants); 2) plant S_1 values may show considerable intraspecific variation; 3) plant S_1 values may or may not show differences in reciprocal crosses.

Considering the first point, since plant S_1 values may be low,

Table 10.10. Summary of breeding experiments in sugar maple, *Acer saccharum*. (From Gabriel 1967.)

Tree	Cross-Pollination Seed Set (S_1)	Self-Pollination Seed Set (S)	S/S_1
M6	0.412	0.206	0.50
H573	0.409	0.271	0.663
M13	0.405	0.183	0.452
M4	0.275	0.081	0.295
H509	0.252	0.13	0.516

Table 10.11. Survivorship of progeny resulting from outcrossing (S_1) and phenotypic criteria used to determine viability.

Species	S_1	Phenotype	Reference
Homo sapiens	0.9	various criteria	For review, see Vogel and Motulsky 1986
Drosophila willistoni	0.843	egg to adult viability	Malagolowkin-Cohen et al. 1964
Tribolium castaneum	0.756	egg to adult viability	Levene et al. 1965
T. confusum	0.763	egg to adult viability	Levene et al. 1965
Phlox drummondii (annual)	0.876	seed abortion	Levin 1984
Ulmus americana	0.55	seed set	Lester 1971
	0.68	seed germination	
	$(0.55)(0.68) = 0.374$	general viability	
Acer saccharum	0.349	seed set	Gabriel 1967
Larix laricina	0.389	sound seed	Park and Fowler 1982
	0.750	seedling survival	
	$(0.389)(0.750) = 0.291$	general viability	
Picea glauca	0.322, 0.507	sound seed	Coles and Fowler 1976
P. mariana	0.63, 0.70	seed germination	Park and Fowler 1984
	0.71, 0.80	seedling survival	
	$(0.70)(0.80) = 0.56$	general viability	
P. omorika	0.334	full seeds	Langner 1959
	0.497	seed germination	
	$(0.334)(0.497) = 0.166$	general viability	
Pinus resinosa	0.72	full seed	Fowler 1965a, b, c
Pseudotsuga menziesii	0.685	sound seeds	Sorensen 1969

lethal equivalents may seriously underestimate genetic loads. The total genetic damage is the sum of B and the genetic component of A and, therefore, lies between B and $B + A$ (Vogel and Motulsky 1986). In organisms where S_1 values are high, A values are low and lethal equivalents (B) may be a useful metric. If S_1 values are low, then A values are high and, consequently, lethal equivalents (B) may seriously underestimate total genetic damage since it lies between B and $B + A$. The crucial question is, of course, what proportion of the low S_1 values in many plants is due to genetics, the segregation of lethals expressed in the haploid gametophytes as well as dominant lethals expressed in the embryos or endosperm (in

Table 10.12. Details of breeding experiments with sugar maple. Trees from Vermont designated by H and trees from Massachusetts by M. The diagonal arrows indicate reciprocal crosses and the offspring marked with an asterisk are the results of selfing. For each cross or self the number of flowers pollinated and the percent of the pollinated flowers setting seed is given. (From Gabriel 1967.)

| | | MALES | | | | | |
		M4	M6	M13	M1	M5	M7
	M4	148*	62	76	52	98	76
		8.1%	46.8%	18.4%	36.5%	19.4%	26%
FEMALES	M6	96	68*	108	80	96	42
		38.5%	20.6%	46.3%	45%	45.8%	16.7%
	M13	100	100	104*	96	100	100
		39%	42%	18.3%	44.8%	38%	39%

		H509	H573	M11	H507
	H509	100*	92	92	86
FEMALES		13%	18.5%	29.4%	28.0%
	H573	42	118*	100	100
		28.6%	27.1%	40.0%	47.0%

angiosperms), and what proportion is due to environmental influences. Two points are suggestive that genetic influences may be important. The high levels of intraspecific variation for S_1 values within tree populations may evidence genetic rather than environmental causes, as one would have expected environmental effects to show greater variability between populations than between individuals in the same population. The second point concerns the results of reciprocal crosses.

One would expect that environmental effects may be more important for the female rather than the male parent (since zygote and embryo development occur on the female). If S_1 values from reciprocal crosses are similar, this is strong evidence for genetic rather than environmental causes. Where reciprocal crosses differ in S_1 values, little can be concluded since either environmental variables or genetics (the segregation of lethals expressed only in either the mega- or microgametophyte) can be invoked. In the sugar maple study, three out of four reciprocal crosses gave similar S_1 values and, thus, support the hypothesis that the low S_1 values have genetic causes at least for those six trees.

Because of the variable and often low S_1 values found in plants, in addition to reporting lethal equivalents, S_1 values should also be reported. The data in table 10.11 show an inverse relationship between life span and S_1 values. If the S_1 depression has a significant genetic component, then it may represent the somatic accumulation of recessive mutants expressed in the gametophytes and/or dominant mutants expressed during embryogenesis or endosperm development. One would then predict an inverse correlation between offspring survivorship (in crosses as well as in selfs) and the age of the parent (see chapter 11 for further discussion).

CHAPTER ELEVEN

Significance of Mutation

Although it is almost axiomatic that mutation is fundamental for evolutionary change in plants (or any organism), it does not necessarily follow that the only important (or even most important) aspect of mutation in plant biology is evolution. It is possible to construct an argument stating that the most important aspect of mutation is the selection for developmental and genetic systems with maximum homeostasis. Such systems best resist the disruptive effects of mutation and, therefore, stasis is promoted rather than evolutionary change.

COMPONENTS OF GENOMIC CHANGE

Any assessment of the significance of mutation in plant biology must consider how the various components of a genome may be affected by mutation. In this regard it is convenient to divide the genome into three components: 1) repeated DNAs *sensu lato;* 2) chromosomes; 3) Mendelian genes.

Repeated DNAs

In a classic paper, Dover (1982) summarized the unique characteristics of repeated DNA sequence variation in and between related species of plants and animals. These characteristics are presented in summary form in figure 11.1. A large proportion of eukaryotic genomes consists of multiple copy families of genes and noncoding sequences. These families of repeated sequences often exhibit unexpected sequence homogeneity within and between individuals of a species, yet reveal substantial heterogeneity between related species. Within a species, the individual repeats of a repeated sequence do not appear to accumulate mutations and evolve independently, but rather, when mutants are fixed, they occur in practically all members of the repeat family. There appears to be a molecular mechanism(s) that results in the homogenization of a repeat family.[1] Such concerted evolution is observed for a wide

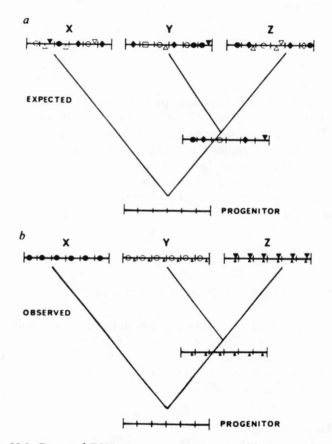

Figure 11.1. Repeated DNA sequence variation in and between related species (X, Y, Z). Families of repeated sequences may be tandem (as shown) or interspersed. If each member of a family evolves independently, one would *expect* considerable within-family variation. What is generally *observed* is a high degree of within-family homogeneity. (From Dover 1982.)

1. Generally some kind of gene conversion (either biased or democratic) is postulated, but the actual molecular mechanisms of the homogenization of a family of repeats are not yet understood (see Flavell 1985 for review).

range of repeated DNA families including coding and noncoding families. The distribution and homogenization of such repeat families is not restricted to the domain of a single chromosome. Repeat chromosomes of the genome also exhibit concerted evolution indicating molecular communication of repeat sequence information between different chromosomes in a genome (Coen et al. 1983).

The evolution of such patterns of variation in repeat families has been the subject of considerable debate. Whether these patterns are a consequence of conventional evolutionary forces (i.e., natural selection, mutation, and genetic drift) or are primarily a consequence of molecular events at the cellular level is the central point. Population fixation of families of repeats as a consequence of cellular processes of homogenization has been called molecular drive (Dover 1982, see table 11.1 for the characteristics of molecular drive). The acceptance of molecular drive as an evolutionary force is in debate (see Dover 1986 for review).

Regardless of the mechanisms involved, the observations concerning sequence spreading and homogenization in repeat families must be considered as aspects of mutational change. As already indicated (chapters 2 and 3), the amount of repeated DNA in plant genomes is high. For repeats 50 bp or longer, more than 75% of the total DNA of genomes in excess of 2 pg DNA consists of repeats. Most of this repeated DNA does not code for proteins or play sequence specific roles (Flavell 1985). In our context then, the critical questions concern the biological consequences of mutation and mutation spreading in this component of the genome.

The major differences at the DNA sequence level that distinguish related species are typically changes in the repeated DNA genome component (e.g., Stein, Thompson, and Belford 1979; see Dover

---

Table 11.1. The essential differences between change in the Mendelian genes (Mendelian genetics) and repeated DNAs (molecular drive) genome components. (From Coen et al. 1983.)

Mendelian Genetics	Molecular Drive
Short timescale	Long timescale
Chromosome independence	Chromosome interdependence
Stochastic	Stochastic and directional
Binomial	Polynomial
Monogenic	Polygenic

1982 for review). It has been hypothesized that the divergence among repeated DNA families constitutes one of the major factors in reducing meiotic chromosome pairing among related species (Dvořák and Appels 1982; Flavell 1982, 1985). If this hypothesis is correct, then mutation and sequence homogenization among families of repeated DNAs may have considerable long-term evolutionary significance.

In many groups of vascular plants, interspecific hybridization and the formation of viable F_1 individuals in nature are not uncommon occurrences (Stebbins 1959; see Grant 1981 for review). Such F_1 hybrids may have either relatively normal chromosome pairing at meiosis and, consequently, be fertile or form some or many univalents at meiosis and, consequently, be sterile. Such hybrid sterility or fertility is not a function of the degree of morphological divergence of the parental species, as there are numerous examples of morphologically very similar species forming sterile hybrids.

If hybrids are fertile, interspecific hybridization often leads to some degree of backcrossing between F_1 and one (or both) parental species. Such a pattern of hybridization and interspecific gene flow is called introgression (Anderson 1949, 1953). On the other hand, if interspecific hybrids are sterile due to a lack of chromosome pairing at meiosis, polyploidy is an escape from such sterility (Stebbins 1971). Interspecific hybridization followed by polyploidy (allopolyploidy) may result in instant speciation with the new species carrying two doses of each original parental genome. Introgression or allopolyploidy are processes that often characterize a phyletic lineage (i.e., a genus, a family, or even a higher category may exhibit patterns of variation primarily determined by polyploidy or introgression). Since each of these processes imposes its own particular set of evolutionary limitations and possibilities, characteristics predisposing a phyletic lineage toward introgression or polyploidy will have very significant long-term evolutionary effects. If, as stated earlier, chromosome pairing is to some extent a function of the degree of divergence found in various families of repeated DNAs, then the forces changing these sequences and the rates of these changes may have profound long-term evolutionary consequence. If such changes are primarily a consequence of molecular drive, then the characteristics of DNA metabolism in relation to sequence spread and homogenization may, in some instances, be the primary determinant affecting long-term patterns of plant evolution.

In addition to the above potential effects of change in repeated sequence families, repeated sequence polymorphisms for ribosomal DNA in plants may be common. Schaal (1985) reported that in

Solidago altissima no two plants within a population were identical for rDNA and that variation also occurred within individuals. Whether such variations are only stochastic noise without phenotypic effects or whether such changes have phenotypic and adaptive significance is undetermined (see Schaal 1985; Walbot and Cullis 1985; Cullis 1986 for review).

Chromosomes

Chromosome mutations, of course, may cause sterility through various meiotic irregularities. The cytological basis of sterility associated with heterozygosity for translocations, inversions and other types of chromosome aberrations as well as sterility due to euploid and aneuploid imbalances has been well documented. The reader is referred to Darlington (1937), Burnham (1962), and Lewis and John (1963) for lucid explanations of these cytological phenomena. It would be unnecessarily repetitive to catalog these effects again. The present discussion will focus on the problem of why some phylogenetic lineages are associated with high levels of chromosome change and other lineages by extreme chromosome conservatism. These cytogenetic differences may reflect, to some degree, the rate at which chromosome mutations occur and the ease with which the chromosome set may be altered without serious effects on viability in these different lineages.

Many researchers have noted that the rates of chromosomal divergence and speciation appear to vary in plants with different growth forms, herbs having the highest rates and trees and shrubs the lowest (Grant 1963; Stebbins 1950, 1971). Assuming that the mean ages of angiosperm genera are 37.4 million years (Myr) for herbs, 67.0 Myr for shrubs, and 74.7 Myr for trees, Levin and Wilson (1976) quantified these rates. When the rate of chromosome diversification was plotted against the rate of increase in species numbers for each angiosperm genus, some interesting correlations were found; for herbs the correlation coefficient (r) was 0.71, for shrubs r = 0.89, and for trees r = 0.15. Thus, in angiospermous trees the rates of chromosome diversification and species numbers were not correlated whereas in herbs and shrubs the correlations were more striking. These relationships in herbs and shrubs may be due to:

    1. Correlative effects—the tempo of both processes is influenced by the same or similar factors

    2. Causative effects—karyological changes determine or enhance the process of speciation.

In general, the second explanation is commonly accepted by biologists. The causative relationships between speciation and karyological change have been summarized by Lewis and John (1963). Chromosome changes may be viewed either as "systems of conservation" or "systems of innovation." In the former the structural or numerical change is favored because it preserves a genotype or component of a genotype from the disruptive effects of recombination. In the latter the chromosome change results in a new adaptive phenotype. To these general explanations are added populational parameters and the ecological characteristics of the various life forms (i.e., herb vs. shrub) (see Levin and Wilson 1976 for review).

The above explanations, although giving broad generalizations concerning groups of different phyletic lineages linked only by life form, fail to explain some of the cytological peculiarities of these lineages. In other words, why have some groups primarily utilized reciprocal translocations as a means of karyological change, whereas in others karyological changes are more likely to involve inversions. Why are Robertsonian fusions or telocentric chromosomes or symmetrical vs. asymmetrical karyotypes or allopolyploidy or aneuploidy, etc., characteristic of some taxonomic groups and not of others? (Stebbins 1971; Jones 1978). The following discussion will develop the theme that such group specific cytogenetic differences may not be a consequence of populational or ecological factors but rather a consequence of DNA related metabolism and the specific topology of chromosomes at the molecular level. These molecular characteristics may differ sufficiently between phyletic lineages to constitute a "chromosome drive" which restricts and, in a way, directs the kinds of cytogenetic changes that may evolve within a group (presumably to fill the above mentioned needs of innovation or conservation). Chromosome drive determines the rate of occurrence of specific chromosome changes, the probability of their acceptance into the genome and whether a genome so changed will still allow a viable organism to develop. Thus, in a broad sense, chromosome drive is an aspect of mutation and mutation buffering.

Since our understanding of eukaryotic DNA metabolism and chromosome organization is yet rudimentary, the characterization of plant lineages with regard to different chromosome drive molecular mechanisms is premature. Perhaps the best course would be to outline some molecular mechanisms that may be involved and then present some relevant plant examples.

Although chromosomes are the bearers of an organism's nuclear genes, chromosome variation (mutations) may have little immediate phenotypic consequence. Ohno calculated that for the haploid set of the mammalian genome with 3.2×10^9 bp of DNA, the average distance between neighboring genes is 35,000 bp. To quote Ohno: "the mammalian euchromatic region can be viewed as a barren stretch of desert in which still functioning genes are scattered as though oases" (1984b:9). He notes that in mammals the genetic distance of one crossover unit covers one million base pairs and may include no more than 30 to 40 still-functioning genes, and that, therefore, there are no closely linked genes *sensu stricto*, "genetic close linkage is largely an illusion." Many plant genomes exceed this mammalian value by ten- or one hundredfold (see table 3.3). Therefore, one might predict that the desertification of portions of chromosomes in some plant groups has progressed even further. Does the extent of this DNA desert and its pattern influence the nature or spectra of karyological changes which a genome may undergo? As was previously discussed, much of plant DNA consists of repeated sequences (presumably such sequences are the main features of Ohno's desert). These sequences homogenize both within chromosomes and between different chromosomes. It is certainly possible that the distribution and amounts of similar sequences either within or between chromosomes could influence the probable directions of chromosome change (Coen et al. 1983; Dover 1982; Flavell 1982).

Genotypes with different propensities for chromosome mutations have been documented in a number of different plant species (Gerstel and Burns 1966; Heneen 1963). For example, in the grass *Elymus farctus* Heneen (1963) discovered an individual in which 78% of the cells showed chromosome aberrations. Mammalian genotypes exhibiting cytogenetic instabilities are also known and in a few instances the molecular basis of the instability has been characterized. In humans, a subset of individuals predisposed to develop cancer are those known to have "chromosome instability syndromes" (German 1972). Such syndromes include ataxia telangiectasia, Bloom's syndrome, and Fanconi anemia, all of which have an elevated frequency of spontaneous chromosome aberrations and whose cells are more sensitive to mutation induction from external agents (Hsu 1983). Bloom's syndrome is attributed to altered DNA ligase I activity, an enzyme which, along with DNA polymerase α, functions during DNA replication (Willis and Lindahl 1987; Chan et al. 1987). Hsu (1983) suggests that populations of humans may vary with respect to the frequencies of individuals with different degrees of instability. There is no reason why similar characteristics may not distinguish plant populations, or species, or even whole lineages. In fact, in plants such genotypes may be less selectively disadvantaged than in animals. In animals, genotypes conferring cytogenetic instabilities are more immediately life-threatening because of their supposed causal relation to cancer. As was discussed in chapter 7, plants are somewhat immune to this malady. The chromosome number instability documented in a clone of *Hymenocallis calathinum* (Amaryllidaceae) is illustrative of the somatic chromosome variation that plants can tolerate. In this clone, which had a spindle abnormality, Snoad (1955) documented from 23 to 83 chromosomes per cell (figure 11.2). Such chromosomal instability would be highly deleterious in a higher animal.

Probably one of the most extreme examples of cytogenetic instability in a plant species is *Scilla autumnalis*, a bulbous, sexual outcrossing member of the family Liliaceae. The range of this species is centered on the Mediterranean, extending from southern England to Israel. Within this single species, Ainsworth, Parker, and Horton (1983) reported a very large array of cytological variability. Eight chromosome races were documented ($2n$ = 10, 12, 14, 26, 28 and 42) with three of the four diploids involved in the generation of polyploids. Within the X = 7 diploids, two karyotypes were documented which differ by 70% of their DNA content. Within races, structural variation for individual chromosomes occurred at a number of levels; individual plants which were chimeras were documented as well as individuals consisting entirely of a single mutant cell type, and populations were found with etensive polymorphisms. In total 109 different structural variants were detected in 1,163 plants, and 37 of the variants were sufficiently common to reach polymorphic proportions. There was no apparent correlation between a specific cytotype and ecology. *S. autumnalis* is able to generate and tolerate large numbers of additional euchromatic segments. Ainsworth, Parker, and Horton speculate that perhaps the segments possess a "drive system" enabling them to sweep through the population and become fixed.

In contrast to *S. autumnalis*, the tribe Aloineae, also of the Liliaceae, is probably one of the most stable groups regarding chromosome number and karyotype in the plant kingdom (Brandham 1983). The three major genera (*Aloë*, *Gasteria*, and *Haworthia*) which comprise several hundred species all have the same basic chromosome number (X = 7), with one long submetacentric, three long acrocentrics and three short acrocentrics, a bimodal karyotype. Brandham (1976) determined the rates of structural chromosome changes in this group. A total of 699 diploid and 333 tetraploid progenies of normal plants were surveyed. These data are presented in table 11.2. In diploids the rate of production of surviving gametes with a single chromosome break is 1 in 1,398 and no gametes survived with two breaks. In contrast, the buffering effect of polyploidy is clear; the frequency of surviving gametes with one or two chro-

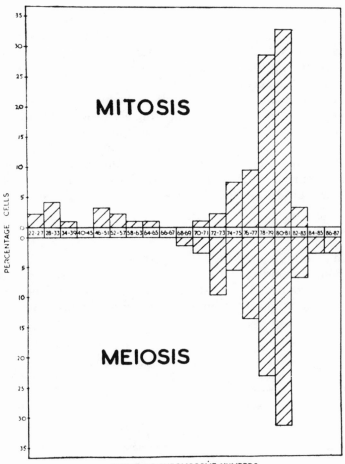

Figure 11.2. Chromosome number instability in *Hymenocallis*. The frequency of different chromosome complements at mitosis (94 cells counted) and meiosis (73 cells counted). The cells with low chromosome numbers are presumably lost through diplontic selection during the growth of the apical meristems and, consequently, are under represented at meiosis. (From Snoad 1955.)

mosome breaks is much higher. From these data one can make a conservative estimate that the rate of formation of gametes with structural changes is ca. 2% per genome per meiosis. To quote Brandham and Johnson (1977), "The Aloineae therefore represent an almost unique situation of a widespread group of many hundreds of species in which interchange hybridity occurs frequently, in polyploids, but in which these aberrations are strongly selected against, maintaining the uniformity of the basic karyotype."

Obviously the majority of plant groups occupy positions in the spectrum between the extreme karyological instability of S. autumnalis and stability of the Aloineae. As stated previously, the mechanisms influencing the rate of the occurrence of specific chromosome changes, the acceptance of specific changes into the genome and whether such altered genomes allow the formation of viable (or sometimes even more adaptive) offspring are essentially unknown. What is established is that such mechanisms of mutation and mutation buffering may vary considerably between different plant groups. Consequently, the significance of chromosome mutations (i.e., whether primarily neutral, disruptive, or occasionally adaptive) will also be a group specific characteristic.

Mendelian Genes

Although in terms of mass the Mendelian genes constitute the smallest component of most plant genomes (10% to 1% or less), generally they are regarded as the most important component. The consequences of mutation in this genome component fall in three cate-

---

Table 11.2. Frequency of chromosomal structural changes in the gametes of diploid and tetraploid *Aloineae*. Only gamete combinations giving rise to viable progeny were counted. (From Brandham 1976.)

	FREQUENCY IN:	
Chromosome change	*1398 Haploid gametes*	*666 Diploid gametes*
Interchange	0	3 (0.45%)
Deletion	1 (0.07%)	21 (3.15%)
Duplication	0	1 (0.15%)
Pericentric Inversion	0	2 (0.3%)
Dicentric	0	1 (0.15%)
Centric Fragment (neo-B)	0	1 (0.15%)

gories: neutral, disadvantageous, and advantageous. Neutral mutations may be very common in organisms and may account for a considerable proportion of the genetic variability detected in populations. The reader is referred to Kimura (1983) for an extensive and excellent discussion. Regarding the relative importance of disadvantageous vs. adaptive mutations, the biological literature is curiously divided; biologists whose main interests are animals tend to emphasize the disadvantageous or genetic load aspects of mutation whereas the adaptive or positive aspects are emphasized by botanists. This altered emphasis is nowhere more apparent than in the differences in coverage given to genetic load and mutational load in evolution and population biology books written by zoologists and botanists (with the notable exception of Solbrig and Solbrig 1979). This difference in emphasis may be in part due to the importance of medically related research in zoology and agricultural research in botany. In the former the relationship of cancers, birth defects, spontaneous abortions, and possibly even aging to mutation and genotype deterioration (Muller 1950; Knudson 1986; Kahn 1985) may bias opinion toward the negative. Plant breeding and crop improvement, with their attendant searching for superior genes and genotypes, may bias botanists toward the positive aspects of mutation.

Mutations may occur in the germline or somatic tissue. As often stated in this book, plants lack a germline. In plants, therefore, mutations may occur in somatic tissue or be generated during meiosis (or in the immediate cell lineages giving rise to meiocytes). It is generally believed that the majority of mutations occurring in plants originated as somatic mutations. Many cultivars that have been selected occurred first as chimeric sports (see Whitham and Slobodchikoff 1981 for review). Of the 8,000 plant varieties cultivated in Europe in 1899, 5,000 originated as somatic mutations (Sutherland and Watkinson 1986).

The relationships between somatic mutation and gametic mutation have been studied mathematically by Slatkin (1984) and Antolin and Strobeck (1985). For mutations that are neutral within the context of the plant (i.e., neutral with regard to developmental selection), somatic mutations will in effect increase the overall mutation rate in the population. This increase will be a function of the number of mitotic divisions between zygote formation and meiosis in those cell lineages giving rise to meiocytes. One would predict, therefore, that the mutation rate (per generation) would be higher in organisms with many mitotic divisions between zygote and meiocyte than in organisms with few such mitotic divisions. In humans this seems to be the case (see Vogel and Motulsky 1986 for review). The female oocyte passes through approximately 22 divisions at the time of birth. From birth to sexual maturity and fertilization, this cell will undergo only two meiotic divisions regardless of the female's age at which fertilization occurs. In contrast, the number of cell divisions in spermatogenesis is a function of the male's age at which sperm are formed. Thus, at puberty the meiocytes have passed through approximately 30 mitotic divisions, at age 28 approximately 380 divisions, and at age 35 approximately 540 divisions. In accordance with expectation, mutation rate is a function of paternal age. Vogel and Motulsky report a fivefold difference in mutation rate between young and old males. In females the mutation rate is constant throughout reproductive life. Similar studies in plants are critically needed. Is mutation rate (and frequency) a function of the individual plant's age? Is mutational load a function of the age distribution of the population of plants being studied? The age of plants in a population (with special regard to at least relative numbers of mitotic divisions between zygote and meiocyte) is a neglected statistic in plant population biology.

Somatic mutations also reduce the average probability of identity by descent in the gamete pool. Thus, somatic mutation may be an important factor in elevating the genetic variation in populations of long-lived plants (e.g., trees, shrubs, and various growth forms with a clonal component, see figure 2.6). For mutations that are not neutral but rather are selectively advantaged within the organism (i.e., ramets or branches that have fixed the mutation are competitively advantaged over nonmutant ramets), the results are dependent on whether selection is hard or soft (Slatkin 1984).

Under soft selection, branches (or ramets) of a genet are in competition with each other for the opportunity to make up the genet's contribution to the gamete pool. Thus, the genet contributes a constant proportion of gametes into the gamete pool depending upon the original zygote genotype from which the genet originated. This constant proportion is partitioned among ramets depending upon their competitive ability. On the other hand, under hard selection the contribution of each ramet to the gamete pool depends only upon its own genotype, as much between ramets of different genets as with sib ramets. Thus, the genet's contribution to the gamete pool is not a constant but rather a function of the individual competitive abilities of the sib ramets. Slatkin (1984) concludes that under the soft selection model, newly arising, advantageous somatic mutations would have a much lower chance of being fixed than a comparable gametic mutation. In contrast, under the hard selection model, strong selection in favor of a somatic mutation could lead to its fixation more readily than for a comparable gametic mutation. But even under the hard selection model and assuming higher somatic mutation rates than gametic mutation rates, "the primary role of somatic mutations is to increase the effectiveness of selection among individuals. Somatic mutations are confined to a single individual for too short a time to have their evolutionary dynamics determined by the process acting within each individual unless selection is very strong" (Slatkin 1984:29).

Recently there have been numbers of papers written concerning the potential importance of somatic mutation in clonal plants with regard to adaptive evolution (Whitham and Slobodchikoff 1981; Gill and Halverson 1984; Silander 1985; Sutherland and Watkinson 1986). What all of these authors acknowledge is the availability of a considerable store of circumstantial evidence in support of their thesis of the evolutionary importance of somatic mutation but very little genetic or experimental evidence.[2]

In addition to random mutations in plants, there is at least one example of stress-induced heritable change. This is the famous case of the induction of heritable changes in *Linum usitatissimum* (flax) (Durrant 1962, 1971). Heritable changes were documented in some flax varieties after they had been grown for a single generation in the appropriate inducing environment. The inducing environments were simply different fertilizer treatments. Stable lines (genotrophs) induced by these environments differed from each other and from the parent in a number of characters. These included plant height, weight, total nuclear DNA, number of genes coding for the 18S ribosomal RNAs (rDNA), 5S RNA (5S DNA) and a number of repetitive sequences (see Cullis 1985 for review). Whether such changes are, in a sense, adaptive responses to the inducing environment or stochastic changes due to the interaction of unstable components of the plant genome to environmental stimuli is unknown. Also

2. In addition to the general thesis of the evolutionary importance of mutation in clonal plants, somatic mutations have been implicated as being important components of a plant's defenses against insect herbivores. Whitham and Slobodchikoff (1981) developed the hypothesis that because of shorter generation times and greater recombination potential, insect herbivores should be able to break the defenses of long-lived plants. Somatic mutations might allow plants to compensate for the recombination advantage of the insects. Although individual trees seem to present a mosaic of environments to herbivorous insects, the extent to which somatic mutation (or recombination) is important is unknown (Whitham 1981, 1983; Gill and Halverson 1984; Whitham, Williams, and Robinson 1984). It is interesting to note that the majority of these insect deterrents are characteristics of the epidermis, a tissue derived from the LI (chapter 5). In dicotyledonous plants the LII typically gives rise to the meiotic cells. If the hypothesis of Whitham and Slobodchikoff is correct, then perhaps the stratified meristem is a means of preserving the genetic integrity of the germ cell lines while allowing at least a portion of the somatic tissue to become genetically unstable.

whether this phenomenon occurs commonly in other plant groups is unknown. Considering the number of fertilizer experiments carried out annually on different crop plants and that the flax example was discovered over twenty years ago, it seems doubtful that genotroph induction is a general phenomenon in plants.

Implicit in the thesis that somatic mutation plays a significant role in adaptive evolution is the occurrence of somatic selection within clonal plants. Well-documented examples of somatic selection are rare. Probably the clearest example is the experiment conducted by Breese, Hayward, and Thomas (1965) with perennial rye grass, *Lolium perenne*. Clones of *L. perenne* from three populations were studied; population 3 was derived from an old pasture with no record of reseeding, population 17 from a pasture reseeded twenty years previously and population 6738 from a commercial variety which was nonpersistent. For each population two types of clones were established, c-clones which were ramets of the original populations and s-clones. The latter were derived from seedlings raised from intrapopulation crosses. Thus, c-clones were asexual derivatives from the original populations and s-clones had a sexual generation interpolated. Somatic selection was carried out on the rate of tillering (or the rate of asexual reproduction) in both s- and c-clones from all three population samples. Although negative results were obtained for all of the c-clones, two of the s-clones showed consistent responses to somatic selection (figure 11.3). It is notable that the clones responding to somatic selection were those that experienced a cycle of sexual reproduction and that had been derived from populations in which asexual reproduction was an important trait. The 6738c- and 6738s-clones showed no responses to somatic selection; both were derived from the nonpersistent rye grass strain.

The genetic or molecular basis of somatic selection in *L. perenne* is unknown. Whether the responses represent selection at the nuclear or cytoplasmic genomic levels or whether the results are due to the selection of persistent epigenetic patterns is not clear. The association with a cycle of sexuality (to generate heterozygosity which is later manifested through mitotic recombination?) is provocative. Clearly this is an area of plant population biology and genetics needing investigation.[3]

3. When population studies of clonal plants have documented intraclonal or intragenotype variation, generally causes other than mutation are invoked. For example, after calculating statistical adjustments for "c-effects" (see Libby and Jund 1962), Harberd (1967) documented intraclonal (or intragenotypic) variation for a number of traits of *Holcus mollis*. To quote Harberd: "Mutations are relatively rare events, so that presumably they could not account for the variability within all three genotypes." It may be that botanists have been preconditioned not to see the effects of somatic mutation.

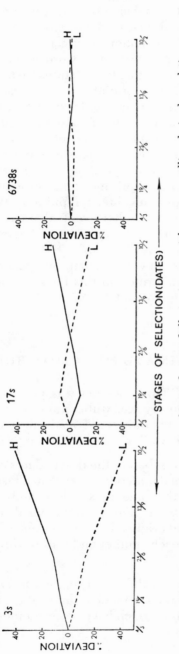

Figure 11.3. Average responses to somatic selection for rate of tillering in perennial ryegrass seedling-derived populations, expressed as percentage deviations from the population mean. (From Breese, Hayward, and Thomas 1965.)

Recently botanists have begun to consider the negative effects of mutation, i.e., mutational load (Wiens et al. 1987). The primary manifestation of mutational load is inbreeding depression. It is generally conceded that inbreeding depression is the only general factor in large outcrossing populations that can prevent the evolution of selfing in most plant species (Charlesworth and Charlesworth 1979). It has been hypothesized that the relationship between inbreeding depression and selfing determines, in large part, the direction of mating system evolution (Lande and Schemske 1985; Schemske and Lande 1985). Thus, accumulated mutational load may be the major determinant in plant mating system evolution, i.e., loadless species may evolve systems of selfing whereas species with high loads must be outcrossing. The critical question then is, what determines or limits load levels? Load levels may be a consequence of historical events at the population level (e.g., rapid reductions in population size, pollinator loss or any other events that may enforce episodes of inbreeding resulting in the purging of deleterious alleles from the gene pool) (Lande and Schemske 1985) as well as the various molecular, metabolic, developmental, ontogenetic, and reproductive characteristics described in this book that may influence the origin and preservation of mutations.

IS SEX THE ULTIMATE MUTATION BUFFER?

"Is sex necessary?" This question was posed and answered by Muller over half a century ago. Muller's answer was an evolutionary one, sexuality facilitated the spread of advantageous mutations in a population. To quote Muller:

> In asexual organisms, before the descendants can acquire a combination of beneficial mutations, these must first have occurred in succession within the same lines of descent. In sexual organisms, however, most of the beneficial mutations that occur simultaneously, or in different original lines of descent, can increase largely independently of each another and diffuse *through* one another, as it were. (1932:121)

Thus, sexual organisms will have faster rates of evolution than asexual organisms since recombination allows the combination of different adaptive mutants which have occurred independently in different lines of descent.

As is the case with many such universal questions, Muller's question, "Is sex necessary?" is not readily amenable to experimental analysis. Thus, answers cannot really be proved or disproved (although some can be shown to be illogical). Consequently, there have been a plethora of answers over the past half century. The reader is referred to Williams (1966, 1975); Felsenstein (1974); Maynard Smith (1978); Bell (1982); and Margolis and Sagan (1986) for reviews and discussions. In the present context, only those aspects of the problem of sexuality dealing with mutation and mutation buffering will be discussed. The majority of thinking about sexuality in these terms begins with an important consequence of asexual reproduction, a phenomenon called "Muller's ratchet." In a classic paper, Muller (1964) considered the problem of mutational load in the absence of recombination. He noted that the rate at which detrimental mutant genes originate would be balanced by the rate of their extinction through "genetic death" in both asexual and sexual species, but that mutant loss would be done in a less flexible manner in the asexual species. Asexual populations have a kind of built-in ratchet mechanism whereby a population can never be selected for fewer mutations than are present in the least-loaded lines in the population. For example, if an asexual population consists of individuals with two, three, four, and five mutations, selection can never result in a population in which individuals with a single mutation predominate (barring, of course, back mutation). If the individuals with two and three mutations are lost or new mutations occur in these lines, then, again, selection cannot restore the population to its previous condition. To quote Muller: ". . . an asexual population incorporates a kind of ratchet mechanism, such that it can never get to contain, in any of its lines, a load of mutations smaller than that already existing in its present least-loaded lines" (1964:8).

In contrast in sexual populations, recombination can result in the formation of genotypes with fewer mutations than the parents; consequently, selection can be effective in reducing mutational loads below the levels found in the least-loaded lines at any point in time (see Haigh 1978 for a theoretical study of the consequences of Muller's ratchet).

Muller's ratchet may also block the spread of favorable mutations in asexual populations. When an advantageous mutation (a → → A) occurs in an asexual population, it will probably occur in an individual that already has accumulated one or more deleterious mutations. If the selective advantage of A is less than the selective disadvantage of the deleterious mutations, then A will not increase in frequency. Conversely, if the selective advantage of A is greater than the disadvantage of the deleterious mutations, A may spread and carry the deleterious mutations to fixation (Manning and Thompson 1984). These conflicting effects may have important consequences for long-lived plants that have the possibility of the fixation of somatic mutations and ramet competition (as well as other forms of developmental selection).

Sexual reproduction is without genetical effect unless recombination occurs. Recombination in eukaryotes is due to two mechanisms: the independent assortment of genes on nonhomologous chromosomes and crossing over. Of course, genes on homologous chromosomes 50 or more map units apart will (because of crossing over) undergo independent assortment, and genes less than 50 map units apart will be linked in their inheritance. Consider the case of independent assortment in a diploid genotype heterozygous at k loci for deleterious mutant allels, where b is the nonmutant allele and B the mutant allele. The distribution and probabilities of the different gamete genotypes is found by expanding the binomial $(b + B)^k$. The value of $B = b = 0.5$ because of Mendel's law of segregation. The frequency of gametes with fewer than k mutants is $(1 - B^k)$. Thus, the majority of gamete genotypes have fewer deleterious mutations than the parent. As the following equation will demonstrate, approximately 50% of the offspring resulting from sexual reproduction will have fewer deleterious mutants than the parent, regardless of whether selfing or outcrossing occurs.

An individual with k mutants forms the following with regard to numbers of mutants per gamete,

$$(b + B)^k = \binom{k}{0}b^k + \binom{k}{1}b^{k-1}B + \binom{k}{2}b^{k-2}B^2 + \cdots + \binom{k}{k}B^k$$

where $\binom{k}{0}b^k$ is the frequency of gametes without mutants, $\binom{k}{1}b^{k-1}B$ the frequency of gametes with one mutant, and so on.

In figure 11.4 are shown the various zygote combinations possible from mating of two individuals with similar k values. The zygotes above the diagonal have fewer mutations than the parents, and the zygotes below the diagonal have more mutations than the parents. The zygotes on the diagonal have the same number of mutants as the parents, the frequency of this class is

$$D = \left[\binom{k}{0}b^k\right]^2 + \left[\binom{k}{1}b^{k-1}B\right]^2 + \left[\binom{k}{2}b^{k-2}B^2\right]^2 + \cdots + \left[\binom{k}{k}B^k\right]^2,$$

since $B = b = 1/2$,

$$D \cong (0.5^{2k})(k + 1)$$

$$D \cong 0, \quad \text{as } k \text{ increases.}$$

The frequency of zygotes (Z) with fewer mutants than the parents is therefore

$$Z = (1 - D)/2$$

$$Z \cong 0.5$$

This result is independent of selfing or crossing since only the number of mutants (k) in the F_1 zygote genotypes is considered, not their identity. Thus, sexuality can reduce the numbers of deleterious mutants below the parental levels with a 50% probability (it can also increase the number above the parental levels with a 50% probability). Is this characteristic, the capacity to reduce mutational load in a proportion of the offspring, sufficient for the evolution and maintenance of sexuality? The answer seems to be yes or no, depending upon how one calculates selection against individuals with multiple nonallelic deleterious mutants.

Since individual mutant alleles are in very low frequency under the mutational load model (chapter 10), the manner in which they reduce the viability of heterozygotes is more important than their homozygous effects. If s is the selective coefficient against a heterozygous genotype with a single deleterious mutation and $(1 - s)$ is the viability of this genotype, then each additional mutation may

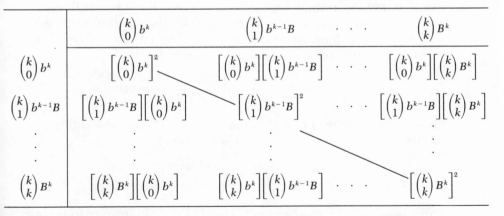

Figure 11.4. Various zygoter combinations from the mating of two parents with the same k values.

reduce viability in a multiplicative manner. Thus if $k = 2$, hetero-zygote viability is $(1 - s)(1 - s)$, consequently heterozygote viabil-ity is $(1 - s)^k$. Using such an exponential decrease in viability with increasing numbers of deleterious mutants, Maynard Smith (1978) calculated that when the population is small and when mutations are mildly deleterious, so s is in general small, and if the rate of mutation is high, then under these conditions sex is at an advantage. For large populations, Muller's ratchet will not operate; thus, whether a population is sexual or asexual has no influence on selec-tion against deleterious mutation.

Kondrashov (1982, 1984) questioned whether the multiplicative reduction in viability due to increasing numbers of deleterious mu-tations was a biologically valid assumption. Decreases in the fitness of heterozygotes with different numbers of deleterious mutations may follow a threshold model, so that only individuals having fewer deleterious mutations than some critical number are viable. Using this assumption, Kondrashov (1982, 1984) concluded that selection against deleterious mutations is the main factor maintaining sexual reproduction regardless of the population size. Under the threshold selection model, all individuals in an asexual population at equilib-rium will have exactly $k - 1$ mutations after selection. If mutations occur independently at all loci, the probability of offspring with $k - 1$ mutations (i.e., no new mutations) is given by the zero term of a Poisson distribution whose mean is U, the mutation rate. Thus, the probability of viable progeny from asexual parents is $l_a = e^{-U}$.

A sexual population under the threshold selection model will have the following characteristics. If λ is the average number of mutations per genome after recombination, then the frequencies of individuals with i mutations in their genome is given by a Poisson distribution whose mean is λ. The viable offspring in this distribution are those with $k - 1$ mutations, the mean number of mutations per genome in this population of viable offspring is (table 11.3)

$$u = \sum_{i=1}^{k-1} \frac{\lambda^i}{(i - 1)!} \left[\sum_{i=0}^{k-1} \frac{\lambda^i}{i!} \right]^{-1}.$$

Then in an equilibrium population $\lambda - u = U$. The probability of survival of the progeny in a sexual population is

$$l_s = e^{-\lambda} \sum_{i=0}^{k-1} \frac{\lambda^i}{i!}.$$

The advantage of sexual reproduction is l_s/l_a. The results in table 11.4 document the clear advantage of sexual populations with regard to fitness under the threshold selection model. These results apply to diploids as well as haploids, since the small viability declines of heterozygotes are more significant than the large viability declines associated with rare homozygotes (individual mutant alleles are assumed to be very rare whereas the overall mutation rate for mutant alleles is summed over many loci and high, U). Kondrashov (1984)

Table 11.3. Viable offspring resulting from sexual reproduction under the threshold selection model. Only genotypes with $k - 1$ mutants are viable. The average number of mutants in a viable offspring is u. (Modified from Kondrashov 1982.)

$$\text{Total offspring} = T = e^{-\lambda} \sum_{i=0}^{\infty} \frac{\lambda^i}{i!}$$

$$\text{Viable offspring} = V = e^{-\lambda} \sum_{i=0}^{k-1} \frac{\lambda^i}{i!}, \qquad \frac{V}{e^{-\lambda}} = Y$$

$$V + (T - V) = 1$$

Mutations	Probability	Weighted Product
0	$\dfrac{e^{-\lambda}}{V}$	0
1	$\dfrac{e^{-\lambda}\lambda}{V}$	$\dfrac{\lambda}{Y}$
2	$\dfrac{e^{-\lambda}\dfrac{\lambda^2}{2!}}{V}$	$\dfrac{\lambda^2}{Y}$
3	$\dfrac{e^{-\lambda}\dfrac{\lambda^3}{3!}}{V}$	$\dfrac{\dfrac{\lambda^3}{2!}}{Y}$
$\vdots$	$\vdots$	$\vdots$
$k - 1$	$\dfrac{e^{-\lambda}\dfrac{\lambda^{k-1}}{(k-1)!}}{V}$	$\dfrac{\dfrac{\lambda^{k-1}}{(k-2)!}}{Y}$

$$u = \frac{0 + \lambda + \lambda^2 + \dfrac{\lambda^3}{2!} + \cdots + \dfrac{\lambda^{k-1}}{(k-2)!}}{Y} = \frac{\displaystyle\sum_{i=1}^{k-1} \frac{\lambda^i}{i!}}{\displaystyle\sum_{i=0}^{k-1} \frac{\lambda^i}{i!}}$$

extended this reasoning to show that under the threshold selection model increased recombination is an effective mutation buffer.[4] He suggested that mutation buffering may be an answer to the question: Why does the genotype not congeal? (Turner 1967). Thus, mutation buffering should lead to a decrease in linkage and increase in the

Table 11.4. Average number of mutations per individual in a sexual population (u), average fitness of a sexual population (l_s), average fitness of an asexual population (l_a), and the relative advantage of sexual reproduction $\left(\dfrac{l_s}{l_a}\right)$ for different mutation rates (U) and selection thresholds, i.e., genotypes with $k - 1$ mutants are viable. (From Kondrashov 1982.)

	k	
	5	20
U	u	
0.5	2.271	13.06
2.0	3.010	15.22
	l_s	
0.5	0.8521	0.9399
2.0	0.4387	0.7180
	l_a	
0.5	0.6065	
2.0	0.1353	
	$\dfrac{l_s}{l_a}$	
0.5	1.405	1.550
2.0	3.2413	5.305

4. Murray and Szostak (1985) suggested that meiotic recombination may have been initially selected to decrease nondisjunction. Crossovers between paired homologs may provide the necessary flexible interconnections for more error-free chromosome disjunctions in the first meiotic division.

The mechanisms involved in homologous chromosome pairing may also increase the frequency of gene conversions. Gene conversion is the nonreciprocal transfer of DNA sequences from one gene to a related gene elsewhere in the genome; consequently, it is a kind of mutational event. Smithies and Powers (1986) suggest that gene conversions are the consequence of a general mechanism whereby DNA strand invasions enable chromosomes to find their homologs during meiosis. Thus, some aspects of meiosis may enhance genomic stability whereas other aspects promote instability.

number of linkage groups (each of which contains fewer genes). If Kondrashov is correct, then perhaps the question is not why a genome does not consist of a single chromosome but rather what restricts the number of chromosomes a genome may have? (See also Bernstein 1977 and Bernstein et al. 1985 for other aspects of the mutation/recombination problem.) In plants practically nothing is known about the viabilities of genotypes heterozygous for different numbers of nonallelic deleterious mutations or genomic mutation rates for deleterious mutations; thus, there is little hard evidence with which to test these theories.

TO CONCLUDE

An understanding of the consequences of mutation in plants must consider every level of plant organization, i.e., the molecular, cellular, meristematic, histological, the mechanisms of development of pattern and form, the different ontogenies of different organs and species (and how these developmental mechanisms and ontogenies influence diplontic selection), the various reproductive characteristics that may allow soft selection, the mutational load characteristics of populations, as well as the components of genome organization and recombination. All of these features must be integrated before one can begin to appreciate the significance of mutation in the biology of plants. In figure 11.5 the relationships between these variables are graphically shown. The central point is the chemostat equation outlined in chapter 2 (figure 2.6),

$$n^*/n = (\lambda/\gamma)t + K$$

where, in the present case, n^* is the number of mutant meiocytes, n the total population of meiocytes (both n^* and n are on a per plant basis and the fraction n^*/n is the mutation frequency), λ is the mutation rate per meristematic initial whose cell lineages give rise ultimately to the meiocyte (λ = mutations per cell per cell generation and is valid regardless of whether the meristem is stochastic or structured), γ is the cell generation time, t is the age of the organism and K is a constant. In light of the discussions in the previous chapters, this equation should be rewritten as

$$n^*/n = (\lambda f/\gamma)t + K$$

where f is the probability of mutation fixation and the product λf is the overall mutation rate (b). The overall mutation rate is a function of the molecular genomic characteristics influencing mutation occurrence (chapters 2 and 3) as well as the developmental parameters affecting mutation fixation (chapters 4, 5, 6, 7, and 8). In figure 11.5 the slopes represent the overall mutation rate (b), the generation length (a) is t/γ, the selective loss of mutants during sexual reproduction is given by (c), this selective loss occurs both through the various mechanisms of soft selection associated with sexual reproduction (chapter 9) as well as the more typical aspects of natural selection (e.g., competition, predation, ecological survival, etc.). The residual mutation frequency that persists from generation to generation represents a balance between the overall mutation rate and selection and is the constant K in the chemostat equation. Changes in generation length, overall mutation rate and selection will change residual mutation frequency (K). The mutational load (chapter 10) is equal to the overall mutation rate per gamete per organism generation (i.e., zygote to meiocyte), in our case approximately $\lambda ft/\gamma$ when γ is the total number of cell generations per

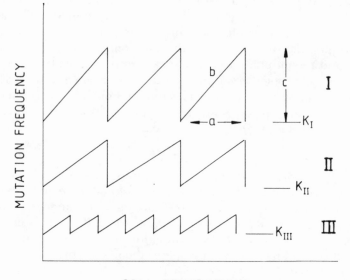

Figure 11.5. Chemostat effect and genetic variability. a: the organism generation in terms of cell generations from zygote to meiocytes; b: mutation rate per cell generation; c: amount of mutations lost through selection; K: the mutational equilibrium between mutation rate and selection. Three combinations of these parameters are shown: I, II, III. See text for further explanation.

organism generation. Mutants that are highly deleterious will have a lower selection-mutation equilibrium frequency than mutants that are less deleterious. Thus, as Haldane (1937) noted, the load (loss of fitness) is related to the mutation rate rather than to the magnitude of the deleterious effect of the mutations. Three idealized situations are depicted in figure 11.5: curve I is the case of high overall mutation rate and long generation length, curve II is the case of low overall mutation rate and long generation length, and curve III represents a reduction in both variables. These curves illustrate how molecular, developmental, and anatomical characteristics of a plant species can influence genetic variation.

PARTHIAN REMARK

The goal of this book has been to provoke the reader to consider the breadth of internal characteristics that can affect the nature of genetic variation in the gene pool as well as influence the patterns of evolutionary change. I trust that the reader has become impressed (as I have) with how much has been learned about the "internal" biology of plants and yet not so impressed that (s)he fails to recognize how little we really know. To me, Emily Dickinson expressed this point perfectly in the following poem.

> But nature is a stranger yet;
>   The ones that cite her most
> Have never passed her haunted house,
>   Nor simplified her ghost.
>
> To pity those that know her not
>   Is helped by the regret
> That those who know her, know her less
>   The nearer her they get.

References

Abrahamson, S., M. A. Bender, A. D. Conger, and S. Wolff. 1973. Uniformity of radiation-induced mutation rates among different species. *Nature* 245:460–461.

Adams, W. T. and R. J. Joly. 1980. Genetics of allozyme variants in loblolly pine. *J. Heredity* 71:33–40.

Ainsworth, C. C., J. S. Parker, and D. M. Horton. 1983. Chromosome variation and evolution in *Scilla autumnalis*. In P. E. Brandham and M. D. Bennett, eds., *Kew Chromosome Conference*, 2:261–268. London: Allen and Unwin.

Albertsen, M. C., T. M. Curry, R. G. Palmer, and C. E. Lamotte. 1983. Genetics and comparative growth morphology of fasciation in soybeans *Glycine max*. *Bot. Gaz.* 144:263–275.

Alff-Steinberger, C. 1987. Codon usage in *Homo sapiens:* Evidence for a coding pattern on the non-coding strand and evolutionary implications of dinucleotide discrimination. *J. theor. Biol.* 124:89–95.

Amiro, B. D. and J. R. Dugle. 1985. Temporal changes in boreal forest tree canopy cover along a gradient of gamma radiation. *Can. J. Bot.* 63:15–20.

Amo Rodriguez, S. Del, and A. Gomez-Pompa. 1976. Variability in *Ambrosia cumanensis* (Compositae). *Syst. Bot.* 1:360–372.

Anderson, E. 1949. *Introgressive Hybridization*. New York: Wiley.

Anderson, E. 1953. Introgressive hybridization. *Biol. Reviews* 28:280–307.

Andersson, E., R. Jansson, and D. Lindgren. 1974. Some results from second-generation crossings involving inbreeding in Norway spruce *(Picea abies)*. *Silvae Genetica* 23:34–43.

Antolin, M. F. and C. Strobeck. 1985. The population genetics of somatic mutation in plants. *Amer. Natur.* 126:52–62.

Apirion, D. and D. Zohary. 1961. Chlorophyll lethal in natural populations of the orchard grass *(Dactylis glomerata* L.). A case of balanced polymorphism in plants. *Genetics* 46:393–399.

Appels, R. and J. Dvorak. 1982a. The wheat ribosomal DNA spacer region: Its structure and variation in populations and among species. *Theor. Appl. Genet.* 63:337–348.

Appels, R. and J. Dvorak. 1982b. Relative rates of divergence of spacer and gene sequences within the rDNA region of species in the *Triticeae:* Implications for the maintenance of homogeneity of a repeated gene family. *Theor. Appl. Genet.* 63:361–365.

Asbeck, F. 1954. Sonnenlicht und Biogenese. *Strahlentherapie* 93:602–609.

Atwood, K. C. 1962. Problems of measurement of mutation rates. In W. J. Schull, *Mutations*, eds., pp. 1–77. Ann Arbor: University of Michigan Press.

Avery, A. G. and A. F. Blakeslee. 1943. Mutation rate in *Datura* seed which had been buried 39 years. *Genetics* 28:69–70.

Ayme, S. and A. Lippmand-Hand. 1982. Maternal age effect in aneuploidy: Does altered embryonic selection play a role? *Amer. J. Hum. Genet.* 34:558–565.

Babcock, E. B. and R. E. Clausen. 1927. *Genetics in Relation to Agriculture.* New York: McGraw Hill.

Balkema, G. H. 1972. Diplontic drift in chimeric plants. *Radiat. Bot.* 12:51–55.

Baltimore, D. 1985. Retroviruses and retrotransposons: The role of reverse transcription in shaping the eukaryotic genome. *Cell* 40:481–482.

Bannan, M. W. 1950. The frequency of anticlinal divisions in fusiform cambial cells of *Chamaecyparis. Amer. J. Bot.* 37:511–519.

Bannan, M. W. and I. L. Bayly. 1956. Cell size and survival in conifer cambium. *Can. J. Bot.* 34:769–776.

Barghoorn, E. S. 1964. Evolution of cambium in geologic time. In M. H. Zimmermann, ed., *The Formation of Wood in Forest Trees*, pp. 3–17. New York: Academic Press.

Barlow, P. W. 1978. The concept of the stem cell in the context of plant growth and development. In B. I. Lord, C. S. Potten, and R. J. Cole, eds., *Stem Cells and Tissue Homeostasis*, pp. 87–113. Cambridge: Cambridge University Press.

Barlow, P. W. and D. J. Carr. 1984. *Positional Controls in Plant Development.* Cambridge: Cambridge University Press.

Bateman, A. J. 1952. Self-incompatibility systems in angiosperms. I. Theory. *Heredity* 6:285–310.

Bateman, A. J. 1975. Simplification of palindromic telomere theory. *Nature* 253:379.

Batschelet, E. 1971. *Introduction to Mathematics for Life Scientists.* Berlin, New York: Springer-Verlag.

Bawa, K. S. and C. J. Webb. 1984. Flower, fruit, and seed abortion in tropical forest trees: Implications for the evolution of paternal and maternal reproductive patterns. *Amer. J. Bot.* 71:736–751.

Bedbrook, J. R., J. Jones, M. O'Dell, R. D. Thompson, and R. B. Flavell. 1980. A molecular description of telomeric heterochromatin in Secale species. *Cell* 19:545–560.

Bell, A. D. 1974. Rhizome organization in relation to vegetative spread in *Medeola virginiana. J. Arnold Arboretum* 55:458–468.

Bell, A. D. and P. B. Tomlinson. 1980. Adaptive architecture in rhizomatous plants. *Bot. J. Linn. Soc.* 80:125–160.

Bell, G. 1982. *The Masterpiece of Nature.* Berkeley: University of California Press.

Bendich, A. J. 1985. Plant mitochondrial DNA: Unusual variation on a common theme. In B. Hohn and E. S. Dennis, eds., *Genetic Flux in Plants*, pp. 111–138. Vienna, New York: Springer-Verlag.

Benne, R. and H. F. Tabak. 1986. Senescence comes of age. *Trends in Genetics* 2:147–148.

Bennett, M. D. 1976. DNA amount, latitude, and crop plant distribution. *Environ. Exp. Bot.* 16:93–108.

Bennett, M. D. 1985. Intraspecific variation in DNA amount and the nucleotypic dimension in plant genetics. In M. Freeling, ed., *Plant Genetics*, pp. 283–302. New York: Liss.

Bennett, M. D. and J. B. Smith. 1976. Nuclear DNA amounts in angiosperms. *Phil. Trans. Roy. Soc. Lond.* B 274:227–274.

Bergann, F. and L. Bergann. 1962. Uber Umschichtungen (Translokationen) an den Sprossscheiteln periklinaler Chimaren. *Der Zuchter* 32:110–119.

Bergann, F. and L. Bergann. 1983. Zur Entwicklungsgeschichte des Angiospermenblattes. 3. Uber unmaskierte Binnenfelder in den Blattspreiten periklinalchimarischer Buntheiten von *Elaeagnus pungens, Coprosma baueri, Ilex aquifolium, Hoya carnosa* und *Nerium oleander. Biol. Zbl.* 102:657–673.

Bernstein, H. 1977. Germ line recombination may be primarily a manifestation of DNA repair processes. *J. theor. Biol.* 69:371–380.

Bernstein, H., H. C. Byerly, F. A. Hopf, and R. E. Michod. 1985. Genetic damage, mutation, and the evolution of sex. *Science* 229:1277–1281.

Bertin, R. I. 1982. Paternity and fruit production in trumpet creeper *(Campsis radicans). Amer. Natur.* 119:694–709.

Bierhorst, D. W. 1971. *Morphology of Vascular Plants.* New York: Macmillan.

Bierhorst, D. W. 1977. On the stem apex, leaf initiation, and early leaf ontogeny in Filicalean ferns. *Amer. J. Bot.* 64:125–152.

Biggs, R. H., S. V. Kossuth, and A. H. Teramura. 1981. Response of 19 cultivars of soybeans to ultraviolet-B irradiance. *Physiol. Plant.* 53:19–26.

Bird, A. P. 1980. DNA methylation and the frequency of CpG in animal DNA. *Nucl. Acids Res.* 8:1499–1504.

Bird, A. P. 1983. DNA modification. In N. Maclean, S. P. Gregory, and R. A. Flavell, eds., *Eukaryotic Genes: Their Structure, Activity, and Regulation,* pp. 53–68. London: Butterworths.

Birky, C. W., Jr. 1987. Evolution and variation in plant chloroplast and mitochondrial genomes. In L. D. Gottlieb and S. K. Jain, eds., *Plant Evolutionary Biology.* London: Chapman and Hall (in press).

Birky, C. W., Jr., T. Maruyama, and P. Fuerst. 1983. An approach to population and evolutionary genetic theory for genes in mitochondria and chloroplasts, and some results. *Genetics* 103:513–527.

Blackburn, E. H. 1984. Telomeres: Do the ends justify the means? *Cell* 37:7–8.

Blackburn, E. H. and J. W. Szostak. 1984. The molecular structure of centromeres and telomeres. *Ann. Rev. Biochem.* 53:163–194.

Blackmore, S. and E. Toothill. 1984. *Dictionary of Botany.* Middlesex, England: Penguin Reference.

Blakeslee, A. F., A. D. Bergner, S. Satina, and E. W. Sinnott. 1939. Induction of periclinal chimeras in *Datura stramonium* by colchicine treatment. *Science* 89: 402.

Blau, H. M., G. K. Pavlath, E. C. Hardeman, C-P. Chiu, L. Silberstein, S. G. Webster, S. C. Miller, and C. Webster. 1985. Plasticity of the differentiated state. *Science* 230:758–766.

Blinkenberg, C., H. Brix, M. Schaffalitzky de Muckadell, and H. Vedel. 1958. Controlled pollinations in *Fagus. Silvae Genetica* 7:116–122.

Bloch, R. 1954. Abnormal plant growth. *Brookhaven Symp. Biol.* 6:41–54.

Bloch, R. 1965. Spontaneous and induced abnormal growth in plants. In W. Ruhland, ed., *Handbuch der Pflanzenphysiologie.* Vol. 15, part 2, Differentiation and Development, pp. 156–183. Berlin: Springer-Verlag.

Blomberg, C., J. Johansson, and H. Liljenstrom. 1985. Error propagation in *E. coli* protein synthesis. *J. theor. Biol.* 113:407–423.

Bloom, K., A. Hill, and E. Yeh. 1986. Structural analysis of a yeast centromere. *BioEssays* 4:100–104.

Bodmer, W. F. and L. L. Cavalli-Sforza. 1976. *Genetics, Evolution, and Man.* San Francisco: Freeman.

Boeke, J. D., D. J. Garfinkel, C. A. Styles, and G. R. Fink, 1985. Ty elements transpose through an RNA intermediate. *Cell* 40:491–500.

Boffey, S. A. 1985. The chloroplast division cycle and its relationship to the cell division cycle. In J. A. Bryant and D. Francis, eds., *The Cell Division Cycle in Plants* (Society for Experimental Biology, Seminar Series 26), pp. 233–246. Cambridge: Cambridge University Press.

Bohr, V. A., C. A. Smith, D. S. Okumoto, and P. C. Hanawalt. 1985. DNA repair in an active gene: Removal of pyrimidine dimers from the DHFR gene of CHO cells is much more efficient than in the genome overall. *Cell* 40:359–369.

Börner, T. 1986. Chloroplast control of nuclear gene function. *Endocyt. C. Res.* 3:265–274.

Börner, T., R. R. Mendel, and J. Schiemann. 1986. Nitrate reductase is not accumulated in chloroplast-ribosome-deficient mutants of higher plants. *Planta* 169:202–207.

Bower, F. O. 1926. *The Ferns (Filicales)*, vol. 2. Cambridge: Cambridge University Press.

Bowman, C. M. and T. A. Dyer. 1986. The location and possible evolutionary significance of small dispersed repeats in wheat ctDNA. *Curr. Genet.* 10:931–941.

Bradbeer, J. W., Y. E. Atkinson, T. Börner, and R. Hagemann. 1979. Cytoplasmic synthesis of plastid polypeptides may be controlled by plastid-synthesized RNA. *Nature* 279:816–817.

Bramlett, D. L. and T. W. Popham. 1971. Model relating unsound seed and embryonic lethal alleles in self-pollinated pines. *Silvae Genetica* 20:192–193.

Brandham, P. E. 1976. The frequency of spontaneous structural change. In K. Jones and P. E. Brandham, eds., *Current Chromosome Research*, pp. 77–87. Amsterdam: Elsevier/North Holland Biomedical Press.

Brandham, P. E. 1983. Evolution in a stable chromosome system. In P. E. Brandham and M. D. Bennett, eds., *Kew Chromosome Conference*, 2:251–260. London: Allen and Unwin.

Brandham, P. E. and M. A. T. Johnson. 1977. Population cytology of structural and numerical chromosome variants in the *Aloineae (Liliaceae)*. *Plant Syst. Evol.* 128:105–122.

Breese, E. L., M. D. Hayward, and A. C. Thomas. 1965. Somatic selection in perennial ryegrass. *Heredity* 20:367–379.

Brink, R. A. 1962. Phase change in higher plants and somatic cell heredity. *Q. Rev. Biol.* 37:1–22.

Brink, R. A. and D. C. Cooper. 1936. The mechanism of pollination in alfalfa *(Medicago sativa)*. *Amer. J. Bot.* 23:678–683.

Brisson, N. and D. P. S. Verma. 1982. Soybean leghemoglobin gene family: Normal, pseudo, and truncated genes. *Proc. Natl. Acad. Sci. USA* 79:4055–4059.

Brown, C. L., R. G. McAlpine, and P. P. Kormanik. 1967. Apical dominance and form in woody plants: A reappraisal. *Amer. J. Bot.* 54:153–162.

Brown, W. M. 1983. Evolution of animal mitochondrial DNA. In M. Nei and R. K. Koehn, eds., *Evolution of Genes and Proteins*, pp. 62–88. Sunderland, Mass.: Sinauer.

Brown, W. V., C. Heimsch, and W. H. P. Emery. 1957. The organization of the grass shoot apex and systematics. *Amer. J. Bot.* 44:590–595.

Brues, A. M. 1969. Genetic load and its varieties. *Science* 164:1130–1136.

Bryant, J. A. 1980. Biochemical aspects of DNA replication with particular reference to plants. *Biol. Rev.* 55:237–284.

Bryant, J. A. 1986. Enzymology of nuclear DNA replication in plants. *CRC Critical Reviews in Plant Sciences* 3:169–199.

Buchholz, J. T. 1922. Developmental selection in vascular plants. *Bot. Gaz.* 73:249–286.

Buchholz, J. T. and A. F. Blakeslee. 1930. Pollen tube growth and control of gametophytic selection in cocklebur, a 25-chromosome *Datura. Bot. Gaz.* 90:366–383.

Burnham, C. R. 1962. *Discussions in Cytogenetics.* Minneapolis: Burgess.

Burr, B. and F. A. Burr. 1982. *Ds* controlling elements of maize at the *Shrunken* locus are large and dissimilar insertions. *Cell* 29:977–986.

Burr, B., S. V. Evola, F. A. Burr, and J. S. Beckmann. 1983. The application of restriction fragment length polymorphism to plant breeding. In J. K. Setlow and A. Hollaender, eds., *Genetic Engineering*, 5:45–59. New York: Plenum Press.

Buss, L. W. 1983. Evolution, development, and the units of selection. *Proc. Natl. Acad. Sci. USA* 80:1387–1391.

Buvat, R. 1952. Structure, evolution et functionnement du méristème apical de quelques dicotyledones. *Ann. Sci. Nat. Bot. Ser.* 13:199–300.

Cairns, J. 1975. Mutation, selection and the natural history of cancer. *Nature* 255:197–200.

Cairns, J. 1978. *Cancer: Science and Society.* San Francisco: Freeman.

Caldwell, M. M. 1968. Solar ultraviolet radiation as an ecological factor for alpine plants. *Ecological Monographs* 38:243–268.

Caldwell, M. M. 1971. Solar UV irradiation and the growth and development of higher plants. In A. C. Giese, ed., *Photophysiology*, 6:131–177. New York: Academic Press.

Caldwell, M. M. 1979. Plant life and ultraviolet radiation: Some perspective in the history of the Earth's UV climate. *BioScience* 29:520–525.

Caldwell, M. M., R. Robberecht, and W. D. Billings. 1980. A steep latitudinal gradient of solar ultraviolet-B radiation in the arctic-alpine life zone. *Ecology* 61:600–611.

Callen, D. F. 1982. Metabolism of chemicals to mutagens by higher plants and fungi. In E. J. Klekowski, Jr., ed., *Environmental Mutagenesis, Carcinogenesis, and Plant Biology*, 1:33–65. New York: Praeger.

Carpenter, R., C. Martin, and E. S. Coen. 1987. Comparison of genetic behaviour of the transposable element *Tam3* in two unlinked pigment loci in *Antirrhinum majus. Mol. Gen. Genet.* 207:82–89.

Carteledge, J. L. and A. F. Blakeslee. 1934. Mutation rate increased by aging seeds as shown by pollen abortion. *Proc. Natl. Acad. Sci. USA* 20:103–110.

Carteledge, J. L. and A. F. Blakeslee. 1935. Mutation rate from old *Datura* seeds. *Science* 81:492–493.

Carteledge, J. L., M. J. Murray, and A. F. Blakeslee. 1935. Increased mutation rate from aged *Datura* pollen. *Proc. Natl. Acad. Sci. USA* 21:597–600.

Casper, B. B. and D. Wiens. 1981. Fixed rates of random ovule abortion in *Cryptantha flava* (Boraginaceae) and its possible relation to seed dispersal. *Ecology* 62:866–869.

Cavalier-Smith, T. 1974. Palindromic base sequences and replication of eukaryote chromosome ends. *Nature* 250:467–470.

Cavalier-Smith, T. 1978. Nuclear volume control by nucleoskeletal DNA, selection for cell volume and cell growth rate, and the solution of the DNA C-value paradox. *J. Cell Sci.* 34:247–278.

Cavalier-Smith, T., ed. 1985. *The Evolution of Genome Size.* New York: Wiley.

Chamberlain, C. J. 1935. *Gymnosperms: Structure and Evolution.* Chicago: University of Chicago Press.

Chan, J. Y. H., F. F. Becker, J. German, and J. H. Ray. 1987. Altered DNA ligase I activity in Bloom's syndrome cells. *Nature* 325:357–359.

Chandler, V. L. and V. Walbot. 1986. DNA modification of a maize transposable element correlates with loss of activity. *Proc. Natl. Acad. Sci. USA* 83:1767–1771.

Charlesworth, B. and D. Charlesworth. 1979. The evolutionary genetics of sexual systems in flowering plants. *Proc. Roy. Soc. London B* 205:513–530.

Charnov, E. L. 1979. Simultaneous hermaphroditism and sexual selection. *Proc. Natl. Acad. Sci. USA* 76:2480–2484.

Cheadle, V. I. 1937. Secondary growth by means of a thickening ring in certain monocotyledons. *Bot. Gaz.* 98:535–555.

Chenou, E., J. Kuligowski, and M. Ferrand. 1986. Alterations de l'embryogenese de *Marsilea vestita* provoquees par un faible abaissement de témpérature et phenomenes de recuperation après retour aux conditions initiales. *Can. J. Bot.* 64:784–792.

Chomet, P. S., S. Wessler, and S. L. Dellaporta. 1987. Inactivation of the maize transposable element *Activator* (*Ac*) is associated with its DNA modification. *The EMBO Journal* 6:295–302.

Cichan, M. A. 1985a. Vascular cambium and wood development in Carboniferous plants. I. Lepidodendrales. *Amer. J. Bot.* 72:1163–1176.

Cichan, M. A. 1985b. Vascular cambium and wood development in Carboniferous plants. II. *Sphenophyllum plurifoliatum* Williamson and Scott (Sphenophyllales). *Bot. Gaz.* 146:395–403.

Cichan, M. A. 1986. Vascular cambium and wood development in Carboniferous plants. III. *Arthropitys* (Equisetales; Calamitaceae). *Can. J. Bot.* 64:688–695.

Clark, A. M. 1982. Endogenous mutagens in green plants. In E. J. Klekowski, Jr., ed., *Environmental Mutagenesis, Carcinogenesis, and Plant Biology*, 1:97–132. New York: Praeger.

Clayton, D. A. 1982. Replication of animal mitochondrial DNA. *Cell* 28:693–705.

Cleland, R. E. 1972. *Oenothera, Cytogenetics, and Evolution*. London: Academic Press.

Clowes, F. A. L. 1961. *Apical Meristems*. Oxford: Blackwell.

Coe, E. H. Jr. and M. G. Neuffer. 1978. Embryo cells and their destinies in the corn plant. In S. Subtelny and I. M. Sussex, eds., *The Clonal Basis of Development*, pp. 113–129. New York: Academic Press.

Coen, E. S. and G. A. Dover. 1982. Multiple Pol I initiation sequences in rDNA spaces of *Drosophila melanogaster*. *Nucleic Acid Res.* 10:7017–7026.

Coen, E. S., R. Carpenter, and C. Martin. 1986. Transposable elements generate novel spatial patterns of gene expression in *Antirrhinum majus*. *Cell* 47:285–296.

Coen, E., T. Strachan, S. Brown, and G. Dover. 1983. On the limited independence of chromosome evolution. In P. E. Brandham and M. D. Bennett, eds., *Kew Chromosome Conference II*, pp. 295–303. London: Allen and Unwin.

Coen, E. S., J. M. Thoday, and G. Dover. 1982. Rate of turnover of structural variants in the rDNA gene family of *Drosophila melanogaster*. *Nature* 295:564–568.

Coles, J. F. and D. P. Fowler. 1976. Inbreeding in neighboring trees in two white spruce populations. *Silvae Genetica* 25:29–34.

Conway, E. 1957. Spore production in bracken (*Pteridium aquilinum* (L.) Kuhn). *J. Ecol.* 45:273–284.

Cooper, D. C., R. A. Brink, and H. R. Albrecht. 1937. Embryo mortality in relation to seed formation in alfalfa (*Medicago sativa*). *Amer. J. Bot.* 24:203–213.

Coulondre, C., J. H. Miller, P. J. Farabaugh, and W. Gilbert. 1978. Molecular basis of base substitution hotspots in *Escherichia coli*. *Nature* 274:775–780.

Coulter, J. M. and C. J. Chamberlain. 1912. *Morphology of Angiosperms* (*Morphology of Spermatophytes*, part 2). New York: Appleton.

Cousens, M. I. 1979. Gametophyte ontogeny, sex expression, and genetic load as measures of population divergence in *Blechnum spicant*. *Amer. J. Bot.* 66:116–132.

Cox, E. C. 1976. Bacterial mutator genes and the control of spontaneous mutation. *Ann. Rev. Genet.* 10:135–156.

Cox, E. C. and T. C. Gibson. 1974. Selection for high mutation rates in chemostats. *Genetics* 77:169–184.

Cram, W. H. 1955. Self-compatibility of *Caragna arborescens* Lam. *Can. J. Bot.* 33:149–155.

Crist, K. C. and D. R. Farrar. 1983. Genetic load and long-distance dispersal in *Asplenium platyneuron. Can. J. Bot.* 61:1809–1814.

Crosland, M. W. J. and R. H. Crozier. 1986. *Myrmecia pilosula*, an ant with only one pair of chromosomes. *Science* 231:1278.

Crow, J. F. 1958. Some possibilities for measuring selection intensities in man. *Human Biology* 30:1–13.

Crow, J. F. and M. Kimura. 1970. *An Introduction to Population Genetics Theory.* New York: Harper and Row.

Crumpacker, D. W. 1967. Genetic loads in maize (*Zea mays* L.) and other cross-fertilized plants and animals. *Evol. Biol.* 1:306–424.

Cullis, C. A. 1985. Sequence variation and stress. In B. Hohn and E. S. Dennis, eds., *Genetic Flux in Plants*, pp. 157–168. Vienna, New York: Springer-Verlag.

Cullis, C. A. 1986. Phenotypic consequences of environmentally induced changes in plant DNA. *Trends in Genetics* 2:307–309.

Cullis, C. A. and W. Cleary. 1985. Fluidity of the flax genome. In M. Freeling, ed., *Plant Genetics*, pp. 303–310. New York: Liss.

Cullis, C. A. and W. Cleary. 1986. Rapidly varying DNA sequences in flax. *Can J. Genet. Cytol.* 28:252–259.

Cullman, G. and J. M. Labouygues. 1984. The mathematical logic of life. *Origins of Life* 14:747–755.

Cumbie, B. G. 1963. The vascular cambium and xylem development in *Hibiscus lasiocarpus. Amer. J. Bot.* 50:944–951.

Cumbie, B. G. 1969. Developmental changes in the vascular cambium of *Polygonum lapathifolium. Amer. J. Bot.* 56:139–146.

D'Amato, F. 1977. *Nuclear Cytology in Relation to Development.* Cambridge: Cambridge University Press.

D'Amato, F. and O. Hoffmann-Ostenhof. 1956. Metabolism and spontaneous mutations in plants. *Advances in Genetics* 3:1–28.

Dancis, B. M. and G. P. Holmquist. 1979. Telomere replication and fusion in eukaryotes. *J. theor. Biol.* 78:211–224.

Darlington, C. D. 1937. *Recent Advances in Cytology*, 2d ed. London: J. and A. Churchill.

Darwin, C. 1883. *The Effects of Cross- and Self-fertilization in the Vegetable Kingdom.* New York: Appleton.

Davidson, D. 1960. Meristem initial cells in irradiated roots of *Vicia faba. Ann. Bot. London.* 24:287–295.

Davis, G. L. 1966. *Systematic Embryology of the Angiosperms.* New York: Wiley.

Dellaporta, S. L. and P. S. Chomet. 1985. The activation of maize controlling elements. In B. Hohn and E. S. Dennis, eds., *Genetic Flux in Plants*, pp. 169–216. Vienna, New York: Springer-Verlag.

De Nettancourt, D. 1977. *Incompatibility in Angiosperms.* Berlin: Springer-Verlag.

Denhardt, D. T. and E. A. Faust. 1985. Eukaryotic DNA replication. *BioEssays* 2:148–154.

DePamphilis, M. L. and P. M. Wassarman. 1980. Replication of eukaryotic chromosomes: A close-up of the replication fork. *Ann. Rev. Biochem.* 49:627–666.

Dermen, H. 1947. Periclinal cytochimeras and histogenesis in cranberry. *Amer. J. Bot.* 34:32–43.

Dermen, H. 1951. Ontogeny of tissues in stem and leaf of cytochimeral apples. *Amer. J. Bot.* 38:753–760.

Dermen, H. 1953. Periclinal cytochimeras and origin of tissues in stem and leaf of peach. *Amer. J. Bot.* 40:154–168.

Dermen, H. 1969. Directional cell division in shoot apices. *Cytologia* 34:541–558.

Diers, L. 1971. Übertragung von Plastiden durch den Pollen bei *Antirrhinum majus*. II. Der Einfluss verschiedener Temperaturen auf die Zahl der Schecken. *Mol. Gen. Genet.* 113:150–153.

DiNardo, S., K. Voelkel, and R. Sternglanz. 1984. DNA topoisomerase II mutant of *Saccharomyces cerevisiae:* Topoisomerase II is required for segregation of daughter molecules at the termination of DNA replication. *Proc. Natl. Acad. Sci.* 81:2616–2620.

Dinter-Gottlieb, G. 1986. Viroids and virusoids are related to group I introns. *Proc. Natl. Acad. Sci. USA* 83:6250–6254.

Dobzhansky, Th. 1955. A review of some fundamental concepts and problems of population genetics. *Cold Spring Harbor Symp. Quant. Biol.* 20:1–15.

Dobzhansky, Th. 1970. *Genetics of the Evolutionary Process.* New York: Columbia University Press.

Dobzhansky, Th., F. J. Ayala, G. L. Stebbins, and J. W. Valentine. 1977. *Evolution.* San Francisco: Freeman.

Doolittle, R. F. 1981. Similar amino acid sequences: Chance or common ancestry? *Science* 214:149–159.

Doolittle, W. F. and C. Sapienza. 1980. Selfish genes, the phenotype paradigm, and genome evolution. *Nature* 284:601–603.

Döring, H.-P. 1985. Plant transposable elements. *BioEssays* 3:164–171.

Döring, H.-P. and P. Starlinger. 1984. Barbara McClintock's controlling elements: Now at the DNA level. *Cell* 39:253–259.

Döring, H.-P. and P. Starlinger. 1986. Molecular genetics of transposable elements in plants. *Ann. Rev. Genet.* 20:175–200.

Döring, H.-P., R. Garber, B. Nelson, and E. Tillmann. 1985. Transposable elements Ds and chromosomal rearrangements. In M. Freeling, ed., *Plant Genetics*, pp. 355–367. New York: Liss.

Dorne, A.-J., J.-P. Carde, J. Joyard, T. Börner, and R. Douce. 1982. Polar lipid composition of a plastid ribosome-deficient barley mutant. *Plant Physiol.* 69:1467–1470.

Dourado, A. M. and E. H. Roberts. 1984. Phenotypic mutations induced during storage in barley and pea seeds. *Ann. Bot.* 54:781–790.

Dover, G. 1980. Ignorant DNA? *Nature* 285:618–620.

Dover, G. 1982. Molecular drive: A cohesive mode of species evolution. *Nature* 299:111–117.

Dover, G. A. 1986. Molecular drive in multigene families: How biological novelties arise, spread, and are assimilated. *Trends in Genetics* 2:159–165.

Dover G. A. and R. B. Flavell. 1984. Molecular coevolution: DNA divergence and the maintenance of function. *Cell* 38:622–623.

Drake, J. W. 1969. Comparative rates of spontaneous mutation. *Nature* 221:1132.

Drake, J. W., B. W. Glickman, and L. S. Ripley. 1983. Updating the theory of mutation. *American Scientist* 71:621–630.

Drouin, G. and G. A. Dover. 1987. A plant processed pseudogene. *Nature* 328:557–558.

Drumm-Herrel, H. and H. Mohr. 1981. A novel effect of UV-B in a higher plant *(Sorghum vulgare)*. *Photochemistry and Photobiology* 33:391–398.

Dufton, M. J. 1983. The significance of redundancy in the genetic code. *J. theor. Biol.* 102:521–526.

Dufton, M. J. 1986. Genetic code redundancy and its differential influence on the evolution of protein interiors versus exteriors. *J. theor. Biol.* 122:231–236.

Dugle, J. R. and J. L. Hawkins. 1985. Leaf development and morphology in ash: Influence of gamma radiation. *Can. J. Bot.* 63:1458–1468.

Dugle, J. R. and K. R. Mayoh. 1984. Responses of 56 naturally-growing shrub taxa to chronic gamma irradiation. *Environ. Exp. Bot.* 24:267–276.

Dunham, V. L. and J. A. Bryant. 1985. Enzymic controls of DNA replication. In J. A. Bryant and D. Francis, eds., *The Cell Division Cycle in Plants* (Society for Experimental Biology Seminar Series 26), pp. 37–59. Cambridge: Cambridge University Press.

Dunham, V. L. and J. A. Bryant. 1986. DNA polymerase activities in healthy and cauliflower mosaic virus-infected turnip *(Brassica rapa)* plants. *Ann. Bot. London.* 57:81–89.

Durrant, A. 1962. The environmental induction of heritable changes in *Linum*. *Heredity* 17:27–61.

Durrant, A. 1971. Induction and growth of flax genotrophs. *Heredity* 27:277–298.

Dvořák, J. and R. Appels. 1982. Chromosome and nucleotide sequence differentiation in genomes of polyploid *Triticum* species. *Theor. Appl. Genet.* 63:349–360.

Dyer, A. F. 1976. Modification and errors of mitotic cell division in relation to differentiation. In M. M. Yeoman, ed., *Cell Division in Higher Plants*, pp. 199–249. London: Academic Press.

Earnshaw, W. C. and M. S. Heck. 1985. Localization of topoisomerase II in mitotic chromosomes. *J. Cell Biol.* 100:1716–1725.

Earnshaw, W. C. and U. K. Laemmli. 1983. Architecture of metaphase chromosomes and chromosome scaffolds. *J. Cell Biol.* 96:84–93.

Edwards, K. L. and B. G. Pickard. In press. Detection and transduction of physical stimuli in plants. In H. Greppin, B. Millet, and E. Wagner, eds., *The Cell Surface and Signal Transduction*, pp. Berlin: Springer-Verlag.

Eiche, V. 1955. Spontaneous chlorophyll mutations in scots pine *(Pinus silvestris* L.). *Meddelanden Fran Statens Skogsforskningsinstitut* 45(13):1–69.

Eigen, M. and P. Schuster. 1977. The hypercycle: A principle of natural self-organization. Part A: Emergence of the hypercycle. *Naturwissenschaften* 64:541–565.

Eldridge, K. G. and A. R. Griffin. 1983. Selfing effects in *Eucalyptus regnans*. *Silvae Genetica* 32:216–221.

Epp, M. D. 1974. The homozygous genetic load in mutagenized populations of *Oenothera hookeri* T. and G. *Mutat. Res.* 22:39–46.

Eriksson, G., I. Ekberg, L. Ehrenberg, and B. Bevilacqua. 1966. Genetic changes induced by semiacute γ-irradiation of pollen mother cells in *Larix leptolepis* (Sieb et Zucc.) Gord. *Hereditas* 55:213–226.

Eriksson, G., B. Schelander, and V. Akebrand. 1973. Inbreeding depression in an old experimental plantation of *Picea abies*. *Hereditas.* 73:185–194.

Esau, K. 1960. *Anatomy of Seed Plants*. New York: Wiley.

Esau, K. 1976. *Anatomy of Seed Plants*. 2d ed. New York: Wiley.

Esau, K. 1977. *Anatomy of Seed Plants*. 2d ed. New York: Wiley.

Evans, D. A. and E. F. Paddock. 1979. Mitotic crossing-over in higher plants. In W. R. Sharp, P. O. Larsen, E. F. Paddock, and V. Raghaven, eds., *Plant Cell Tissue Culture: Principles and Applications*, pp. 315–351. Columbus: Ohio State University Press.

Evans, D. A., W. R. Sharp, and H. P. Medina-Filho. 1984. Somaclonal and gametoclonal variation. *Amer. J. Bot.* 71:759–774.

Evans, H. J. and A. H. Sparrow. 1961. Nuclear factors affecting radiosensitivity. II. Dependence on nuclear and chromosome structure and organization. *Brookhaven Symp. Biol.* 14:101–127.

Fahn, A. 1974. *Plant Anatomy*. 2d ed. New York: Pergamon Press.

Falke, L., K. L. Edwards, S. Misler, and B. G. Pickard. 1986. A mechanotransductive ion channel in patches from cultured tobacco cell plasmalemma. *Plant Physiol. Suppl.* 80:9.

Fedoroff, N. V. 1983. Controlling elements in maize. In J. A. Shapiro, ed., *Mobile Genetic Elements*, pp. 1–63. New York: Academic Press.

Felsenstein, J. 1974. The evolutionary advantage of recombination. *Genetics* 78:737–756.

Fersht, A. R., J. W. Knill-Jones, and W.-C. Tsui. 1982. Kinetic basis of spontaneous mutation: Misinsertion frequencies, proofreading specificities, and cost of proofreading by DNA polymerases of *Escherichia coli*. *J. Mol. Biol.* 156:37–51.

Figureau, A. and M. Pouzet. 1984. Genetic code and optimal resistance to the effects of mutations. *Origins of Life* 14:579–588.

Finnegan, D. J. 1983. Retroviruses and transposable elements—which came first? *Nature* 302:105–106.

Flavell, A. J. 1984. Role of reverse transcription in the generation of extrachromosomal *copia* mobile genetic elements. *Nature* 310:514–516.

Flavell, R. B. 1980. The molecular characterization and organization of plant chromosomal DNA sequences. *Ann. Rev. Plant Physiol.* 31:569–596.

Flavell, R. B. 1982. Amplification deletion and rearrangement: major sources of variation during species divergence. In G. A. Dover and R. B. Flavell, eds., *Genome Evolution*, pp. 301–324. London: Academic Press.

Flavell, R. B. 1985. Repeated sequences and genome change. In B. Hohn and E. S. Dennis, eds., *Genetic Flux in Plants*, pp. 139–156. Wien, New York: Springer-Verlag.

Flavell, R. B. 1986. The structure and control of expression of ribosomal RNA genes. In B. J. Miflin, ed., *Oxford Surveys of Plant Molecular and Cell Biology*, 3:251–274. Oxford: Oxford University Press.

Flavell, R. B., M. O'Dell, D. B. Smith, and W. F. Thompson. 1985. Chromosome architecture: The distribution of recombination sites, the structure of ribosomal DNA loci, and the multiplicity of sequences containing inverted repeats. In L. van Vloten-Doting, G. S. P. Groot, and T. C. Hall, eds., *Molecular Form and Function of the Plant Genome*, pp. 1–14. New York and London: Plenum Press.

Foard, D. E. 1971. The initial protrusion of a leaf primordium can form without concurrent periclinal cell divisions. *Can. J. Bot.* 49:1601–1603.

Foard, D. E., A. H. Haber, and T. N. Fishman. 1965. Initiation of lateral root primordia without completion of mitosis and without cytokinesis in uniseriate pericycle. *Amer. J. Bot.* 52:580–590.

Foster, A. S. 1940. Further studies on zonal structure and growth of the shoot apex of *Cycas revoluta. Amer. J. Bot.* 27:487–501.

Foster, A. S. and E. M. Gifford Jr. 1959. *Comparative Morphology of Vascular Plants.* San Francisco: Freeman.

Fowler, D. P. 1964. Pre-germination selection against a deleterious mutant in red pine. *For. Sci.* 10:335–336.

Fowler, D. P. 1965a. Effects of inbreeding in red pine, *Pinus resinosa* Ait. II. Pollination studies. *Silvae Genetica* 14:12–23.

Fowler, D. P. 1965b. Effects of inbreeding in red pine, *Pinus resinosa* Ait. III. Factors affecting natural selfing. *Silvae Genetica* 14:37–45.

Fowler, D. P. 1965c. Effects of inbreeding in red pine, *Pinus resinosa* Ait. IV. Comparison with other northeastern *Pinus* species. *Silvae Genetica* 14:76–81.

Franklin, E. C. 1969. Mutant forms founded by self-pollination in loblolly pine. *J. Heredity* 60:315–320.

Franklin, E. C. 1970. Survey of mutant forms and inbreeding depression in species of the family Pinaceae. *USDA Forest Service Research Paper SE* 61:1–21.

Franklin, E. C. 1972. Genetic load in loblolly pine. *Amer. Natur.* 106:262–265.

Freeling, M. 1984. Plant transposable elements and insertion sequences. *Ann. Rev. Plant Physiol.* 35:277–298.

Freeling, M. and S. Hake. 1985. Developmental genetics of mutants that specify knotted leaves in maize. *Genetics* 111:617–634.

Fukshansky, L. and E. Wagner. 1985. Anwendung von Variationsprinzipien zur Beschreibung der hierarchischen Organisation des Energiestoffwechsels. *Ber. Deutsch Bot. Ges.* 98:25–34.

Furner, I. J., G. A. Huffman, R. M. Amasino, D. J. Garfinkel, M. P. Gordon, and E. W. Nester. 1986. An *Agrobacterium* transformation in the evolution of the genus *Nicotiana. Nature* 319:422–427.

Furnier, G. R., P. Knowles, M. A. Aleksiuk, and B. P. Dancik. 1986. Inheritance and linkage of allozymes in seed tissue of whitebark pine. *Can. J. Genet. Cytol.* 28:601–604.

Gabriel, W. J. 1967. Reproductive behavior in sugar maple: Self-compatibility, cross-compatibility, agamospermy, and agamocarpy. *Silvae Genetica* 16:165–168.

Gallant, J. A. and J. Prothero. 1980. Testing models of error propagation. *J. theor. Biol.* 83:561–578.

Ganders, F. R. 1972. Heterozygosity for recessive lethals in homosporous fern populations: *Thelypteris palustris* and *Onoclea sensibilis. Bot. J. Linn. Soc.* 65:211–221.

Gasser, S. M., T. Laroche, J. Falquet, E. Boy de la Tour, and U. K. Laemmli. 1986. Metaphase chromosome structure. Involvement of topoisomerase II. *J. Mol. Biol.* 188:613–629.

Gaul, H. 1965. Selection in M⟨d1⟩ generation after mutagenic treatment of barley seeds. In J. Veleminsky and T. Gichner, eds., *Induction of Mutations and the Mutation Process*, pp. 62–71. Prague: Czechoslovak Academy of Sciences.

Geburek, Th. 1986. Some results of inbreeding depression in serbian spruce (*Picea omorika* (Panc.) Purk.). *Silvae Genetica* 35:169–172.

German, J. 1972. Genes which increase chromosomal instability in somatic cells and predispose to cancer. *Prog. Med. Genet.* 8:61–101.

Gerstel, D. U. and J. A. Burns. 1966. Chromosomes of unusual length in hybrids between two species of *Nicotiana*. In C. D. Darlington and K. R. Lewis, eds., *Chromosomes Today*, 1:41–56. London: Oliver and Boyd.

Gifford, E. M., Jr. 1954. The shoot apex in angiosperms. *Bot. Rev.* 20:477–529.

Gifford, E. M., Jr. and G. E. Corson. 1971. The shoot apex in seed plants. *Bot. Rev.* 37:143–229.

Gifford, E. M. and E. Kurth. 1983. Quantitative studies of the vegetative shoot apex of *Equisetum scirpoides. Amer. J. Bot.* 70:74–79.

Giles, N. 1940. Spontaneous chromosome aberrations in *Tradescantia. Genetics* 25:69–87.

Gill, D. E. and T. G. Halverson. 1984. Fitness variation among branches within trees. In B. Shorrocks, ed., *Evolutionary Ecology*, pp. 105–116. Oxford: Blackwell Scientific Publications.

Glickman, B. W. and M. Radman. 1980. *Escherichia coli* mutator mutants deficient in methylation-instructed DNA mismatch correction. *Proc. Natl. Acad. Sci. USA* 77:1063–1067.

Goebl, M. G. and T. D. Petes. 1986. Most of the yeast genomic sequences are not essential for cell growth and division. *Cell* 46:983–992.

Good, C. W. and T. N. Taylor. 1972. The ontogeny of Carboniferous articulates: The apex of *Sphenophyllum. Amer. J. Bot.* 59:617–626.

Goodman, M. F. and E. W. Branscomb. 1986. DNA replication fidelity and base mispairing mutagenesis. In T. B. L. Kirkwood, R. F. Rosenberger, and D. J. Galas, eds., *Accuracy in Molecular Processes*, pp. 191–232. New York: Chapman and Hall.

Gorter, C. J. 1965. Origin of fasciation. In W. Ruhland, ed., *Handbuch der Pflanzenphysiologie*. Vol. 15, part 2, Differentiation and Development, pp. 330–351. Berlin: Springer-Verlag.

Gottlieb, L. D. 1984. Genetics and morphological evolution in plants. *Amer. Natur.* 123:681–709.

Grant, V. 1963. *The Origin of Adaptations*. New York and London: Columbia University Press.

Grant, V. 1975. *Genetics of Flowering Plants*. New York and London: Columbia University Press.

Grant, V. 1981. *Plant Speciation*. 2d ed. New York: Columbia University Press.

Gray, A. 1887. *The Elements of Botany (Gray's Lessons in Botany)*. New York: American Book Company.

Green, P. B. 1980. Organogenesis—a biophysical view. *Ann. Rev. Plant Physiol.* 31:51–82.

Green, P. B. 1985. Surface of the shoot apex: A reinforcement-field theory for phyllotaxis. *J. Cell Sci. Suppl.* 2:181–201.

Greenblatt, I. M. 1984. A chromosome replication pattern deduced from pericarp phenotypes resulting from movements of the transposable element, Modulator, in maize. *Genetics* 108:471–485.

Greenblatt, I. M. and R. A. Brink. 1962. Twin mutations in medium variegated pericarp maize. *Genetics* 47:489–501.

Griffin, A. R. and D. Lindgren. 1985. Effect of inbreeding on production of filled seed in *Pinus radiata*—experimental results and a model of gene action. *Theor. Appl. Genet.* 71:334–343.

Griffin, A. R., G. F. Moran, and Y. J. Fripp. 1987. Preferential outcrossing in *Eucalyptus regnans* F. Muell. *Aust. J. Bot.* 35: (in press).

Griffith, M. M. 1952. The structure and growth of the shoot apex in *Araucaria*. *Amer. J. Bot.* 39:253–263.

Grodzinsky, D. B. and I. M. Gudkov. 1982. Heterogeneity of meristems: The basis of higher plant reliability. Academy of Sciences of the Ukranian S.S.R., *Physiology and Biochemistry of Cultivated Plants* 14:107–118 (translated from Russian).

Gruenbaum, Y., T. Naveh-Many, H. Cedar, and A. Razin. 1981. Sequence specificity of methylation in higher plant DNA. *Nature* 292:860–862.

Guédes, M. 1982. A simpler morphological system of tree and shrub architecture. *Phytomorphology* 32:1–14.

Guharay, F. and F. Sachs. 1984. Stretch-activated single ion channel currents in tissue-cultured embryonic chick skeletal muscle. *J. Physiol.* 352:685–701.

Guharay, F. and F. Sachs. 1985. Mechanotransducer ion channels in chick skeletal muscle: the effects of extracellular pH. *J. Physiol.* 363:119–134.

Gunckel, J. E. and A. H. Sparrow. 1954. Aberrant growth in plants induced by ionizing radiation. *Brookhaven Symp. Biol.* 6:252–279.

Gunckel, J. E. and A. H. Sparrow. 1961. Ionizing radiations: Biochemical, physiological, and morphological aspects of their effects on plants. In W. Ruhland, ed., *Handbuch der Pflanzenphysiologie*, 16:555–611. Berlin: Springer-Verlag.

Gunning, B. E. S. and R. L. Overall. 1983. Plasmodesmata and cell-to-cell transport in plants. *BioScience* 33:260–265.

Gunning, B. E. S. and A. W. Robards, eds. 1976. *Intercellular Communication in Plants: Studies on Plasmodesmata*. Berlin: Springer-Verlag.

Haber, A. H. 1962. Nonessentiality of concurrent cell divisions for degree of polarization of leaf growth. I. Studies with radiation-induced mitotic inhibition. *Amer. J. Bot.* 49:583–589.

Haber, A. H. and D. E. Foard. 1963. Nonessentiality of concurrent cell divisions for degree of polarization of leaf growth. II. Evidence from untreated plants and from chemically induced changes of the degree of polarization. *Amer. J. Bot.* 50:937–944.

Haber, A. H. and D. E. Foard. 1964. Interpretations concerning cell division and growth. *Proceedings of the 5th International Conference on Plant Growth Regulators, Gif-sur-Yvette, France, Regulateurs Naturels de la Croissance Vegetale* 123:491–503.

Haber, A. H., T. J. Long, and D. E. Foard. 1964. Is final size determined by rate and duration of growth? *Nature* 201:479–480.

Hadidi, A. 1986. Relationship of viroids and certain other plant pathogenic nucleic acids to group I and II introns. *Plant Molecular Biology* 7:129–142.

Hadorn, E. 1948. Gene action in growth and differentiation of lethal mutants of *Drosophila. Symp. of the Society for Experimental Biology.* 2:177–195.

Haigh, J. 1978. The accumulation of deleterious genes in a population—Muller's Ratchet. *Theor. Pop. Biol.* 14:251–267.

Hake, S. and M. Freeling. 1986. Analysis of genetic mosaics shows that the extra epidermal cell divisions in *Knotted* mutant maize plants are induced by adjacent mesophyll cells. *Nature* 320:621–623.

Haldane, J. B. S. 1937. The effect of variation on fitness. *Amer. Natur.* 71:337–349.

Hallé, F., R. A. A. Oldeman, and P. B. Tomlinson. 1978. *Tropical Trees and Forests: An Achitectural Analysis.* Heidelberg: Springer-Verlag.

Hannan, M. A., J. Calkins, and W. L. Lasswell. 1980. Recombinagenic and mutagenic effects of sunlamp (UV-B) irradiation in *Saccharomyces cerevisiae. Mol. Gen. Genet.* 177:577–580.

Hara, N. 1971. Structure of the vegetative shoot apex of *Clethra barbinervis* III. Longisectional view, summary analysis and discussion. *Bot. Mag.* Tokyo 84:283–292.

Hara, N. 1975. Structure of the vegetative shoot apex of *Cassiope lycopodioides. Bot. Mag.* Tokyo 88:89–101.

Harberd, D. J. 1967. Observations on natural clones in *Holcus mollis. New Phytol.* 66:401–408.

Hardwick, R. C. 1986. Physiological consequences of modular growth in plants. *Phil. Trans. Roy. Soc. B London* 313:161–173.

Harper, J. L. 1977. *The Population Biology of Plants.* London: Academic Press.

Harris, H., D. A. Hopkinson, and Y. H. Edwards. 1977. Polymorphism and the subunit structure of enzymes: A contribution to the neutralist-selectionist controversy. *Proc. Natl. Acad. Sci. USA* 74:698–701.

Harte, C. and A. Maek. 1976. Genabhangigkeit des Wachstums pflanzlicher Meristeme untersucht an der Entwicklung isolierter Wurzeln verschiedener Genotypen von *Antirrhinum majus* L. *Biol. Zbl.* 95:267–299.

Hartman, P. E. 1980. Bacterial mutagenesis: Review of new insights. *Environmental Mutagenesis* 2:3–16.

Hébant, C., R. Hébant-Mauri, and J. Barthonnet. 1978. Evidence for division and polarity in apical cells of bryophytes and pteridophytes. *Planta* 138:49–52.

Hedrick, P. W. 1987. Genetic load and the mating system in homosporous ferns. *Evolution* 41:1282–1289.

Heidecker, G. and J. Messing. 1986. Structural analysis of plant genes. *Ann. Rev. Plant Physiol.* 37:439–466.

Hejnowicz, Z., J. Nakielski, and K. Hejnowicz. 1984. Modeling of spatial variations of growth within apical domes by means of the growth tensor. II. Growth specified on dome surface. *Acta Societatis Botanicorum Poloniae* 53:301–316.

Heneen, W. K. 1963. Extensive chromosome breakage occurring spontaneously in a certain individual of *Elymus farctus* (= *Agropyron junceum*). *Hereditas* 49:1–32.

Hepler, P. K. and R. O. Wayne. 1985. Calcium and plant development. *Ann. Rev. Plant Physiol.* 36:397–439.

Hiatt, H. H., J. D. Watson, and J. A. Winsten. 1977. *Origins of Human Cancer.* Cold Spring Harbor Conference on Cell Proliferation. New York: Cold Spring Harbor.

Hieter, P., C. Mann, M. Snyder, and R. W. Davis. 1985. Mitotic stability of yeast chromosomes: A colony color assay that measures nondisjunction and chromosome loss. *Cell* 40:381–392.

Hill, J. B., L. O. Overholts, and H. W. Popp. 1950. *Botany: A Textbook for Colleges.* New York: McGraw-Hill.

Hillier, H. G. 1972. *Hillier's Manual of Trees and Shrubs.* New York: A. S. Barnes.

Hinegardner, R. 1976. Evolution of genome size. In F. J. Ayala, ed., *Molecular Evolution,* pp. 179–199. Sunderland, Mass.: Sinauer.

Hoffmann, G. W. 1974. On the origin of the genetic code and the stability of the translation apparatus. *J. Mol. Biol.* 86:349–362.

Hohn, T., B. Hohn, and P. Pfeiffer. 1985. Reverse transcription in CaMV. *Trends Biochem. Sci.* 10:205–209.

Holbrook-Walker, S. G. and R. M. Lloyd. 1973. Reproductive biology and gametophyte morphology of the Hawaiian fern genus *Sadleria* (Blechnaceae) relative to habitat diversity and propensity for colonization. *Bot. J. Linn. Soc.* 67:157–174.

Holliday, R. and G. M. Tarrant. 1972. Altered enzymes in ageing human fibroblasts. *Nature* 238:26–30.

Holm, C., T. Goto, J. C. Wang, and D. Botstein. 1985. DNA topoisomerase II is required at the time of mitosis in yeast. *Cell* 41:553–563.

Holsinger, K. E. and M. W. Feldman. 1983. Modifiers of mutation rate: Evolutionary optimum with complete selfing. *Proc. Natl. Acad. Sci. USA* 80:6732–6734.

Honda, H., P. B. Tomlinson, and J. B. Fisher. 1981. Computer simulation of branch interaction and regulation by unequal flow rates in botanical trees. *Amer. J. Bot.* 68:569–585.

Honda, H., P. B. Tomlinson, and J. B. Fisher. 1982. Two geometrical models of branching of botanical trees. *Ann. Bot.* 49:1–11.

Horn, H. S. 1971. *The Adaptive Geometry of Trees.* Princeton, N.J.: Princeton University Press.

Howland, G. P. 1975. Dark-repair of ultraviolet-induced pyrimidine dimers in the DNA of wild carrot protoplasts. *Nature* 254:160–161.

Howland, G. P., R. W. Hart, and M. L. Yette. 1975. Repair of DNA strand breaks after gamma-irradiation of protoplasts isolated from cultured wild carrot cells. *Mutat. Res.* 27:81–87.

Hsu, T. C. 1983. Genetic instability in the human population: A working hypothesis. *Hereditas* 98:1–9.

Huxley, J. S. 1932. *Problems of Relative Growth.* London: Methuen.

Iwasa, Y., D. Cohen, and J. A. Leon. 1984. Tree height and crown shape, as results of competitive games. *J. theor. Biol.* 112:279–297.

Jabs, E. W., S. F. Wolf, and B. R. Migeon. 1984. Characterization of a cloned DNA sequence that is present at centromeres of all human autosomes and the X chromosome and shows polymorphic variation. *Proc. Natl. Acad. Sci. USA* 81:4884–4888.

Jacquard, A. 1974. *The Genetic Structure of Populations*. Berlin: Springer-Verlag.

Janzen, D. H. 1977. Variation in seed size within a crop of a Costa Rican *Mucuna andreana* (Leguminosae). *Amer. J. Bot.* 64:347–349.

Jaynes, R. A. 1968. Self incompatibility and inbreeding depression in three laurel *(Kalmia)* species. *Proc. Amer. Soc. Hort. Sci.* 93:618–622.

Jaynes, R. A. 1969. Chromosome counts of *Kalmia* species and revaluation of *K. polifolia* var. *microphylla*. *Rhodora* 71:280–284.

Johansen, D. A. 1950. *Plant Embryology*. Waltham, Mass.: Chronica Botanica.

Johns, M. A., J. Mottinger, and M. Freeling. 1985. A low copy number, *copia*-like transposon in maize. *The EMBO Journal* 4:1093–1102.

Johns, M. A., J. N. Strommer, and M. Freeling. 1983. Exceptionally high levels of restriction site polymorphism in DNA near the maize *Adh1* gene. *Genetics* 105:733–743.

Johnson, L. C. and W. B. Critchfield. 1974. A white-pollen variant of bristle-cone pine. *J. Hered.* 65:123.

Johnson, M. A. 1951. The shoot apex in gymnosperms. *Phytomorphology* 1:188–204.

Johnsson, H. 1976. Contributions to the genetics of empty grains in the seed of pine *(Pinus sylvestris)*. *Silvae Genetica* 25:10–15.

Johri, M. M. and E. H. Coe, Jr. 1982. Genetic approaches to meristem organization. In W. F. Sheridan, ed., *Maize for Biological Research*, pp. 301–310. Charlottesville, Va.: Plant Molecular Biology Association.

Johri, M. M. and E. H. Coe Jr. 1983. Clonal analysis of corn plant development. I. The development of the tassel and the ear shoot. *Dev. Biol.* 97:154–172.

Jones, D. F. 1928. *Selective Fertilization*. Chicago: Chicago University Press.

Jones, D. F. 1935. The similarity between fasciations in plants and tumors in animals and their genetic basis. *Science* 81:75–76.

Jones, G. N. and O. Tippo. 1952. John Theodore Buchholz, 1888–1951. *Phytomorphology* 2:181–185.

Jones, K. 1978. Aspects of chromosome evolution in higher plants. In H. Woolhouse, ed., *Advances in Botanical Research*, 6:119–194. London: Academic Press.

Jones, R. N. and H. Rees. 1982. *B Chromosomes*. London: Academic Press.

Jørgensen, J. H. and H. P. Jensen. 1986. The spontaneous chlorophyll mutation frequency in barley. *Hereditas* 105:71–72.

Kahn, C. 1985. *Beyond the Helix*. New York: Times Books.

Kay, H. E. M. 1965. How many cell-generations? *The Lancet* 2:418–419.

Kehr, A. E. 1965. The growth and development of spontaneous plant tumors. In W. Ruhland, ed., *Handbuch der Pflanzenphysiologie*. Vol. 15, part 2, Differentiation and Development, pp. 184–196. Berlin: Springer-Verlag.

Kemble, R. J., S. Gabay-Laughnan, and J. R. Laughnan. 1985. Movement of genetic information between plant organelles: Mitochondria-nuclei. In B. Hohn and E. S. Dennis, *Genetic Flux in Plants*, pp. 79–87. Vienna, New York: Springer-Verlag.

Kemble, R. J., R. J. Mans, S. Gabay-Laughnan, and J. R. Laughnan. 1983. Sequences homologous to episomal mitochondrial DNAs in the maize nuclear genome. *Nature* 304:744–747.

Kenrick, J., V. Kaul, and E. G. Williams. 1986. Self-incompatibility in *Acacia retinodes*: Site of pollen-tube arrest is the nucellus. *Planta* 169:245–250.

Kimura, M. 1983. *The Neutral Theory of Molecular Evolution*. Cambridge: Cambridge University Press.

King, J. L. and T. H. Jukes. 1969. Non-Darwinian evolution. *Science* 164:788–798.

Kirk, J. T. O. and R. A. E. Tilney-Bassett. 1978. *The Plastids.* New York: Elsevier/North-Holland Biomedical Press.

Kirkwood, T. B. L. 1977. Evolution of ageing. *Nature* 270:301–304.

Kirkwood, T. B. L. 1980. Error propagation in intracellular information transfer. *J. theor. Biol.* 82:363–382.

Kirkwood, T. B. L. and R. Holliday. 1975. The stability of the translation apparatus. *J. Mol. Biol.* 97:257–265.

Klekowski, E. J., Jr. 1970. Populational and genetic studies of a homosporous fern—*Osmunda regalis. Amer. J. Bot.* 57:1122–1138.

Klekowski, E. J., Jr. 1972. Evidence against genetic self-incompatibility in the homosporous fern *Pteridium aquilinum. Evolution* 26:66–73.

Klekowski, E. J., Jr. 1973. Genetic load in *Osmunda regalis* populations. *Amer. J. Bot.* 60:146–154.

Klekowski, E. J., Jr. 1982. Genetic load and soft selection in ferns. *Heredity* 49:191–197.

Klekowski, E. J., Jr. 1984. Mutational load in clonal plants: A study of two fern species. *Evolution* 38:417–426.

Klekowski, E. J., Jr. 1987. Mechanisms that maintain the genetic integrity of plants. *Proceedings of the XIV International Botanical Congress,* (in press).

Klekowski, E. J., Jr. and N. Kazarinova-Fukshansky. 1984a. Shoot apical meristems and mutation: Fixation of selectively neutral cell genotypes. *Amer. J. Bot.* 71:22–27.

Klekowski, E. J., Jr. and N. Kazarinova-Fukshansky. 1984b. Shoot apical meristems and mutation: Selective loss of disadvantageous cell genotypes. *Amer. J. Bot.* 71:28–34.

Klekowski, E. J., Jr., N. Kazarinova-Fukshansky, and H. Mohr. 1985. Shoot apical meristems and mutation: Stratified meristems and angiosperm evolution. *Amer. J. Bot.* 72:1788–1800.

Klekowski, E. J., Jr., H. Mohr, and N. Kazarinova-Fukshansky. 1986. Mutation, apical meristems, and developmental selection in plants. In J. P. Gustafson, G. L. Stebbins, and F. J. Ayala, eds., *Genetics, Development, and Evolution.* New York: Plenum Press.

Klekowski, E. J. and D. E. Levin. 1979. Mutagens in a river heavily polluted with paper recycling wastes: Results of field and laboratory mutagen assays. *Environmental Mutagenesis* 1:209–220.

Knudson, A. G. Jr. 1986. Genetics of human cancer. *Ann. Rev. Genet.* 20:231–251.

Kny, L. 1875. Die Entwickelung der Parkeriaceen dargestellt an *Ceratopteris thalictroides* Brongn. *Nova Acta* 37:5–81.

Kondrashov, A. S. 1982. Selection against harmful mutations in large sexual and asexual populations. *Genet. Res., Camb.* 40:325–332.

Kondrashov, A. S. 1984. Deleterious mutations as an evolutionary factor. I. The advantage of recombination. *Genet. Res., Camb.* 44:199–217.

Korn, R. W. 1982. Positional specificity within plant cells. *J. theor. Biol.* 95:543–568.

Koski, V. 1973. On self-pollination, genetic load, and subsequent inbreeding in some conifers. *Comm. Inst. For. Fenn.* 78.10:1–40.

Kraszewska, E. K., C. A. Bjerknes, S. S. Lamm, and J. Van't Hof. 1985. Extrachromosomal DNA of pea-root *(Pisum sativum)* has repeated sequences and ribosomal genes. *Plant Molecular Biology* 5:353–361.

Krimer, D. B. and J. Van't Hof. 1983. Extrachromosomal DNA of pea *(Pisum sativum)* root-tip cells replicates by strand displacement. *Proc. Natl. Acad. Sci. USA* 80:1933–1937.

Kuhlemeier, C., P. J. Green, and N.-H. Chua. 1987. Regulation of gene expression in higher plants. *Ann. Rev. Plant Physiol.* 38:221–257.

Kuligowski, J., M. Ferrand, E. Chenou, and Y. Tourte. 1985. Recuperation d'une cellule apicale de tige après alteration de l'embryogenese par la colchicine chez le *Marsilea vestita*. *Cytobios* 42:157–170.

Kuligowski-Andres, J. and Y. Tourte. 1979. La détermination des cellules apicales chez une Pteridophyte; role particulier du genome paternel. *Bull. Soc. bot. Fr., Lettres bot.* 126:491–505.

Kuligowski-Andres, J., Y. Tourte, and M. Faivre-Baron. 1979. La differenciation cellulaire de l'embryon: genome paternel et cellules organogenes chez une Pteridophyte. *C. R. Acad. Sc. Paris* 289:1093–1096.

Kunkel, T. A. 1985. The mutational specificity of DNA polymerases-α and -γ during *in vitro* DNA synthesis. *J. Biol. Chem.* 260:12866–12874.

Kuser, J. 1983. Inbreeding depression in *Metasequoia*. *J. Arnold Arboretum* 64:475–481.

Labouygues, J. M. and A. Figureau. 1984. The logic of the genetic code: Synonyms and optimality against effects of mutations. *Origins of Life* 14:685–692.

Lande, R. and D. W. Schemske. 1985. The evolution of self-fertilization and inbreeding depression in plants. I. Genetic models. *Evolution* 39:24–40.

Langenauer, H. and E. L. Davis. 1973. *Helianthus annuus* responses to acute x-irradiation. I. Damage and recovery in the vegetative apex and effects on development. *Bot. Gaz.* 134:301–316.

Langner, W. 1959. Selbsfertilitat und Inzucht bei *Picea omorika* (Pancic) Purkyne. *Silvae Genetica* 8:84–93.

Langridge, J. 1958. A hypothesis of developmental selection exemplified by lethal and semi-lethal mutants of *Arabidopsis*. *Aust. J. Biol. Sci.* 11:58–68.

Langridge, J. 1974. Mutation spectra and the neutrality of mutations. *Aust. J. Biol. Sci.* 27:309–319.

Lawrence, G. H. M. 1951. *Taxonomy of Vascular Plants*. New York: Macmillan.

Lee, T. D. 1984. Patterns of fruit maturation: A gametophyte competition hypothesis. *Amer. Natur.* 123:427–432.

Lee, T. D. and F. A. Bazzaz. 1982. Regulation of fruit and seed production in an annual legume, *Cassia fasciculata*. *Ecology* 63:1363–1373.

Leigh, E. G., Jr. 1970. Natural selection and mutability. *Amer. Natur.* 104:301–305.

Leigh, E. G., Jr. 1973. The evolution of mutation rates. *Genetics Suppl.* 73:1–18.

Lester, D. T. 1971. Self-compatibility and inbreeding depression in American elm. *For. Sci.* 17:321–322.

Levene, H., I. M. Lerner, A. Sokoloff, F. K. Ho, and I. R. Franklin. 1965. Genetic load in *Tribolium*. *Proc. Natl. Acad. Sci. USA* 53:1042–1050.

Levin, D. A. 1984. Inbreeding depression and proximity-dependent crossing success in *Phlox drummondii*. *Evolution* 38:116–127.

Levin, D. A. and A. C. Wilson. 1976. Rates of evolution in seed plants: Net increase in diversity of chromosome numbers and species numbers through time. *Proc. Natl. Acad. Sci. USA* 73:2086–2090.

Levings, C. S. III. 1983. The plant mitochondrial genome and its mutants. *Cell* 32:659–661.

Lewis, D. 1954. Comparative incompatibility in angiosperms and fungi. *Advances in Genetics* 6:235–285.

Lewis, K. R. and B. John. 1963. *Chromosome Marker*. London: J. and A. Churchill.

Li, W.-H, C.-I. Wu, and C.-C. Luo. 1984. Nonrandomness of point mutation as reflected in nucleotide substitutions in pseudogenes and its evolutionary implications. *J. Mol. Evol.* 21:58–71.

Li, W.-H., C.-C. Luo, and C.-I. Wu. 1985. Evolution of DNA sequences. In R. J. Mac-Intyre, ed., *Molecular Evolutionary Genetics*, pp. 1–94. New York: Plenum Press.

Li, S. L. and G. P. Rédei. 1969. Estimation of mutation rate in autogamous diploids. *Radiat. Bot.* 9:125–131.

Libby, W. J. and E. Jund. 1962. Variance associated with cloning. *Heredity* 17:533–540.

Libby, W. J., B. G. McCutchan, and C. I. Millar. 1981. Inbreeding depression in selfs of redwood. *Silvae Genetica* 30:15–25.

Lichtenstein, C. 1986. A bizarre vegetal bestiality. *Nature* 322:682–683.

Lima-de-Faria, A. 1983. *Molecular Evolution and Organization of the Chromosome.* Amsterdam: Elsevier.

Lindgren, D. 1975. The relationship between self-fertilization, empty seeds and seeds originating from selfing as a consequence of polyembryony. *Studia Forestalia Suecica* 126:1–24.

Lindoo, S. J. and M. M. Caldwell. 1978. Ultraviolet-B radiation-induced inhibition of leaf expansion and promotion of anthocyanin production. *Plant Physiol.* 61:278–282.

Lintilhac, P. M. 1984. Positional controls in meristem development: a caveat and an alternative. In P. W. Barlow and D. J. Carr, eds., *Positional Controls in Plant Development*, pp. 83–105. Cambridge: Cambridge University Press.

Lintilhac, P. M. and P. B. Green. 1976. Patterns of microfibrillar order in a dormant fern apex. *Amer. J. Bot.* 63:726–728.

Litvak, S. and M. Castroviejo. 1985. Plant DNA polymerases. *Plant Molecular Biology* 4:311–314.

Litvak, S., L. Keclard-Christophe, M. Echeverria, and M. Castroviejo. 1983. DNA polymerase and DNA synthesis in wheat embryo mitochondria. In O. Cifferi and L. Dure III, eds., *Structure and Function of Plant Genomes*, pp. 381–385. New York: Plenum Press.

Lloyd, D. G. 1982. Selection of combined versus separate sexes in seed plants. *Amer. Natur.* 120:571–585.

Lloyd, R. M. 1974. Systematics of the genus *Ceratopteris* Brongn. (Parkeriaceae) II. Taxonomy. *Brittonia* 26:139–160.

Lloyd, R. M. 1980. Reproductive biology and gametophyte morphology of New World populations of *Acrostichum aureum. Amer. Fern J.* 70:99–110.

Lloyd, R. M. and T. L. Gregg. 1975. Reproductive biology and gametophyte morphology of *Acrostichum danaeifolium* from Mexico. *Amer. Fern J.* 65:105–120.

Lloyd, R. M. and E. J. Klekowski, Jr. 1970. Spore germination and viability in Pteridophyta: Evolutionary significance of chlorophyllous spores. *Biotropica* 2:129–137.

Lloyd, R. M. and T. R. Warne. 1978. The absence of genetic load in a morphologically variable sexual species, *Ceratopteris thalictroides* (Parkeriaceae). *Syst. Bot.* 3:20–36.

Lockhart, J. A. and U. Brodführer Franzgrote. 1961. The effects of ultraviolet radiation on plants. In W. Ruhland, ed., *Handbuch der Pflanzenphysiologie*, 16:532–554. Berlin: Springer-Verlag.

Loeb, L. A. 1985. Apurinic sites as mutagenic intermediates. *Cell* 40:483–484.

Loeb, L. A. and T. A. Kunkel. 1982. Fidelity of DNA synthesis. *Ann. Rev. Biochem.* 51:429–457.

Lonsdale, D. M. 1985. Movement of genetic material between the chloroplast and mitochondrion in higher plants. In B. Hohn and E. S. Dennis, eds., *Genetic Flux in Plants*, pp. 51–60. Vienna, New York: Springer-Verlag.

Lyman, J. C. and N. C. Ellstrand. 1984. Clonal diversity in *Taraxacum officinale* (Compositae) an apomict. *Heredity* 53:1–10.

Lyrene, P. 1983. Inbreeding depression in rabbiteye blueberries. *HortScience* 18:226–227.

McAlpin, B. W. and R. A. White. 1974. Shoot organization in the filicales: The promeristem. *Amer. J. Bot.* 61:562–579.

McClintock, B. 1934. The relationship of a particular chromosomal element to the development of the nucleoli in *Zea mays. Zeit Zellforsch. mik Anat.* 21:294–328.

McClintock, B. 1941. The stability of broken ends of chromosomes in *Zea mays. Genetics* 26:234–282.

McClintock, B. 1984. The significance of responses of the genome to challenge. *Science* 226:792–801.

MacKay, T. F. C. 1985. Transposable element-induced response to artificial selection in *Drosophila melanogaster. Genetics* 111:351–374.

Maheshwari, P. 1950. *An Introduction to the Embryology of Angiosperms.* New York: McGraw-Hill.

Maillette, L. 1982a. Structural dynamics of silver birch. I. The fates of buds. *J. Appl. Ecol.* 19:203–218.

Maillette, L. 1982b. Structural dynamics of silver birch. II. A matrix model of the bud population. *J. Appl. Ecol.* 19:219–238.

Malogolowkin-Cohen, Ch., H. Levene, N. P. Dobzhansky, and A. Solima Simmons. 1964. Inbreeding and the mutational and balanced loads in natural populations of *Drosophila willistoni. Genetics* 50:1299–1311.

Manning, J. T. and D. J. Thompson. 1984. Muller's ratchet and the accumulation of favourable mutations. *Acta Biotheoretica* 33:219–225.

Marcotrigiano, M. 1986. Experimentally synthesized plant chimeras. 3. Qualitative and quantitative characteristics of the flowers of interspecific *Nicotiana* chimeras. *Ann. Bot.* London 57:435–442.

Margolis, L. and D. Sagan. 1986. *Origins of Sex: Three Billion Years of Genetic Recombination.* New Haven: Yale University Press.

Marx, G. A. 1983. Developmental mutants in some annual seed plants. *Ann. Rev. Plant Physiol.* 34:389–417.

Masters, M. T. 1869. *Vegetable Teratology.* London: Robert Hardwicke.

Masuyama, S. 1979. Reproductive biology of the fern *Phegopteris decursive-pinnata.* I. The dissimilar mating systems of diploids and tetraploids. *Bot. Mag.* Tokyo 92:275–289.

Masuyama, S. 1986. Reproductive biology of the fern *Phegopteris decursive-pinnata.* II. Genetic analyses of self-sterility in diploids. *Bot. Mag.* Tokyo 99:107–121.

Matheson, A. C. 1980. Unexpectedly high frequencies of outcrossed seedlings among offspring from mixtures of self and cross pollen in *Pinus radiata* D. Don. *Aust. For. Res.* 10:21–27.

Mauseth, J. D. 1978a. An investigation of the morphogenetic mechanisms which control the development of zonation in seedling shoot apical meristems. *Amer. J. Bot.* 65:158–167.

Mauseth, J. D. 1978b. An investigation of the phylogenetic and ontogenetic variability of shoot apical meristems in the Cactaceae. *Amer. J. Bot.* 65:326–333.

Mauseth, J. D. 1979. Cytokinin-elicited formation of the pith-rib meristem and other effects of growth regulators on the morphogenesis of *Echinocereus* (Cactaceae) seedling shoot apical meristems. *Amer. J. Bot.* 66:446–451.

Mauseth, J. D. and K. J. Niklas. 1979. Constancy of relative volumes of zones in shoot apical meristems in Cactaceae: implications concerning meristem size, shape, and metabolism. *Amer. J. Bot.* 66:933–939.

Maynard Smith, J. 1978. *The Evolution of Sex.* Cambridge: Cambridge University Press.

Mayr, E. 1963. *Animal Species and Evolution.* Cambridge: Harvard University Press.

Meinhardt, H. 1984. Models of pattern formation and their application to plant development. In P. W. Barlow and D. J. Carr, eds., *Positional Controls in Plant Development,* pp. 1–32. Cambridge: Cambridge University Press.

Meinke, D. W. 1985. Embryo-lethal mutants of *Arabidopsis thaliana:* analysis of mutants with a wide range of lethal phases. *Theor. Appl. Genet.* 69:543–552.

Meinke, D. W. and I. M. Sussex. 1979a. Embryo-lethal mutants of *Arabidopsis thaliana:* A model system for genetic analysis of plant embryo development. *Dev. Biol.* 72:50–61.

Meinke, D. W. and I. M. Sussex. 1979b. Isolation and characterization of six embryo-lethal mutants of *Arabidopsis thaliana. Dev. Biol.* 72:62–72.

Meins, F., Jr. 1983. Heritable variation in plant cell culture. *Ann. Rev. Plant Physiol.* 34:327–346.

Meins, F., Jr. 1985. Cell heritable changes during development. In M. Freeling, ed., *Plant Genetics,* pp. 45–59. New York: Liss.

Meins, F., Jr. and A. N. Binns. 1977. Epigenetic variation of cultured somatic cells: Evidence for gradual changes in the requirement for factors promoting cell division. *Proc. Natl. Acad. Sci. USA* 74:2928–2932.

Meins, F., Jr. and A. N. Binns. 1979. Cell determination in plant development. *Bio-Science* 29:221–225.

Melchior, R. C. and J. W. Hall. 1961. A calamitean shoot apex from the Pennsylvanian of Iowa. *Amer. J. Bot.* 48:811–815.

Mergen, F., J. Burley, and G. M. Furnival. 1965. Embryo and seedling development in *Picea glauca* (Moench) Voss after self-, cross-, and wind-pollination. *Silvae Genetica* 14:188–194.

Michaelis, P. 1955a. Uber Gesetzmassigkeiten der Plasmon-Unkombination und uber eine Methode zur Trennung einer Plastiden-, Chondriosomen-, resp. Sphaerosomen-, (Mikrosomen)-, und einer Zytoplasmavererbung. *Cytologica* 20:315–338.

Michaelis, P. 1955b. Modellversuche zur Plastiden- und Plasmavererbung. *Der Zuchter* 25:209–221.

Michaelis, P. 1967. The investigation of plasmone segregation by the pattern-analysis. *Nucleus* 10:1–14.

Miflin, B. J. 1974. The location of nitrite reductase and other enzymes related to amino acid biosynthesis in the plastids of roots and leaves. *Plant Physiol.* 54:550–555.

Miflin, B. J. and P. J. Lea. 1977. Amino acid metabolism. *Ann. Rev. Plant Physiol.* 28:299–329.

Mohr, H. and H. Drumm-Herrel. 1983. Coaction between phytochrome and blue/UV light in anthocyanin synthesis in seedlings. *Physiol. Plant.* 58:408–414.

Mohr, H., H. Drumm-Herrel, and R. Oelmüller. 1984. Coaction of phytochrome and blue/UV light photoreceptors. In H. Senger, ed., *Blue Light Effects in Biological Systems,* pp. 6–19. Berlin: Springer-Verlag.

Moran, G. F. and A. R. Griffin. 1985. Non-random contribution of pollen in polycrosses of *Pinus radiata* D. Don. *Silvae Genetica* 34:117–121.

Moreau, P. and R. Devoret. 1977. Potential carcinogens tested by induction and mutagenesis of prophage in *Escherichia coli* K_12. In H. H. Hiatt, J. D. Watson, and J. A. Winsten, eds., *Origins of Human Cancer* (Cold Spring Harbor Conference on Cell Proliferation), 4:1451–1472. New York: Cold Spring Harbor.

Morton, N. E., J. F. Crow, and H. J. Muller. 1956. An estimate of the mutational damage in man from data on consanguineous marriages. *Proc. Natl. Acad. Sci. USA* 42:855–863.

Mottinger, J. P., M. A. Johns, and M. Freeling. 1984. Mutations of the Adhl gene in maize following infection with barley stripe mosaic virus. *Mol. Gen. Genetics.* 195:367–369.

Mulcahy, D. L. 1979. The rise of the angiosperms: a genecological factor. *Science* 206:20–23.

Mulcahy, D. L., G. Bergamini Mulcahy, and E. Ottaviano. 1986. *Biotechnology and Ecology of Pollen.* New York: Springer-Verlag.

Mulcahy, D. L. and S. M. Kaplan. 1979. Mendelian ratios despite nonrandom fertilization? *Amer. Natur.* 113:419–425.

Muller, H. J. 1932. Some genetic aspects of sex. *Amer. Natur.* 66:118–138.

Muller, H. J. 1950. Our load of mutations. *Am. J. Hum. Genet.* 2:111–176.

Muller, H. J. 1964. The relation of recombination to mutational advance. *Mutat. Res.* 1:2–9.

Murray, A. W. and J. W. Szostak. 1985. Chromosome segregation in mitosis and meiosis. *Ann. Rev. Cell Biol.* 1:289–315.

Murray, M. G., D. L. Peters, and W. F. Thompson. 1981. Ancient repeated sequences in the pea and mung bean genomes and implications for genome evolution. *J. Mol. Evol.* 17:31–42.

Nagl, W., J. Pohl, and A. Radler. 1985. The DNA endoreduplication cycles. In J. A. Bryant and D. Francis, eds., *The Cell Division Cycle in Plants* (Society for Experimental Biology, Seminar Series 26), pp. 217–232 Cambridge: Cambridge University Press.

Nagley, P. 1987. The naked truth about eukaryotic DNA polymerase errors. *Trends in Genetics* 3:31–32.

Navashin, M. 1934. Chromosomal alterations caused by hybridization and their bearing upon certain genetic problems. *Cytologia* 5:169–203.

Neilson-Jones, W. 1969. *Plant Chimeras.* 2d ed. London: Methuen.

Newman, I. V. 1965. Pattern in the meristems of vascular plants. III. Pursuing the pattern in the apical meristem where no cell is a permanent cell. *Bot. J. Linn. Soc.* 59:185–214.

Niklas, K. J. 1986. Computer-simulated plant evolution. *Scientific American* 254:78–86.

Ninio, J. 1983. *Molecular Approaches to Evolution.* Princeton, N.J.: Princeton University Press.

Nitsch, J. P. 1952. Plant hormones and the development of fruits. *Q. Rev. Biol.* 27:33–57.

Nitsch, J. P. 1963. Fruit development. In P. Maheshwari, ed., *Recent Advances in the Embryology of Angiosperms,* pp. 361–394. Ranchi, India: The Catholic Press, University of Delhi, International Society of Plant Morphologists.

Nordenskiöld, H. 1961. Tetrad analysis and the course of meiosis in three hybrids of *Luzula campestris. Hereditas* 47:203–238.

Novick, A. and L. Szilard. 1950. Experiments with the chemostat on spontaneous mutations of bacteria. *Proc. Natl. Acad. Sci. USA* 36:708–719.

Novick, A. and L. Szilard, 1951. Genetic mechanisms in bacteria and bacterial viruses. I. Experiments on spontaneous and chemically induced mutations of bacteria growing in the chemostat. *Cold Spring Harbor Symp. on Quantitative Biology* 16:337–343.

Oelmüller, R. and H. Mohr. 1986. Photooxidative destruction of chloroplasts and its consequences for expression of nuclear genes. *Planta* 167:106–113.

Ohlrogge, J. B. 1982. Fatty acid synthetase: plants and bacteria have similar organization. *Trends Biochem. Sci.* 7:386–387.

Ohnishi, O. 1982. Population genetics of cultivated common buckwheat, *Fagopyrum esculentum* Moench. I. Frequency of chlorophyll-deficient mutants in Japanese populations. *Jpn. J. Genet.* 57:623–639.

Ohnishi, O. 1985. Population genetics of cultivated common buckwheat, *Fagopyrum esculentum* Moench. IV. Allozyme variability in Nepali and Kashmirian populations. *Jpn. J. Genet.* 60:293–305.

Ohnishi, O. and T. Nagakubo. 1982. Population genetics of cultivated common buckwheat, *Fagopyrum esculentum* Moench. II. Frequency of dwarf mutants in Japanese populations. *Jpn. J. Genet.* 57:641–650.

Ohno, S. 1984a. Repeats of base oligomers as the primordial coding sequences of the primeval earth and their vestiges in modern genes. *J. Mol. Evol.* 20:313–321.

Ohno, S. 1984b. Conservation of linkage relationships between genes as the underlying theme of karyological evolution in mammals. In A. K. Sharma and A. Sharma, eds., *Chromosomes in Evolution of Eukaryotic Groups*, 2:1–11. Boca Raton, Fl: CRC Press.

Ohno, S. 1985. Immortal genes. *Trends in Genetics* 1:196–200.

Ohta, T. and G. A. Dover. 1983. Population genetics of multigene families that are dispersed into two or more chromosomes. *Proc. Natl. Acad. Sci. USA* 80:4079–4083.

Oinonen, E. 1967. Sporal regeneration of bracken *(Pteridium aquilinum)* in Finland in the light of the dimensions and the age of its clones. *Acta For. Fenn.* 83:1–96.

Orgel, L. E. 1963. The maintenance of the accuracy of protein synthesis and its relevance to ageing. *Proc. Natl. Acad. Sci. USA* 49:517–521.

Orgel, L. E. 1970. The maintenance of the accuracy of protein synthesis and its relevance to ageing: A correction. *Proc. Natl. Acad. Sci. USA* 67:1476.

Orgel, L. E. and F. H. C. Crick. 1980. Selfish DNA: The ultimate parasite. *Nature* 284:604–607.

Orr-Ewing, A. L. 1957. A cytological study of the effects of self-pollination on *Pseudotsuga menziesii* (Mirb.) Franco. *Silvae Genetica* 6:179–185.

Östergren, G. 1945. Parasitic nature of extra fragment chromosomes. *Bot. Notiser* 2:157–163.

Ottaviano, E., M. Sari Gorla, and D. L. Mulcahy. 1980. Pollen tube growth rate in *Zea mays:* Implications for genetic improvement of crops. *Science* 210:437–438.

Palmer, J. D. 1983. Chloroplast DNA exists in two orientations. *Nature* 301:92–93.

Palmer, J. D. 1985. Comparative organization of chloroplast genomes. *Ann. Rev. Genet.* 19:325–354.

Palmer, J. D. 1986. Evolution of chloroplast and mitochondrial DNA in plants and algae. In R. J. MacIntyre, ed., *Molecular Evolutionary Genetics*, pp. 131–240. New York: Plenum Press.

Palmer, J. D. and W. F. Thompson. 1982. Chloroplast DNA rearrangements are more frequent when a large inverted repeat sequence is lost. *Cell* 29:537–550.

Palmer, J. D., J. M. Nugent, and L. A. Herbon. 1987. Unusual structure of geranium chloroplast DNA: A triple-sized inverted repeat, extensive gene duplications, multiple inversions, and two repeat families. *Proc. Natl. Acad. Sci. USA* 84:769–773.

Pandey, K. K. 1986. Gene transfer through the use of sublethally irradiated pollen: the theory of chromosome repair and possible implication of DNA repair enzymes. *Heredity* 57:37–46.

Paolillo, D. J., Jr. and E. M. Gifford, Jr. 1961. Plastochronic changes and the concept of apical initials in *Ephedra altissima. Amer. J. Bot.* 48:8–16.

Park, Y. S. and D. P. Fowler. 1982. Effects of inbreeding and genetic variances in a natural population of tamarack (*Larix laricina* (Du Roi) K. Koch) in eastern Canada. *Silvae Genetica* 31:21–26.

Park, Y. S. and D. P. Fowler. 1984. Inbreeding in black spruce (*Picea mariana* (Mill.) B.S.P.): self-fertility, genetic load, and performance. *Can. J. For. Res.* 14:17–21.

Pate, D. W. 1983. Possible role of ultraviolet radiation in evolution of *Cannabis* chemotypes. *Economic Botany* 37:396–405.

Peterson, P. A. 1985. Transposon-induced events at gene loci. *BioEssays* 3:199–204.

Pfeiffer, P. and T. Hohn. 1983. Involvement of reverse transcription in the replication of cauliflower mosaic virus: A detailed model and test of some aspects. *Cell* 33:781–789.

Philipson, W. R., J. M. Ward, and B. G. Butterfield. 1971. *The Vascular Cambium*. London: Chapman and Hall.

Poethig, R. S. 1984a. Patterns and problems in angiosperm leaf morphogenesis. In G. M. Malacinski and S. V. Bryant, eds., *Pattern Formation*, pp. 413–432. New York: Macmillan.

Poethig, R. S. 1984b. Cellular parameters of leaf morphogenesis in maize and tobacco. In R. A. White and W. C. Dickinson, eds., *Contemporary Problems in Plant Anatomy*, pp. 235–259. New York: Academic Press.

Poethig, R. S. and I. M. Sussex. 1985a. The developmental morphology and growth dynamics of the tobacco leaf. *Planta* 165:158–169.

Poethig, R. S. and I. M. Sussex. 1985b. The cellular parameters of leaf development in tobacco: a clonal analysis. *Planta* 165:170–184.

Pohlheim, F. 1971a. Untersuchungen zur Sprossvariation der Cupressaceae. 1. Nachweis immerspaltender Periklinalchimaren. *Flora* (Jena) 160:264–293.

Pohlheim, F. 1971b. Untersuchungen zur Sprossvariation der Cupressaceae. 3. Quantitative Auswertung des Scheckungsmusters immerspaltender Periklinalchimaren. *Flora* (Jena) 160:360–372.

Pohlheim, F. 1971c. *Spiraea bumalda* "Anthony Waterer" und *Mentha arvensis* "Variegata"—zwei immerspaltende Periklinalchimaren unter den Angiosperm. *Biol. Zbl.* 90:295–319.

Pohlheim, F. 1973. Untersuchungen zur periklinalchimarischen Konstitution von *Pelargonium zonale* 'freak of nature'. *Flora* 162:284–294.

Pohlheim, F. 1980. zur Sprossvariation bei den Cupressaceae. *Wiss. Z. d. Humboldt-Univ. zu Berlin, Math.-Nat. R.* 29:295–306.

Pohlheim, F. 1981. Induced mutations for investigation of histogenetic processes as the basis for optimal mutant selection. *Proc. Intern. Symp. on Induced Mutations as a Tool for Crop Plant Improvement, International Atomic Energy Agency* 251/40:489–495.

Pohlheim, F. 1983. Vergleichende Untersuchungen zur Anderung der Richtung von Zellteilungen in Blattepidermen. *Biol. Zbl.* 102:323–336.

Pohlheim, F. and M. Kaufhold. 1985. On the formation of variegation patterns in *Filipendula ulmaria* "Aureo-Variegata" through changes in the plane of cell division in the epidermisses of young leaves. *Flora* 177:167–174.

Popham, R. A. 1951. Principal types of vegetative shoot apex organization in vascular plants. *Ohio J. Sci.* 51:249–270.

Poulson, D. F. 1945. Chromosomal control of embryogenesis in *Drosophila*. *Amer. Natur.* 79:340–363.

Pratt, C., D. K. Ourecky, and J. Einset. 1967. Variation in apple cytochimeras. *Amer. J. Bot.* 54:1295–1301.

Pressing, J. and D. C. Reanney. 1984. Divided genomes and intrinsic noise. *J. Mol. Evol.* 20:135–146.

Price, H. J. 1976. Evolution of DNA content in higher plants. *Bot. Rev.* 42:27–52.

Radman, M. 1974. Phenomenology of an inducible mutagenic DNA repair pathway in *Escherichia coli* SOS repair hypothesis. In L. Prakash, F. Sherman, M. W. Miller, C. W. Lawrence, and H. W. Taber, eds., *Molecular and Environmental Aspects of Mutagenesis*, pp. 129–140. Springfield, Ill.: C. C. Thomas.

Razin, A. and A. D. Riggs. 1980. DNA methylation and gene function. *Science* 210:604–610.

Reanney, D. C., D. G. MacPhee, and J. Pressing. 1983. Intrinsic noise and the design of the genetic machinery. *Aust. J. Biol. Sci.* 36:77–90.

Reeder, R. H. 1984. Enhancers and ribosomal gene spacers. *Cell* 38:349–351.

Reeder, R. H. and J. G. Roan. 1984. The mechanism of nucleolar dominance in *Xenopus* hybrids. *Cell* 38:39–44.

Reiss, T., R. Bergfeld, G. Link, W. Thien, and H. Mohr. 1983. Photooxidative destruction of chloroplasts and its consequences for cytosolic enzyme levels and plant development. *Planta* 159:518–528.

Relichova, J. 1977. Distribution of mutations in the apical meristem of the shoot in *Arabidopsis thaliana* (L.) Heynh. *Folia* 18:5–109.

Robberecht, R., C. C. Caldwell, and W. D. Billings. 1980. Leaf ultraviolet optical properties along a latitudinal gradient in the arctic-alpine life zone. *Ecology* 61:612–619.

Romberger, J. A. 1963. *Meristems, Growth, and Development in Woody Plants.* U.S. Department of Agriculture, Forest Service Technical Bulletin No. 1293.

Rosenberger, R. F. and T. B. L. Kirkwood. 1986. Errors and the integrity of genetic information transfer. In T. B. L. Kirkwood, R. F. Rosenberger, and D. J. Galas, eds., *Accuracy in Molecular Processes*, pp. 17–35. New York: Chapman and Hall.

Rouffa, A. S. and J. E. Gunckel. 1951. A comparative study of vegetative shoot apices in the Rosaceae. *Amer. J. Bot.* 38:290–300.

Rowlands, D. G. 1958. The nature of the breeding system in the field bean (*V. faba* L.) and its relationship to breeding for yield. *Heredity* 12:113–126.

Rowlands, D. G. 1960a., Fertility studies in the field bean (*Vicia faba* L.). I. Cross and self-fertility. *Heredity* 15:161–173.

Rowlands, D. G. 1960b. Fertility studies in the field bean (*Vicia faba* L.). II. Inbreeding. *Heredity* 16:497–508.

Rowlands, D. G. 1963. Fertility studies in the broad bean (*Vicia faba* L.). *Heredity* 19:271–277.

Rudin, D. and I. Ekberg. 1978. Linkage studies in *Pinus sylvestris* L. using macro gametophyte allozymes. *Silvae Genetica* 27:1–12.

Ruth, J., E. J. Klekowski, Jr., and O. L. Stein. 1985. Impermanent initials of the shoot apex and diplontic selection in a juniper chimera. *Amer. J. Bot.* 72:1127–1135.

Rutishauser, A. 1956. Chromosome distribution and spontaneous chromosome breakage in *Trillium grandiflorum. Heredity* 10:367–407.

Rutishauser, A. 1969. *Embryologie und Fortpflanzungsbiologie der Angiospermen.* Vienna, New York: Springer-Verlag.

Rutishauser, A. and L. F. La Cour. 1956a. Spontaneous chromosome breakage in endosperm. *Nature* 177:324–325.

Rutishauser, A. and L. F. La Cour. 1956b. Spontaneous chromosome breakage in hybrid endosperms. *Chromosome* 8:317–340.

Sachs, T. 1975. Plant tumors resulting from unregulated hormone synthesis. *J. theor. Biol.* 55:445–453.

Sachs, T. 1982. A morphogenetic basis for plant morphology. *Acta Biotheoretica* 31A:118–131.

Saedler, H. and P. Nevers. 1985. Transposition in plants: a molecular model. *The EMBO Journal* 4:585–590.

Saghai-Maroof, M. A., K. M. Soliman, R. A. Jorgensen, and R. W. Allard. 1984. Ribosomal DNA spacer-length polymorphisms in barley: Mendelian inheritance, chromosomal location, and population dynamics. *Proc. Natl. Acad. Sci. USA* 81:8014–8018.

Salser, W. 1978. Globin mRNA sequences: Analysis of base pairing and evolutionary implications. *Cold Spring Harbor Symp. on Quantitative Biology* 42:985–1002.

Sandermann, H., Jr. 1982. Metabolism of environmental chemicals: A comparison of plant and liver enzyme systems. In E. J. Klekowski, Jr., ed., *Environmental Mutagenesis, Carcinogenesis, and Plant Biology*, 1:1–32. New York: Praeger.

Satina, S. 1945. Periclinal chimeras in *Datura* in relation to the development and structure of the ovule. *Amer. J. Bot.* 32:71–81.

Satina, S., A. F. Blakeslee, and A. G. Avery. 1940. Demonstration of the three germ layers in the shoot apex of *Datura* by means of induced polyploidy in periclinal chimeras. *Amer. J. Bot.* 27:895–905.

Saus, G. L. and R. M. Lloyd. 1976. Experimental studies on mating systems and genetic load in *Onoclea sensibilis* L. (Aspleniaceae: Athyrioideae). *Bot. J. Linn. Soc.* 72:101–113.

Schaal, B. A. 1985. Genetic variation in plant populations: From demography to DNA. In J. Haeck and J. W. Woldendorp, eds., *Structure and Functioning of Plant Populations*, 2:321–341, Amsterdam: North Holland.

Schaffalitzky de Muckadell, J. 1954. Juvenile stages in woody plants. *Physiol. Plant.* 7:782–796.

Scheel, D. and H. Sandermann. 1981. Metabolism of 2,4-dichlorophenoxyacetic acid in cell suspension cultures of soybean (*Glycine max* L.) and wheat (*Triticum aestivum* L.). II. Evidence for incorporation into lignin. *Planta* 152:253–258.

Schemske, D. W. and R. Lande. 1985. The evolution of self-fertilization and inbreeding depression in plants. II. Empirical observations. *Evolution* 39:41–52.

Schensted, I. V. 1958. Model of subnuclear segregation in the macronucleus of ciliates. *Amer. Natur.* 92:161–170.

Schieder, O. 1980. Somatic hybrids between a herbaceous and two tree *Datura-species*. *Z. Pflanzenphysiol.* 98:119–127.

Schimke, R. T., S. W. Sherwood, and A. B. Hill. 1986. Overreplication of DNA and the rapid generation of genomic change. In H. Gershowitz, D. L. Rucknagel, and R. E. Tashian, eds., *Evolutionary Perspectives and the New Genetics*, pp. 109–117. New York: Liss.

Schlarbaum, S. E. and T. Tsuchiya. 1984. A chromosome study of coast redwood, *Sequoia sempervirens* (D. Don) Endl.). *Silvae Genetica* 33:56–62.

Schmalhausen, I. I. 1949. *Factors of Evolution: The Theory of Stabilizing Selection*. Philadelphia: Blakiston.

Schmitt, D. and T. O. Perry. 1964. Self-sterility in sweetgum. *For. Sci.* 10:302–305.

Schneller, J. J. 1979. Biosystematic investigations on the lady fern (*Athyrium filix-femina*). *Plant Syst. Evol.* 132:255–277.

Schuster, W. and A. Brennicke. 1987. Plastid, nuclear, and reverse transcriptase sequences in the mitochondrial genome of *Oenothera*: Is genetic information transferred between organelles via RNA? *The EMBO Journal* 6:2857–2863.

Schwartz, D. and C. E. Bay. 1956. Further studies on the reversal in the seedling height dose curve at very high levels of ionizing radiations. *Amer. Natur.* 90:323–327.

Schwarz-Sommer, Z., A. Gierl, H. Cuypers, P. A. Peterson, and H. Saedler. 1985. Plant transposable elements generate the DNA sequence diversity needed in evolution. *The EMBO Journal* 4:591–597.

Scott, N. S. and J. V. Possingham. 1983. Changes in chloroplast DNA levels during growth of spinach leaves. *J. Exp. Bot.* 34:1756–1767.

Seavey, S. R. and K. S. Bawa. 1986. Late-acting self-incompatibility in angiosperms. *Bot. Rev.* 52:195–219.

Sederoff, R. R. and C. S. Levings, III. 1985. Supernumerary DNAs in plant mitochondria.

In B. Hohn and E. S. Dennis, eds., *Genetic Flux in Plants*, pp.91–109. Vienna, New York: Springer-Verlag.

Shampay, J., J. W. Szostak, and E. H. Blackburn. 1984. DNA sequences of telomeres maintained in yeast. *Nature* 310:154–157.

Shannon, C. E. 1949. The mathematical theory of communication. In C. E. Shannon and W. Weaver, eds., *The Mathematical Theory of Communication*, pp. 3–91. Urbana: University of Illinois Press.

Shapiro, H. S. 1968. Distribution of purines and pyrimidines in deoxyribosenucleic acids. In H. A. Sober, eds., *Handbook of Biochemistry*, pp. H39–H48. Cleveland: Chemical Rubber Co.

Shapiro, J. A. 1979. Molecular model for the transposition and replication of bacteriophage Mu and other transposable elements. *Proc. Natl. Acad. Sci. USA* 76:1933–1937.

Shapiro, J. A. 1983. *Mobile Genetic Elements*. New York: Academic Press.

Shepard, J. F., D. Bidney, and E. Shahin. 1980. Potato protoplasts in crop improvement. *Science* 208:17–24.

Shepherd, N. S., Z. Schwarz-Sommer, J. Blumberg vel Spaive, M. Gupta, U. Wienand, and H. Saedler. 1984. Similarity of the *Cin1* repetitive family of *Zea mays* to eukaryotic transposable elements. *Nature* 307:185–187.

Shiba, T. and K. Saigo. 1983. Retrovirus-like particles containing RNA homologous to the transposable element *copia* in *Drosophila melanogaster*. *Nature* 302:119–124.

Shul'Ga, V. V. 1979. On the nature of dwarfish trees and "witch's broom" shoots of *Pinus sylvestris* L. *Lesovedenie* 3:82–86.

Silander, J. A. Jr. 1985. Microevolution in clonal plants. In J. B. C. Jackson, L. W. Buss, and R. E. Cook, eds., *Population Biology and Evolution of Clonal Organisms*, pp. 107–152, New Haven: Yale University Press.

Simmons, M. J. and J. F. Crow. 1977. Mutations affecting fitness in *Drosophila* populations. *Ann. Rev. Genet.* 11:49–78.

Sims, L. E. and H. J. Price. 1985. Nuclear DNA content variation in *Helianthus* (Asteraceae). *Amer. J. Bot.* 72:1213–1219.

Sinnott, E. W. 1960. *Plant Morphogenesis*. New York: McGraw-Hill.

Slatkin, M. 1984. Somatic mutations as an evolutionary force. In P. J. Greenwood, P. H. Harvey and M. Slatkin, eds., *Evolution* pp. 19–30. Cambridge: Cambridge University Press.

Smith, G. P. 1976. Evolution of repeated DNA sequences by unequal crossover. *Science* 191:528–535.

Smith, H. H. 1972. Plant genetic tumors. *Progr. exp. Tumor Res.* 15:138–164.

Smithies, O. and P. A. Powers. 1986. Gene conversions and their relation to homologous chromosome pairing. *Phil. Trans. Roy. Soc. B Lond.* B312:291–302.

Smoot, E. L. and T. N. Taylor. 1986. Evidence of simple polyembryony in Permian seeds from Antarctica. *Amer. J. Bot.* 73:1079–1081.

Snoad, B. 1955. Somatic instability of chromosome number in *Hymenocallis calathinum*. *Heredity* 9:129–134.

Snoad, B. and P. Matthews. 1969. Neoplasms of the pea pod. In C. D. Darlington and K. R. Lewis, eds., *Chromosomes Today*, 2:126–131. New York: Plenum.

Snow, A. A. 1986. Pollination dynamics in *Epilobium canum* (Onagraceae): Consequences for gametophytic selection. *Amer. J. Bot.* 73:139–151.

Sokal, R. R. and F. J. Rohlf. 1969. *Biometry*. San Francisco: Freeman.

Solbrig, O. T. and D. J. Solbrig. 1979. *Introduction to Population Biology and Evolution*. Reading, Mass.: Addison-Wesley.

Soll, D. R. 1986. The regulation of cellular differentiation in the dimorphic yeast *Candida albicans*. *BioEssays* 5:5–11.

Soma, K. and E. Ball. 1964. Studies of the surface growth of the shoot apex of *Lupinus albus*. *Brookhaven Symp. Biol.* 16:13–45.

Sommer, H., R. Carpenter, B. J. Harrison, and H. Saedler. 1985. The transposable element Tam3 of *Antirrhinum majus* generates a novel type of sequence alterations upon excision. *Mol. Gen. Genet.* 199:225–231.

Sommer, H., E. Krebbers, R. Piotrowiak, U. Bonas, R. Hehl, and H. Saedler. 1985. Transposable elements of *Antirrhinum majus*. In M. Freeling, ed., *Plant Genetics*, pp. 433–444. New York: Liss.

Sorensen, F. 1967. Linkage between marker genes and embryonic lethal factors may cause disturbed segregation ratios. *Silvae Genetica* 16:132–134.

Sorensen, F. 1969. Embryonic genetic load in coastal Douglas-fir, *Pseudotsuga Menziesii* var. *Menziesii*. *Amer. Natur.* 103:389–398.

Sorensen, F. C. 1982. The roles of polyembryony and embryo viability in the genetic system of conifers. *Evolution* 36:725–733.

Sorensen, F. C., J. F. Franklin, and R. Woollard. 1976. Self-pollination effects on seed and seedling traits in noble fir. *For. Sci.* 22:155–159.

Soyfer, V. N. 1983. Influence of physiological conditions on DNA repair and mutagenesis in higher plants. *Physiol. Plant.* 58:373–380.

Sparrow, A. H., H. J. Price, and A. G. Underbrink. 1972. A survey of DNA content per cell and per chromosome of prokaryotic and eukaryotic organisms: Some evolutionary considerations. In H. H. Smith, ed., *Evolution of Genetic Systems*, pp. 451–494. New York: Gordon and Breach.

Sparrow, A. H., A. F. Rogers, and S. S. Schwemmer. 1968. Radiosensitivity studies with woody plants. I. Acute gamma irradiation survival data for 28 species and predictions for 190 species. *Radiat. Bot.* 8:149–186.

Sparrow, A. H., L. A. Schairer, and R. C. Sparrow. 1963. A relationship between estimated interphase chromosome volumes and relative radiosensitivities. *Genetics* 48:911–912.

Sparrow, R. C. and A. H. Sparrow. 1965. Relative radiosensitivities of woody and herbaceous spermatophytes. *Science* 147:1449–1451.

Sprague, G. F., W. A. Russell, and L. H. Penny. 1960. Mutations affecting quantitative traits in the selfed progeny of doubled monoploid maize stocks. *Genetics* 45:855–866.

Stachel, S. E., B. Timmerman, and P. Zambryski. 1986. Generation of single-stranded T-DNA molecules during the initial stages of T-DNA transfer from *Agrobacterium tumefaciens* to plant cells. *Nature* 322:706–712.

Stadler, L. J. 1942. Some observations on gene variability and spontaneous mutation. The Spragg Memorial Lectures on Plant Breeding (Third Series) East Lansing, Michigan State College.

Stanley, R. G. and H. F. Linskens. 1974. *Pollen*. New York: Springer-Verlag.

Stebbins, G. L. 1950. *Variation and Evolution in Plants*. New York: Columbia University Press.

Stebbins, G. L. 1959. The role of hybridization in evolution. *Proc. Amer. Phil. Soc.* 103:231–251.

Stebbins, G. L. 1966. Chromosomal variation and evolution. *Science* 152:1463–1469.

Stebbins, G. L. 1971. *Chromosomal Evolution in Higher Plants*. London: Arnold.

Stebbins, G. L. and E. Yagil. 1966. The morphogenetic effects of the hooded gene in barley. I. The course of development in hooded and awned genotypes. *Genetics* 54:727–741.

Steeves, T. A. and I. M. Sussex. 1972. *Patterns in Plant Development*. Englewood Cliffs, N.J.: Prentice-Hall.

Stein, D. B., W. F. Thompson, and H. S. Belford. 1979. Studies on DNA sequences in the Osmundaceae. *J. Mol. Evol.* 13:215–232.

Stein, O. L., A. H. Sparrow, and L. A. Schairer. 1959. Leaf tumors induced in *Graptopetalum paraguayense* by gamma radiation. *Cancer Research* 19:746–748.

Stein, O. L. and D. M. Steffensen. 1959. The activity of x-rayed apical meristems: A genetic and morphogenetic analysis in *Zea mays. Z. für Vererbungslehre* 90:483–502.

Steinhauer, D. A. and J. J. Holland. 1986. Direct method for quantitation of extreme polymerase error frequencies at selected single base sites in viral RNA. *J. Virology* 57:219–228.

Steinmetz, V. and E. Wellmann. 1986. The role of solar UV-B in growth regulation of cress *Lepidium sativum* L. seedlings. *Photochem. Photobiol.* 43:189–193.

Stephens, S. G. 1944a. The genetic organization of leaf-shape development in the genus *Gossypium. J. Genetics* 46:28–51.

Stephens, S. G. 1944b. The modifier concept: A developmental analysis of leaf-shape "modification" in New World cottons. *J. Genetics* 46:331–344.

Stephens, S. G. 1944c. Canalization of gene action in the *Gossypium* leaf-shape system and its bearing on certain evolutionary mechanisms. *J. Genetics* 46:345–357.

Stephenson, A. G. 1981. Flower and fruit abortion: Proximate causes and ultimate functions. *Ann. Rev. Ecol. Syst.* 12:253–279.

Stephenson, A. G. and J. A. Winsor. 1986. *Lotus corniculatus* regulates offspring quality through selective fruit abortion. *Evolution* 40:453–458.

Stephenson, A. G., J. A. Winsor, and L. E. Davis. 1986. Effects of pollen load size on fruit maturation and sporophyte quality in zucchini. In D. L. Mulcahy, G. Bergamini Mulcahy, and E. Ottaviano, eds., *Biotechnology and Ecology of Pollen*, pp. 429–434. New York: Springer-Verlag.

Stern, D. B. and J. D. Palmer. 1984. Extensive and widespread homologies between mitochondrial DNA and chloroplast DNA in plants. *Proc. Natl. Acad. Sci. USA* 81:1946–1950.

Stevenson, D. W. 1980. Radial growth in the Cycadales. *Amer. J. Bot.* 67:465–475.

Stevenson, D. W. and J. B. Fisher. 1980. The developmental relationship between primary and secondary thickening growth in *Cordyline* (Agavaceae). *Bot. Gaz.* 141:264–268.

Stewart, R. N. 1978. Ontogeny of the primary body in chimeral forms of higher plants. In S. Subtelny and I. M. Sussex, eds., *The Clonal Basis of Development*, pp. 131–160. New York: Academic Press.

Stewart, R. N. and H. Dermen. 1970. Determination of number and mitotic activity of shoot apical initial cells by analysis of mericlinal chimeras. *Amer. J. Bot.* 57:816–826.

Stewart, R. N. and H. Dermen. 1975. Flexibility in ontogeny as shown by the contribution of the shoot apical layers to leaves of periclinal chimeras. *Amer. J. Bot.* 62:935–947.

Stewart, R. N. and H. Dermen. 1979. Ontogeny in monocotyledons as revealed by studies of the developmental anatomy of periclinal chloroplast chimeras. *Amer. J. Bot.* 66:47–58.

Stewart, R. N., P. Semeniuk, and H. Dermen. 1974. Competition and accommodation between apical layers and their derivatives in the ontogeny of chimeral shoots of *Pelargonium X hortorum. Amer. J. Bot.* 61:54–67.

Stewart, S. C. and D. J. Schoen. 1986. Segregation at enzyme loci in megagametophytes of white spruce, *Picea glauca. Can. J. Genet. Cytol.* 28:149–153.

Stumpf, P. K. 1980. Biosynthesis of saturated and unsaturated fatty acids. In P. K. Stumpf and E. E. Conn, eds., *The Biochemistry of Plants.* Vol. 4, *Lipids: Structure and Function*, pp. 177–204. New York: Academic Press.

Sussex, I. and D. Rosenthal. 1973. Differential ^{3}H-Thymidine labeling of nuclei in the shoot apical meristem of *Nicotiana*. *Bot. Gaz.* 134:295–301.

Sutherland, W. J. and A. R. Watkinson. 1986. Do plants evolve differently? *Nature* 320:305.

Swanson, C. P., T. Merz, and W. J. Young. 1981. *Cytogenetics*. 2d ed. Englewood Cliffs, N.J.: Prentice-Hall.

Sybenga, J. 1972. *General Genetics*. New York: American Elsevier.

Tanksley, S. D., D. Zamir, and C. M. Rick. 1981. Evidence for extensive overlap of sporophytic and gametophytic gene expression in *Lycopersicon esculentum*. *Science* 213:453–455.

Tautz, D., M. Trick, and G. A. Dover. 1986. Cryptic simplicity in DNA is a major source of genetic variation. *Nature* 322:652–656.

Taylor, L. P. and V. Walbot. 1985. A deletion adjacent to the maize transposable element Mu-1 accompanies loss of Adh1 expression. *The EMBO Journal* 4:869–876.

Taylor, T. N. 1981. *Paleobotany, An Introduction to Fossil Plant Biology*. New York: McGraw-Hill.

Tazima, Y. 1982. Mutagenic and carcinogenic mycotoxins. In E. J. Klekowski, Jr., ed., *Environmental Mutagenesis, Carcinogenesis, and Plant Biology*, 1:67–95. New York: Praeger.

Teramura, A. H. 1983. Effects of ultraviolet-B radiation on the growth and yield of crop plants. *Physiol. Plant.* 58:415–427.

Thielke, C. 1951. Uber die Moglichkeiten der Periklinalchimarenbildung bei Grasern. *Planta* 39:402–430.

Thielke, C. 1954. Die histologische Struktur des Sprossvegetationskegels einiger Commelinaceen unter Berucksichtigung panaschierter Formen. *Planta* 44:18–74.

Thielke, C. 1955. Die Struktur des Vegetationskegels einer sektorial panaschierten *Hemerocallis fulva* L. *Ber. dtsch. bot. Ges.* 68:233–238.

Thomas, H. and D. Grierson (eds.). 1987. *Developmental Mutants in Higher Plants*. Cambridge: Cambridge University Press.

Thomashow, M. F., S. Hugly, W. G. Buchholz, and L. S. Thomashow. 1986. Molecular basis for the auxin-independent phenotype of crown gall tumor tissues. *Science* 231:616–618.

Tilney-Bassett, R. A. E. 1963. The structure of periclinal chimeras. *Heredity* 18:265–285.

Tilney-Bassett, R. A. E. 1986. *Plant Chimeras*. London: Arnold.

Timmis, J. N. and N. S. Scott. 1983. Sequence homology between spinach nuclear and chloroplast genomes. *Nature*. 305:65–67.

Timmis, J. N. and N. S. Scott. 1985. Movement of genetic information between the chloroplast and nucleus. In B. Hohn and E. S. Dennis, eds., *Genetic Flux in Plants*, pp. 61–78. Vienna, New York: Springer-Verlag.

Tomlinson, P. B. 1961. *Anatomy of the Monocotyledons II. The Palmae*. Oxford: Oxford University Press.

Tomlinson, P. B. 1982. Chance and design in the construction of plants. In R. Sattler, ed., *Axioms and Principles of Plant Construction*. *Acta Biotheoretica*, vol. 31A. The Hague, Boston, London: Martinus Nijhoff/Junk.

Tomlinson, P. B. 1986. *The Botany of Mangroves*. London, New York, Cambridge University Press.

Tourte, Y., J. Kuligowski-Andres, and C. Barbier-Ramond. 1980. Comportement differentiel des chromatines paternelles et maternelles au cours de l'embryogenese d'une fougere: Le Marsilea. *European Journal of Cell Biology* 21:28–36.

Trewavas, A. 1982. Possible control points in plant development. In H. Smith and D.

Grierson, eds., *The Molecular Biology of Plant Development*, pp. 7–27. Berkeley: University of California Press.

Tryon, R. M. 1941. Revision of the genus *Pteridium*. *Rhodora* 43:1–70.

Tschermak-Woess, E. 1956. Karyologische Pflanzenanatomie. *Protoplasma* 46:798–834.

Tucker, E. B. 1982. Translocation in the staminal hairs of *Setcreasea purpurea*. I. A study of cell ultrastructure and cell-to-cell passage of molecular probes. *Protoplasma* 113:193–201.

Tucker, S. C. 1962. Ontogeny and phyllotaxis of the terminal vegetative shoots of *Michelia fuscata*. *Amer. J. Bot.* 49:722–737.

Turner, J. R. G. 1967. Why does the genotype not congeal? *Evolution* 21:645–656.

Ullu, E. and C. Tschudi. 1984. *Alu* sequences are processed 7SL RNA genes. *Nature* 312:171–172.

Van't Hof, J. 1985. Control points within the cell cycle. In J. A. Bryant and D. Francis, eds., *The Cell Division Cycle in Plants* (Society for Experimental Biology, Seminar Series 26), pp. 1–13. Cambridge: Cambridge University Press.

Van't Hof, J. and C. A. Bjerknes. 1979. Chromosomal DNA replication in higher plants. *BioScience* 29:18–22.

Van't Hof, J. and C. A. Bjerknes. 1981. Similar replicon properties of higher plant cells with different S periods and genome sizes. *Exp. Cell Res.* 136:461–466.

Van't Hof, J., C. A. Bjerknes, and N. C. Delihas. 1983. Excision and replication of extrachromosomal DNA of Pea *(Pisum sativum)*. *Molecular and Cellular Biology* 3:172–181.

Van't Hof, J., A. Kuniyuki, and C. A. Bjerknes. 1978. The size and number of replicon families of chromosomal DNA of *Arabidopsis thaliana*. *Chromosoma* (Berlin) 68:269–285.

Van Den Ende, H. 1976. *Sexual Interactions in Plants*. New York: Academic Press.

Van Groenendael, J. M. 1985. Teratology and metameric plant construction. *New Phytol.* 99:171–178.

Vanin, E. F. 1984. Processed pseudogenes characteristics and evolution. *Biochim. Biophys. Acta* 782:231–241.

Van Valen, L. 1973. A new evolutionary law. *Evol. Theory* 1:1–30.

Varmus, H. E. 1982. Form and function of retroviral proviruses. *Science* 216:812–820.

Varshavsky, A. 1981. On the possibility of metabolic control of replicon "misfiring": Relationship to emergence of malignant phenotypes in mammalian cell lineages. *Proc. Natl. Acad. Sci. USA* 78:3673–3677.

Veleminsky, J. and T. Gichner. 1978. DNA repair in mutagen-induced higher plants. *Mutat. Res.* 55:71–84.

Vig, B. K. 1982. Somatic crossing-over in higher plants. In E. J. Klekowski, Jr., ed., *Environmental Mutagenesis, Carcinogenesis, and Plant Biology*, 2:25–54. New York: Praeger.

Vogel, F. and A. G. Motulsky. 1986. *Human Genetics*. 2d ed. Berlin: Springer-Verlag.

Wagner, M. 1986. A consideration of the origin of processed pseudogenes. *Trends in Genetics* 2:134–137.

Walbot, V. 1985. On the life strategies of plants and animals. *Trends in Genetics* 1:165–169.

Walbot, V. 1986. Inheritance of mutator activity in *Zea mays* as assayed by somatic instability of the *bz2-mu1* allele. *Genetics* 114:1293–1312.

Walbot, V. and E. H. Coe, Jr. 1979. Nuclear gene *iojap* conditions a programmed change to ribosome-less plastids in *Zea mays*. *Proc. Natl. Acad. Sci. USA* 76:2760–2764.

Walbot, V. and C. A. Cullis. 1983. The plasticity of the plant genome—is it a requirement for success? *Plant Mol. Biol. Rep.* 1:3–11.

Walbot, V. and C. A. Cullis. 1985. Rapid genomic change in higher plants. *Ann. Rev. Plant Physiol.* 36:367–396.

Wallace, B. 1968. *Topics in Population Genetics.* New York: Norton.

Wallace, B. 1970. *Genetic Load.* Englewood Cliffs, N.J.: Prentice-Hall.

Wallace, B. 1981. *Basic Population Genetics.* New York: Columbia University Press.

Wallace, B. and C. Madden. 1953. The frequencies of sub- and supervitals in experimental populations of *Drosophila melanogaster. Genetics* 38:456–470.

Wallace, H. and W. H. R. Langridge. 1971. Differential amphiplasty and the control of ribosomal RNA synthesis. *Heredity* 27:1–13.

Walmsley, R. W., C. S. M. Chan, B.-K. Tye, and T. D. Petes. 1984. Unusual DNA sequences associated with the ends of yeast chromosomes. *Nature* 310:157–160.

Ward, B. L., R. S. Anderson, and A. J. Bendich. 1981. The mitochondrial genome is large and variable in a family of plants (Cucurbitaceae). *Cell* 25:793–803.

Wareing, P. F. 1959. Problems of juvenility and flowering in trees. *J. Linn. Soc. London* 56:282–289.

Warne, T. R. and R. M. Lloyd. 1981. Inbreeding and homozygosity in the fern, *Ceratopteris pteridoides* (Hooker) Hieronymus (Parkeriaceae). *Bot. J. Linn. Soc.* 83:1–13.

Watson, J. D. 1972. Origin of concatemeric T7 DNA. *Nature New Biol.* 239:197–201.

Watson, J. D., N. H. Hopkins, J. W. Roberts, J. A. Steitz, and A. M. Weiner. 1987. *Molecular Biology of the Gene.* 4th ed. Vol. 1. Reading, Mass.: Benjamin Cummings.

Watt, A. S. 1976. The ecological status of bracken. *Bot. J. Linn. Soc.* 73:217–239.

Waxman, S. 1975. Witches'-brooms' sources of new and interesting dwarf forms of *Picea, Pinus* and *Tsuga* species. *Acta Horticulturae* 54:25–32.

Weiling, F. 1962. Lasst sich die Zahl der Initialzellen der embryonalen Sprossanlage hoherer Pflanzen mit Hilfe von Rontgenmutanten ermitteln? *Ber Deutsch Bot. Ges.* 75:211–217.

Weiner, A. M., P. L. Deininger, and A. Efstratiadis. 1986. Nonviral retroposons: Genes, pseudogenes, and transposable elements generated by the reverse flow of genetic information. *Ann. Rev. Biochem.* 55:631–661.

Weismann, A. 1892. *Das Keimplasma. Eine Theorie der Vererbung.* Jena, Germany: Fischer.

White, J. 1979. The plant as a metapopulation. *Ann. Rev. Ecol. Syst.* 10:109–145.

White, M. J. D. 1973. *Animal Cytology and Evolution.* 3d ed. Cambridge: Cambridge University Press.

White, O. E. 1948. Fasciation. *Bot. Rev.* 14:319–358.

White, P. R. and A. C. Braun. 1941. Crown gall production by bacteria-free tumor tissues. *Science* 94:239–241.

Whitehouse, H. L. K. 1950. Multiple-allelomorph incompatibility of pollen and style in the evolution of the angiosperms. *Ann. Bot.* 14:199–216.

Whitham, T. G. 1981. Individual trees as heterogeneous environments: Adaptation to herbivory or epigenetic noise? In R. F. Denno and H. Dingle, eds., *Insect Life History Patterns: Habitat and Geographic Variation,* pp. 9–27. New York: Springer-Verlag.

Whitham, T. G. 1983. Host manipulation of parasites: Within-plant variation as a defense against rapidly evolving pests. In R. F. Denno and M. S. McClure, eds., *Variable Plants and Herbivores in Natural and Managed Systems,* pp. 15–41. New York: Academic Press.

Whitham, T. G. and S. Mopper. 1985. Chronic herbivory: Impacts on architecture and sex expression of pinyon pine. *Science* 228:1089–1091.

Whitham, T. G. and C. N. Slobodchikoff. 1981. Evolution by individuals, plant-herbivore interactions, and mosaics of genetic variability: The adaptive significance of somatic mutations in plants. *Oecologia* (Berl) 49:287–292.

Whitham, T. G., A. G. Williams, and A. M. Robinson. 1984. The variation principle: Individual plants as temporal and spatial mosaics of resistance to rapidly evolving pests. In P. W. Price, C. N. Slobodchikoff, and W. S. Gaud, eds., *A New Ecology: Novel Approaches to Interactive Systems*, pp. 16–51. New York: Wiley.

Whyte, L. L. 1960. Developmental selection of mutations. *Science* 132:954, 1694–1695.

Whyte, L. L. 1964. Internal factors in evolution. *Acta Biotheoretica* 17:33–48.

Whyte, L. L. 1965. *Internal Factors in Evolution*. New York: Braziller.

Wiens, D. 1984. Ovule survivorship, brood size, life history, breeding systems, and reproductive success in plants. *Oecologia* (Berl) 64:47–53.

Wiens, D., C. L. Calvin, C. A. Wilson, C. I. Davern, D. Frank, and S. R. Seavey. 1987. Reproductive success, spontaneous embryo abortion, and genetic load in flowering plants. *Oecologia* (Berl) 71:501–509.

Wilcox, M. D. 1983. Inbreeding depression and genetic variances estimated from self- and cross-pollinated families of *Pinus radiata*. *Silvae Genetica* 32:89–95.

Wilkie, D. 1956. Incompatibility in bracken. *Heredity* 10:247–256.

Williams, G. C. 1966. *Adaptation and Natural Selection*. Princeton, N.J.: Princeton University Press.

Williams, G. C. 1975. *Sex and Evolution*. Princeton, N.J.: Princeton University Press.

Williams, R. F. 1975. *The Shoot Apex and Leaf Growth*. Cambridge: Cambridge University Press.

Willis, A. E. and T. Lindahl. 1987. DNA ligase I deficiency in Bloom's syndrome. *Nature* 325:355–357.

Willson, M. F. and N. Burley. 1983. *Mate Choice in Plants*. Princeton, N.J.: Princeton University Press.

Wilson, B. F. 1966. Development of the shoot system of *Acer rubrum* L. *Harvard Forest Paper* 14:1–21.

Wolpert, L. 1969. Positional information and the spatial pattern of cellular differentiation. *J. theor. Biol.* 25:1–47.

Wolpert, L. 1971. Positional information and pattern formation. In A. A. Moscona and A. Monroy, eds., *Current Topics in Developmental Biology*, 6:183–224. New York: Academic Press.

Wolpert, L. 1978. Pattern formation in biological development. *Scientific American* 239:154–164.

Woodwell, G. M. 1967. Radiation and the patterns of nature. *Science* 156:461–470.

Wright, S. 1977. *Evolution and the Genetics of Populations*. Vol. 3. Chicago: University of Chicago Press.

Yagil, E. and G. L. Stebbins. 1969. The morphogenetic effects of the hooded gene in barley. II. Cytological and environmental factors affecting gene expression. *Genetics* 62:307–319.

Yamagishi, H. 1986. Role of mammalian circular DNA in cellular differentiation. *BioEssays* 4:218–221.

Ying. C. C. 1978. Performance of white spruce [*Picea glauca* (Moench) Voss] progenies after selfing. *Silvae Genetica* 27:214–215.

Zamir, D. and Y. Tadmor. 1986. Unequal segregation of nuclear genes in plants. *Bot. Gaz.* 147:355–358.

Zimmermann, F. K. 1971. Induction of mitotic gene conversion by mutagens. *Mutat. Res.* 11:327–337.

Zouros, E. 1976. Hybrid molecules and the superiority of the heterozygote. *Nature* 262:227–229.

Zuckerkandl, E. and L. Pauling. 1965. Evolutionary divergence and convergence in proteins. In V. Bryson and H. J. Vogel, eds., *Evolving Genes and Proteins*, pp. 97–166. New York and London: Academic Press.

Author Index

Abrahamson, S., 43
Adams, W. T., 275
Ainsworth, C. C., 300
Akebrand, V., 271, 272
Albertsen, M. C., 172
Albrecht, H. R., 287
Aleksiuk, M. A., 275
Alff-Steinberger, C., 185
Allard, R. W., 188
Amasino, R. M., 27
Amiro, B. D., 48, 54
Amo Rodriguez, S. Del, 33
Anderson, E., 296
Anderson, R. S., 72
Andersson, E., 256
Antolin, M. F., 303
Apirion, D., 281
Appels, R., 188, 296
Asbeck, F., 178
Atkinson, Y. E., 76
Atwood, K. C., 30
Avery, A. G., 40, 114, 195
Ayala, F. J., 32
Ayme, S., 233

Babcock, E. B., 2
Balkema, G. H., 86
Ball, E., 96, 120
Baltimore, D., 23
Bannan, M. W., 164, 165
Barbier-Ramond, C., 90, 92
Barghoorn, E. S., 159
Barlow, P. W., 119, 175, 210
Barthonnet, J., 90
Bateman, A. J., 56, 284

Batschelet, E., 208
Bawa, K. S., 238, 284, 285
Bay, C. E., 205, 206
Bayly, I. L., 165
Bazzaz, F. A., 238, 242
Becker, F. F., 299
Beckmann, J. S., 32
Bedbrook, J. R., 27, 59
Belford, H. S., 295
Bell, A. D., 158, 161, 162, 309
Bell, G., 54
Bender, A. D., 43
Bendich, A. J., 72, 73
Benne, R., 15
Bennett, M. D., 43, 59
Bergamini Mulcahy, G., 232
Bergann, F., 114-16, 124, 126
Bergann, L., 114-16, 124, 126
Bergfeld, R., 76
Bergner, A. D., 195
Bernstein, H., 315
Bertin, R. I., 237
Bevilacqua, B., 30
Bidney, D., 34
Bierhorst, D. W., 89, 90, 93, 96
Biggs, R. H., 179
Billings, W. D., 177-79
Binns, A. N., 202
Bird, A. P., 185
Birky, C. W., Jr., 81, 83
Bjerknes, C. A., 65-67
Blackburn, E. H., 56, 58, 59
Blakeslee, A. F., 40, 114, 195, 230-32
Blau, H. M., 201
Blinkenberg, C., 279

Bloch, R., 170, 171
Blomberg, C., 14
Bloom, K., 54
Blumberg vel Spaive, J., 22
Bodmer, W. F., 285, 287
Boeke, J. D., 22
Boffey, S. A., 77, 81
Bohr, V. A., 64
Bonas, U., 23
Börner, T., 75, 76
Botstein, D., 65
Bower, F. O., 158
Bowman, C. M., 73
Boy de la Tour, E., 65
Bradbeer, J. W., 76
Bramlett, D. L., 273, 275
Brandham, P. E., 300, 302
Branscomb, E. W., 63
Braun, A. C., 171
Breese, E. L., 306, 307
Brennicke, A., 23, 71
Brinikh, L. I., 97
Brink, R. A., 44, 201, 202, 287
Brisson, N., 26
Brix, H., 279
Brodführer Franzgrote, U., 177, 178
Brown, C. L., 153
Brown, S., 295, 299
Brown, W. M., 74
Brown, W. V., 110
Brues, A. M., 253
Bryant, J. A., 57, 60, 62, 63
Buchholz, J. T., 3, 6, 222, 225-27, 230-33, 243
Buchholz, W. G., 171
Buckley, D. P., 270
Burley, J., 271, 285, 286
Burley, N., 225, 237, 244
Burnham, C. R., 297
Burns, J. A., 299
Burr, B., 21, 32
Burr, F. A., 21, 32
Buss, L. W., 2
Butterfield, B. G., 159, 163, 164
Buvat, R., 117
Byerly, H. C., 315

Cairns, J., 6, 92, 134, 169
Caldwell, M. M., 177-79
Calkins, J., 177

Callen, D. F., 176
Calvin, C. L., 279, 308
Carde, J.-P., 76
Carpenter, R., 19, 21, 45
Carr, D. J., 210
Carteledge, J. L., 40
Casper, B. B., 279
Castroviejo, M., 62, 63
Cavalier-Smith, T., 43, 56, 58
Cavalli-Sforza, L. L., 285, 287
Cedar, H., 186
Chamberlain, C. J., 226, 229, 235, 237
Chan, C. S. M., 56
Chan, J. Y., 299
Chandler, V. L., 17
Charlesworth, B., 308
Charlesworth, D., 308
Charnov, E. L., 238
Cheadle, V. I., 159
Chenou, E., 204, 205
Chiu, C-P., 201
Chomet, P. S., 17, 44, 45
Chua, N.-H., 216
Cichan, M. A., 159, 163
Clark, A. M., 176
Clausen, R. E., 2
Clayton, D. A., 74
Cleary, W., 43
Cleland, R. E., 288
Clowes, F. A. L., 96, 119, 136-38
Coe, E. H., Jr., 76, 140, 142, 143
Coen, E. S., 19, 20, 45, 187, 189, 295, 299
Cohen, D., 143
Coles, J. F., 281, 290
Conger, A. D., 43
Conway, E., 263
Cooper, D. C., 287
Corson, G. E., 110
Coulondre, C., 185
Coulter, J. M., 229
Cousens, M. I., 267
Cox, E. C., 35
Cram, W. H., 279
Crick, F. H. C., 37, 43
Crist, K. C., 267
Critchfield, W. B., 178
Crosland, M. W. J., 54
Crow, J. F., 32, 235, 236, 252, 254
Crozier, R. H., 54

Crumpacker, D. W., 32, 276
Cullis, C. A., 2, 28, 43, 297, 305
Cullmann, G., 184
Cumbie, B. G., 163
Curry, T. M., 172
Cuypers, H., 18, 20

D'Amato, F., 39, 42
Dancik, B. P., 275
Dancis, B. M., 58
Darlington, C. D., 297
Darwin, C., 238
Davern, C. I., 279, 308
Davidson, D., 167
Davis, E. L., 117
Davis, G. L., 229
Davis, L. E., 238, 240, 241
Davis, R. W., 54, 55
Deininger, P. L., 23
Delihas, N. C., 67
Dellaporta, S. L., 17, 44, 45
De Nettancourt, D., 284
Denhardt, D. T., 60
De Pamphilis, M. L., 60
Dermen, H., 117, 120, 122, 126, 129,
  197, 198
Devoret, R., 36
Diers, L., 81
DiNardo, S., 65
Dinter-Gottlieb, G., 27
Dobzhanksy, N. P., 290
Dobzhansky, Th., 32, 252, 253, 261
Doolittle, R. F., 183
Doolittle, W. F., 37, 43
Döring, H.-P., 16, 17, 21-23
Dorne, A.-J., 76
Douce, R., 76
Dourado, A. M., 39
Dover, G., 12, 16, 26, 28, 43, 187-89,
  293-95, 299
Draginskaya, L. Ya., 97
Drake, J. W., 12, 13, 60
Drouin, G., 26
Drumm-Herrel, H., 180, 181
Dufton, M. J., 183
Dugle, J. R., 48, 54, 203
Dunham, V. L., 61-63
Durrant, A., 305
Dvořák, J., 188, 296
Dyer, A. F., 74, 167, 168

Earnshaw, W. C., 65
Echeverria, M., 61
Edwards, K. L., 211, 212
Edwards, Y. H., 191
Efstratiadis, A., 23
Ehrenberg, L., 30
Eiche, V., 245
Eigen, M., 8, 9
Einset, J., 117
Ekberg, L., 275
Eldridge, K. G., 277, 279, 280
Ellstrand, N. C., 117
Emery, W. H. P., 110
Epp, M. D., 288
Eriksson, G., 30, 271, 272
Esau, K., 94, 96, 148, 201
Evans, D. A., 34, 168
Evans, H. J., 47-50, 52
Evola, S. V., 32

Fahn, A., 96, 143
Faivre-Baron, M., 90
Falke, L., 211
Falquet, J., 65
Farabaugh, P. J., 185
Farrar, D. R., 267
Faust, E. A., 60
Fedoroff, N. V., 16
Feldman, M. W., 35
Felsenstein, J., 309
Ferrand, M., 204, 205
Fersht, A. R., 11
Figureau, A., 184
Fink, G. R., 22
Finnegan, D. J., 26
Fisher, J. B., 143, 159
Fishman, T. N., 206
Flavell, A. J., 22
Flavell, R. B., 21, 27, 28, 59, 187-89,
  294-96, 299
Foard, D. E., 206-10, 214
Foster, A. S., 146, 195
Fowler, D. P., 247, 271-73, 279, 281,
  290
Frank, D., 279, 308
Franklin, E. C., 271, 272, 275, 279
Franklin, I. R., 290
Franklin, J. F., 272
Freeling, M., 21, 22, 32, 37, 215
Fripp, Y. J., 277

Fuerst, P., 81
Fukshansky, L., 144, 145, 217
Furner, I. J., 27
Furnier, G. R., 275
Furnival, G. M., 271, 285, 286

Gabay-Laughnan, S., 71
Gabriel, W. J., 289-91
Gallant, J. A., 14
Ganders, F. R., 265
Garber, R., 21
Garfinkel, D. J., 22, 27
Gasser, S. M., 65
Gaul, H., 3
Geburek, Th., 273
German, J., 299
Gerstel, D. U., 299
Gibson, T. C., 35
Gichner, T., 64
Gierl, A., 18, 20
Gifford, E. M., Jr., 91, 97, 110, 126, 146
Gilbert, W., 185
Giles, N., 167
Gill, D. E., 305
Glickman, B. W., 12, 13, 63
Goebl, M. G., 192
Gomez-Pompa, A., 33
Good, C. W., 89
Goodman, M. F., 63
Gordon, M. P., 27
Gorter, C. J., 171, 172
Goto, T., 65
Gottlieb, L. D., 170, 215
Grant, V., 168, 214, 296, 297
Gray, A., 152
Green, P. B., 90, 211, 214
Green, P. J., 216
Greenblatt, I. M., 44-46
Gregg, T. L., 266
Grierson, D., 215
Griffin, A. R., 248, 256, 277, 279, 280
Griffith, M. M., 195, 196
Grigorieva, G. A., 97
Grinikh, L. I., 97
Grodzinsky, D. B., 119, 175
Gruenbaum, Y., 186
Gudkov, I. M., 119, 175
Guédes, M., 152
Guharay, F., 211
Gunckel, J. E., 110, 172, 174

Gunning, B. E. S., 193
Gupta, M., 22

Haber, A. H., 206-10, 214
Hackett, G., 148
Hadidi, A., 27
Hadorn, E., 260, 261
Hagemann, R., 76
Haigh, J., 130, 309
Hake, S., 215
Haldane, J. B. S., 317
Hall, J. W., 89
Hallé, F., 148, 152, 162
Halverson, T. G., 305
Hanawalt, P. C., 64
Hannan, M. A., 177
Hara, N., 109-11
Harberd, D. J., 306
Hardeman, E. C., 201
Hardwick, R. C., 169
Harper, J. L., 5, 219, 263, 266
Harris, H., 191
Harrison, B. J., 18, 19
Hart, R. W., 64
Harte, C., 215
Hartman, P. E., 11
Hawkins, J. L., 203
Hayward, M. D., 306, 307
Hébant, C., 90
Hébant-Mauri, R., 90
Heck, M. S., 65
Hedrick, P. W., 258
Hehl, R., 23
Heidecker, G., 189-91
Heimsch, C., 110
Hejnowicz, K., 214
Hejnowicz, Z., 214
Heneen, W. K., 299
Hepler, P. K., 212
Herbon, L. A., 74
Hiatt, H. H., 169, 176
Hieter, P., 54, 55
Hill, A., 54
Hill, A. B., 41, 67, 70
Hill, J. B., 142, 147
Hillier, H. G., 129
Hinegardner, R., 43
Ho, F. K., 290
Hoffmann, G. W., 14
Hoffmann-Ostenhoff, O., 39

Hohn, B., 22, 25
Hohn, T., 22, 25
Holbrook-Walker, S. G., 267
Holland, J. J., 8
Holliday, R., 14, 15
Holm, C., 65
Holmquist, G. P., 55, 58
Holsinger, K. E., 35
Honda, H., 143
Hopf, F. A., 315
Hopkins, N. H., 60
Hopkinson, D. A., 191
Horn, H. S., 143
Horton, D. M., 300
Howland, G. P., 64
Hsu, T. C., 299
Huffman, G. A., 27
Hugly, S., 171
Huxley, J. S., 206

Iwasa, Y., 143

Jabs, E. W., 56
Jacquard, A., 254
Jansson, R., 256
Janzen, D. H., 237, 238
Jaynes, R. A., 278
Jensen, H. P., 32
Johansen, D. A., 229
Johansson, J., 14
John, B., 297, 298
Johns, M. A., 22, 32
Johnson, L. C., 178
Johnson, M. A., 93, 110
Johnson, M. A. T., 302
Johnsson, H., 271
Johri, M. M., 140, 142
Joly, R. J., 275
Jones, D. F., 42, 170, 193, 230
Jones, G. N., 230
Jones, J., 27, 59
Jones, K., 48, 54, 298
Jones, R. N., 67-70
Jørgensen, J. H., 32
Jorgensen, R. A., 188
Joyard, J., 76
Jukes, T. H., 183
Jund, E., 306

Kahn, C., 303

Kaplan, S. M., 248
Kaufhold, M., 200
Kaul, V., 286
Kay, H. E. M., 7, 86, 88, 89, 119
Kazarinova-Fukshansky, N., 87, 98-107, 122, 126-32, 144, 145
Keclard-Christophe, L., 63
Kehr, A. E., 171
Kemble, R. J., 71
Kenrick, R. J., 286
Kimura, M., 191, 236, 252, 254, 303
King, J. L., 183
Kirk, J. T. O., 7, 79, 81, 124, 125, 129
Kirkwood, T. B. L., 14, 15
Klekowski, E. J., Jr., 7, 33, 34, 90, 98-107, 122, 126-33, 144, 145, 249, 259-61, 263, 265, 267, 269, 285
Knill-Jones, J. W., 11
Knowles, P., 275
Knudson, A. G., Jr., 303
Kny, L., 265
Kondrashov, A. S., 312-15
Kormanik, P. P., 153
Korn, R. W., 214
Koski, V., 271
Kossuth, S. V., 179
Kraczewska, E. K., 67
Krebbers, R., 23
Krimer, D. B., 67
Kuhlemeier, C., 216
Kuligowski, J., 204, 205
Kuligowski-Andres, J., 90, 92
Kuniyuki, A., 66
Kunkel, T. A., 60, 62
Kurth, E., 91
Kuser, J., 271

Labouygues, J. M., 184
La Cour, L. F., 167
Laemmli, U. K., 65
Lamm, S. S., 67
Lamotte, C. E., 172
Lande, R., 308
Langenauer, H., 117
Langner, W., 273, 290
Langridge, J., 102, 132, 192, 194
Langridge, W. H. R., 189
Laroche, T., 65
Lasswell, W. L., 177
Laughnan, J. F., 71

Lawrence, G. H. M., 271
Lea, P. J., 76
Lee, T. D., 238, 242
Leigh, E. G., Jr., 35
Leon, J. A., 143
Lerner, I. M., 290
Lester, D. T., 279, 290
Levene, H., 290
Levin, D. A., 276-78, 282, 290, 297, 298
Levin, D. E., 265
Levings, C. S., III, 72
Lewis, D., 284
Lewis, K. R., 297, 298
Li, S. L., 97, 246, 275
Li, W.-H., 184
Libby, W. J., 272, 306
Lichtenstein, C., 171
Liljenstrom, H., 14
Lima-de-Faria, A., 41, 56
Lindahl, T., 299
Lindgren, D., 235, 236, 255, 256, 273
Lindoo, S. J., 179
Link, G., 76
Linskens, H. F., 179
Lintilhac, P. M., 90, 211
Lippmand-Hand, A., 233
Litvak, S., 62, 63
Lloyd, R. M., 261, 265-67
Lockhart, J. A., 177, 178
Loeb, L. A., 12, 60, 62
Long, T. J., 209
Lonsdale, D. M., 71
Luo, C.-C., 184
Lyman, J. C., 117
Lyrene, P., 278

McAlpin, B. W., 89
McAlpine, R. G., 153
McClintock, B., 16, 17, 21, 36, 58, 59, 189
McCutchan, B. G., 272
MacKay, T. F. C., 36
MacPhee, D. G., 8-10, 29, 95
Madden, C., 261
Maek, A., 215
Maheshwari, P., 227, 228, 237
Maillette, L., 149, 154-57, 166
Malogolowkin-Cohen, Ch., 290
Mann, C., 54, 55
Manning, J. T., 310

Mans, R. J., 71
Marcotrigiano, M., 214
Margolis, L., 309
Martin, C., 19, 21, 45
Maruyama, T., 81
Marx, G. A., 215
Masters, M. T., 171
Masuyama, S., 34, 261, 264, 266, 285
Matheson, A. C., 248
Matthews, P., 170
Mauseth, J. D., 108, 216
Maynard Smith, J., 309, 312
Mayoh, K. R., 54
Mayr, E., 96
Medina-Filho, H. P., 34
Meinhardt, H., 210
Meinke, D. W., 102, 204, 231
Meins, F., Jr., 34, 202
Melchior, R. C., 89
Mendel, R. R., 76
Mergen, F., 271, 285, 286
Merz, T., 58
Messing, J., 189-91
Michaelis, P., 77-80, 82
Michod, R. E., 315
Miflin, B. J., 76
Migeon, B. R., 56
Millar, C. I., 272
Miller, J. H., 185
Miller, S. C., 201
Misler, S., 211
Mohr, H., 76, 103, 104, 106, 107, 126-32, 180, 181
Mopper, S., 154
Moran, G. F., 248, 277
Moreau, P., 36
Morton, N. E., 252, 254
Mottinger, J. P., 22
Motulsky, A. G., 289-91, 304
Mulcahy, D. L., 232, 237, 248
Muller, H. J., 130, 252-54, 303, 308, 309
Murray, A. W., 55, 56, 314
Murray, M. G., 43
Murray, M. J., 40

Nagakubo, T., 283
Nagl, W., 53
Nagley, P., 60
Nakielski, J., 214
Navashin, M., 189

Naveh-Many, T., 186
Neilson-Jones, W., 110, 139
Nelson, B., 21
Nester, E. W., 27
Neuffer, M. G., 140, 143
Nevers, P., 18
Newman, I. V., 96, 120
Niklas, K. J., 108, 143
Ninio, J., 9
Nitsch, J. P., 242, 243
Nordenskiöld, H., 50, 51
Novick, A., 37
Nugent, J. M., 74

O'Dell, M., 21, 27, 59
Oelmüller, R., 76, 180, 181
Ohlrogge, J. B., 76
Ohnishi, O., 282, 283
Ohno, S., 186, 299
Ohta, T., 28
Oinonen, E., 263
Okumoto, D. S., 64
Oldeman, R. A. A., 148, 152, 162
Orgel, L. E., 14, 37, 43, 87
Orr-Ewing, A. L., 285
Östergren, G., 68, 70
Ottaviano, E., 232
Ourecky, D. K., 117
Overall, R. L., 193
Overholts, L. O., 142, 147

Paddock, E. F., 168
Palmer, J. D., 71-75
Palmer, R. G., 172
Pandey, K. K., 65
Paolillo, D. J., Jr., 97
Park, Y. S., 271, 279, 290
Parker, J. S., 300
Pate, D. W., 178
Pauling, L., 191
Pavlath, G. K., 201
Penny, L. H., 32
Perry, T. O., 279
Peters, D. L., 43
Peterson, P. A., 16, 18, 20
Petes, T. D., 56, 192
Pfeiffer, P., 22, 25
Philipson, W. R., 159, 163, 164
Pickard, B. G., 211, 212
Piotrowiak, U., 23

Poethig, R. S., 166, 200, 211, 212
Pohl, J., 53
Pohlheim, F., 110, 112-14, 133, 134, 198-200
Popham, R. A., 89, 93, 135
Popham, T. W., 273, 275
Popp, H. W., 142, 147
Possingham, J. V., 77
Poulson, D. F., 260
Pouzet, M., 184
Powers, P. A., 314
Pratt, D., 117
Pressing, J., 8-10, 29, 95
Price, H. J., 28, 43
Prothero, J., 14

Radler, A., 53
Radman, M., 11, 63
Ray, J. H., 299
Razin, A., 185, 186
Reanney, D. C., 8-10, 29, 95
Rédei, G. P., 97, 246, 276
Reeder, R. H., 188, 189
Rees, H., 67-70
Reiss, T., 76
Relichova, J., 97, 122
Rick, C. M., 232
Riggs, A. D., 185
Ripley, L. S., 12, 13
Roan, J. G., 189
Robards, A. W., 193
Robberecht, R., 177-79
Roberts, E. J., 39
Roberts, J. W., 60
Robinson, A. M., 305
Rogers, A. F., 52
Rohlf, F. J., 248, 249
Romberger, J. A., 146, 148
Rosenberger, R. F., 14
Rosenthal, D., 117, 119
Rouffa, A. S., 110
Rowlands, D. G., 286, 287
Rudin, D., 275
Russell, W. A., 32
Ruth, J., 133, 249
Rutishauser, A., 167, 227

Sachs, F., 211
Sachs, T., 171, 204
Saedler, H., 18-20, 22, 23

Sagan, D., 309
Saghai-Maroof, M. A., 188
Saigo, K., 22
Salser, W., 185
Sandermann, H., Jr., 176, 177
Sapienza, C., 37, 43
Sari-Gorla, M., 232
Satina, S., 114, 195, 197
Saus, G. L., 261, 265
Schaal, B. A., 296, 297
Schaffalitzky de Muckadell, J., 201-3, 279
Schairer, L. A., 52, 174
Scheel, D., 177
Schelander, B., 271, 272
Schemske, D. W., 308
Schensted, I. V., 79
Schieder, O., 170
Schiemann, J., 76
Schimke, R. T., 41, 67, 70
Schlarbaum, S. E., 272
Schmalhausen, I. I., 175, 217
Schmitt, D., 279
Schneller, J. J., 260, 266
Schoen, D. J., 275
Schuster, P., 8, 9
Schuster, W., 23, 71
Schwartz, D., 205, 206
Schwarz-Sommer, Z., 18, 20, 22
Schwemmer, S. S., 52
Scott, N. S., 71, 77
Seavey, S. R., 279, 284, 285, 308
Sederoff, R. R., 72
Semeniuk, P., 197, 198
Shahin, E., 34
Shampay, J., 58
Shannon, C. E., 8
Shapiro, H. S., 186
Shapiro, J. A., 17, 18
Sharp, W. R., 34
Shepard, J. F., 34
Shepherd, N. S., 22
Sherwood, S. W., 41, 67, 70
Shevchenko, V. V., 97
Shiba, T., 22
Shul'Ga, V. V., 173
Silander, J. A., Jr., 305
Silberstein, L., 201
Simmons, M. J., 32, 252
Sims, L. E., 28

Sinnott, E. W., 170, 195
Slatkin, M., 303-5
Slobodchikoff, C. N., 2, 33, 106, 303, 305
Smith, C. A., 64
Smith, D. B., 21
Smith, H. H., 170, 174, 175
Smith, J. B., 43
Smithies, O., 314
Smoot, E. L., 235
Snoad, B., 171, 300, 301
Snow, A. A., 232
Snyder, M., 54, 55
Sokal, R. R., 248, 249
Sokoloff, A., 290
Solbrig, D. J., 254, 303
Solbrig, O. T., 254, 303
Soliman, K. M., 188
Solima Simmons, A., 290
Soll, D. R., 212, 213
Soma, K., 96, 120
Sommer, H., 18, 19, 23
Sorensen, F., 249, 254, 271-73, 275, 276, 279, 290
Soyfer, V. N., 63, 64
Sparrow, A. H., 28, 43, 46-49, 52-54, 172, 174
Sparrow, R. C., 49, 53, 54
Sprague, G. F., 32
Stachel, S. E., 171
Stadler, L. J., 31
Stanley, R. G., 179
Starlinger, P., 16, 17
Stebbins, G. L., 32, 54, 215, 216, 296-98
Steeves, T. A., 117
Steffensen, D. M., 140
Stein, D. B., 295
Stein, O. L., 133, 140, 170, 249
Steinhauer, D. A., 8
Steinmetz, V., 179
Steitz, J. A., 60
Stephens, S. G., 206
Stephenson, A. G., 237, 238, 240, 241
Stern, D. B., 71
Sternglanz, R., 65
Steucek, G., 211
Stevenson, D. W., 159
Stewart, R. N., 117, 120, 122, 126, 129, 197, 198
Stewart, S. C., 275

Strachan, T., 295, 299
Strobeck, C., 303
Strommer, J. N., 32
Stumpf, P. K., 76
Styles, C. A., 22
Sussex, I. M., 102, 117, 166, 200
Sutherland, W. J., 303, 305
Swanson, C. P., 58
Sybenga, J., 4
Szilard, L., 37
Szostak, J. W., 56, 58, 59, 314

Tabak, H. F., 15
Tadmor, Y., 248
Tanksley, S. D., 232
Tarrant, G. M., 14, 15
Tautz, D. M., 12
Taylor, L. P., 18
Taylor, T. N., 89, 235, 265
Tazima, Y., 176
Teramura, A. H., 179
Thielke, C., 137-41
Thien, W., 76
Thoday, J. M., 187
Thomas, A. C., 306, 307
Thomas, H., 215
Thomashow, L. S., 171
Thomashow, M. F., 171
Thompson, D. J., 310
Thompson, R. D., 27, 59
Thompson, W. F., 21, 43, 73, 295
Tillman, E., 21
Tilney-Bassett, R. A. E., 7, 79, 81, 112,
    117, 119, 124-26, 129
Timmerman, B., 171
Timmis, J. N., 71
Tippo, O., 230
Tomlinson, P. B., 143, 148, 149, 152,
    158, 159, 161, 162, 266
Tourte, Y., 90, 91, 204
Trewavas, A., 170, 201
Trick, M., 12
Tryon, R. M., 263
Tschermak-Woess, E., 42
Tschudi, C., 23
Tsuchiya, T., 272
Tsui, W.-C., 11
Tsvelev, N. N., 54
Tucker, E. B., 193
Tucker, S. C., 109, 112

Turner, J. R. G., 53, 314
Tye, B.-K., 56

Ullu, E., 23
Underbrink, A. G., 28, 43

Valentine, J. W., 32
Van den Ende, H., 284
Van Groenendael, J. M., 203
Vanin, E. F., 26
Van't Hof, J., 65-67
Van Valen, L., 29
Varmus, H. E., 22
Varshavsky, A., 66, 67, 70
Vedel, H., 279
Veleminsky, J., 64
Verma, D. P. S., 26
Vig, B. K., 4
Voelkel, K., 65
Vogel, F., 289-91, 304

Wagner, E., 217
Wagner, M., 21, 26, 43
Walbot, V., 2, 17, 18, 28, 76, 297
Wallace, B., 32, 221, 252, 254, 261
Wallace, H., 189
Walmsley, R. W., 56
Wang, J. C., 65
Ward, B. L., 72
Ward, J. M., 159, 163, 164
Wareing, P. F., 202
Warne, T. R., 266
Wassarman, P. M., 60
Watkinson, A. R., 303, 305
Watson, J. D., 56, 60, 169, 176
Watt, A. S., 263
Waxman, S., 172, 174
Wayne, R. O., 212
Webb, C. J., 238
Webster, C., 201
Webster, S. G., 201
Weiling, F., 276
Weiner, A. M., 23, 60
Weismann, A., 2
Wellmann, E., 179
Wessler, S., 17
White, J., 5
White, M. J. D., 54
White, O. E., 172
White, P. R., 171

White, R. A., 89
Whitehouse, H. L. K., 285
Whitham, T. G., 2, 33, 106, 154, 303, 305
Whyte, L. L., 3, 6, 249
Wienand, U., 22
Wiens, D., 238, 239, 279, 308
Wilcox, M. D., 271, 273, 274
Wilkie, D., 263, 285
Williams, A. G., 305
Williams, E. G., 286
Williams, G. C., 35, 36, 284, 309
Williams, R. F., 97, 121
Willis, A. E., 299
Willson, M. F., 225, 237, 244
Wilson, A. C., 297, 298
Wilson, B. F., 149, 151, 152, 163
Wilson, C. A., 279, 308
Winsor, J. A., 238, 240, 241
Winsten, J. A., 169, 176
Wolf, S. F., 56

Wolff, S., 43
Wolpert, L., 201, 210
Woodwell, G. M., 50
Woollard, R., 272
Wright, S., 252, 254
Wu, C.-I., 184

Yagil, E., 215, 216
Yamagishi, H., 67
Yeh, E., 54
Yette, M. L., 64
Ying, C. C., 271
Young, W. J., 58

Zambryski, P., 171
Zamir, D., 232, 248
Zhukova, P. G., 54
Zimmermann, F. K., 4
Zohary, D., 281
Zouros, E., 191
Zuckerkandl, E., 191

Subject Index

Abies: balsamea, 203; *procera*, 271, 274
Abnormal growth, 166
Aborted pollen, 277
Acacia retinodes, 286
Acentric chromosomes, 67
Acentric fragments, 48, 54, 56
Acer, 149, 202, 210; *negundo*, 203; *platanoides*, 200, 212; *rubrum*, 149-52; *saccharum*, 289-91
Acrostichum: aureum, 261, 266; *danaeifolium*, 266
Actin, 212
Activator (Ac) in corn, 16, 17
Advantaged mutants, 104
Adventitious buds, 116, 117, 162
Agavaceae, 159
Agenic developmental program, 212; in yeast, 213
Aging, 15
Agraphis, 229
Agrobacterium: rhizogenes, 27; *tumefaciens*, 171
Allium cepa, 57
Allometric constant, 206-8
Allopolyploidy, 296, 298
Allozyme variability, 283
Aloë, 300
Aloineae, 300, 302
Alu DNA in humans, 23
Amaryllidaceae, 300
Amino acid biosynthesis, 76
Anemone virginiana, 57
Aneuploidy, 56, 230, 298; pollen transmission, 231
Angiosperm life cycle, 168, 228

Anneau initial, 117
Annuals, 33
Anthocyanins, 178, 181
Anticlinal divisions, 135, 164
Antirrhinum majus, 17, 18, 26, 45, 81, 215
Apical control, 153
Apical dominance, 153
Apical initial redundancy, 95
Apical meristem: *Cassiope lycopodioides*, 110; *Clethra barbinervis*, 111; gymnosperms, 94; *Michelia fuscata*, 112; *Zebrina pendula*, 138
Apical meristem size increase with aging, 121
Apical meristem topology, 133
Apparent cell number (ACN), 140
Apurinic sites, 11
Arabidopsis thaliana, 28, 65, 66, 97, 121, 231
Araucaria excelsa, 195, 196, 202
Araucariaceae, 110
Artificial chromosomes, 54-56
Asexual populations, 309
Astilbe, 229
Ataxia telangiectasia, 299
Athyrium filix-femina, 258, 260, 266
Autocatalytic activity, 200, 201
Autocatalytic systems, 217
Auxin, 171
Auxotrophs, 217
Axillary buds, 117, 146, 148
Axillary flowers, 240, 241

B-chromosomes, 67-70

Balanced lethal system, 287
Balanced load, 252
Barley, *see Hordeum vulgare*
Barley strip mosaic virus, 22
Beta vulgaris, 57
Betula pendula, 148, 149, 166
Bimodal karyotype, 300
Biological age, 86, 144
Biological code, 182
Biological time, 29, 30, 33, 40, 99
Birth defects, 285
Bloom's syndrome, 299
Branching, 143-58; demography, 146;
    meristems, 100; nomenclature in
    Acer rubrum, 150-51; senescence and
    mutation loading, 145; sympodial vs
    monopodial, 147
Breakage-fusion-bridge cycle, 58, 59
Brookhaven Gamma Forest, 50
Bs 1 insert in corn, 22
Buckwheat, *see Fagopyrum esculentum*,
    282
Bud pollination, 277
Bud populations, 154; production in
    silver birch, 155-57

C-effects, 306
C-value, 66
Cambium, 163; loss of initials, 164
Cancer, 134, 300
Candida albicans, 212, 214
Cannabinol, 178
Cannabis, 178
Capsicum, 248
Caragana arborescens, 279
Carex arenaria, 158, 162
Carotenoids, 178; synthesis, 75
Carpinus, 148
Carrot protoplasts, 64
Caryophyllaceae, 143
Cassiope lycopodioides, 109, 110
Castanea, 148
Casuarina, 229
Cauliflory, 85, 162
Cauliflower mosaic virus, 22
Cell division patterns, tangential and
    logarithmic, 88
Cell displacements, 109, 111, 112, 124-
    26
Cell fitness, 5, 107, 133, 144
Celtis, 148

Centromere, 41, 46; DNA (CEN), 54;
    DNA sequences, 70; position, 45
Ceratopteris, 288; *pteridoides*, 265,
    266; *thalictroides*, 265, 266
Chalcone synthase, 18
Chamaecyparis: lawsoniana, 164; *pisi-*
    fera plumosa argentea, 133; *thy-*
    oides, 164
Chemostat, 37-40; consequences for
    genetic variability, 316; equation, 315
Chemostat-like effect, 39, 271
Chenopodiaceae, 143
Chimeras, 3, 7, 300; classification, 7;
    lengths, 124; stability, 104, 105, 123,
    129-31; *see also* Periclinal chimeras;
    Sectorial chimeras
Chlorophyll deficient mutants, 32, 282,
    283
Chlorophyll formation, 181
Chlorophytum elatum, 140
Chloroplast, 70-83; DNA isomerization,
    75; genome, 70, 72, 74; genome copy
    number, 77; genomic stasis, 74; major
    genome types, 73; random sorting in
    cells, 80, 82
Chromatid segregation, 92
Chromatin, 60, 65
Chromosomal scaffold, 65
Chromosome aberrations (mutations),
    21, 58, 193, 297-302; breakage, 58,
    180; drive, 298; frequency in poly-
    ploids, 302; instability syndromes,
    299, 301; mosaicism, 33; rates of, 167;
    size, 54
Chromosome stability, linear vs circu-
    lar, 55
Cicer, 73
Cin 1 in corn, 22
Circular chromosomes, 54-56
Cleavage polyembryony, 233, 236, 237
Clethra barbinervis, 109, 111
Clonal growth, 219
Clonal plants, 115
Clone longevity, 270
Codons, 182
Codon synonymy, 183
Commelina benghalensis, 138, 139
Commelinaceae, 136-38
Component meristems, 109, 110, 120,
    126, 197
Concerted evolution, 28, 187, 294

Cone abortion, 272
Congealed genotype, 54
Coniferales (conifers), 149, 271
Conjugation reactions, 176
Connations, 171
Convulate embryos, 234
Consanguineous marriages, 252
Copia, 22
Coprosma baueri, 114, 115
Cork cambia, 159, 165
Corpus, *see* Apical meristem
Corylus, 148, 149, 229
Cranberry, 197
Crataegomespilus asniersii, 214
Crepis capillaris, 49, 68
Cross feeding of mutant cells, 102, 193, 217
Crown gall, 171
Cruciferae, 22
Cryptochrome, 181
Cryptomeria, 229
Cucurbita pepo, 238-40
Cucurbitaceae, 72
Cunninghamia, 229
Cupressaceae, 110, 133
Cycas revoluta, 194, 195
Cyperaceae, 48
Cytochrome P-450-dependent reactions, 176
Cytogenetic instabilities, 299, 300
Cytokinin, 171
Cytoplasmic genomes, 70-81
Cytosine methylation, 17
Cytoskeletal organization, 212

Dactylis glomerata, 281
Dark repair (UV), 64, 174
Datura stramonium, 40, 197, 230, 231, 238
Deamination, 185
Delay of fertilization, 243, 244
Deleterious alleles, 251, 252
Deliquescent growth, 153
Delphinium, 229
Density-dependent selection, 194
Depurination, 11
Design for survival, 284
Deterministic growth, 152
Developmental buffering, 203; in embryo of *Marsilea*, 205
Developmental constraints, 108, 272

Developmental-mutants, 215
Developmental selection, 3, 6, 148, 219, 222-24, 227, 251, 310
Developmental-weak-points, 134
Development of animals vs plants, 170
Deviant growth, 166
Deviant Mendelian ratios, 248, 249
Dicentrics, 55, 58
Dicotyledons, 117
Diffuse centromere, 48, 50
Dioryctria abovitella, 154
Diplontic drift, 86
Diplontic selection, 3, 29, 39, 50, 75, 102, 106, 107, 132, 142, 182, 187, 194, 198, 217, 315
Directional selection, 225
Disadvantaged mutations, 104
Disadvantageous mutations, 90, 128
Dispersed repeats, 28
Dissociation (Ds) in corn, 16, 17
Distichous phyllotaxes, 136
DNA amplification, 28; content, 28; ligase I, 299; methylation, 185; per chromosome, 52; polymerases, 60-70, 299; repair, 60, 63, 64; replication, 60; replication fidelity in bacteria, 11; replication fork, 61, 65, 66; slippage, 12; synthesis, 60, 90; topoisomerases, 65
Drosophila, 33, 60, 142, 235, 253, 260, 261; *melanogaster*, 32, 36, 189; *willistoni*, 290
Dryopteris filix-mas, 266
Duplications, 67
Durchbrenner, 261
Dwarfing, 174
Dwarfism, 283

Early-acting lethals, 260
Ecological load, 253
Editing, 11, 62
Elaeagnus pungens, 115, 116
Elymus farctus, 299
Embryo: abortion, 284, 287; development, 243; lethals, 204, 219; sacs, 227, 228
Embryo-lethal mutants, 103, 231
Embryonic competition, 233
Embryonic lethal equivalents, 255
Embryonic lethals, 219, 255
Embryonic load, 272, 278

En (Spm) in corn, 18, 26
Endomitosis, 42
Endonuclear polyploids, 50
Endoreduplication, 11, 42, 67
Endosperm, 167; protein, 189
Enhancer (En) in corn, 18, 26
Environmental mutagens, 175
Ephedra, 97
Ephedraceae, 110
Epidermal cells, 111
Epilobium canum, 232
Equisetum, 90, 143
Ericaceae, 278
Error catastrophe, 8, 14, 15
Error rate, 8-11
Escherichia coli, 11, 36
Etiolation, 181
Eucalyptus regnans, 277, 279, 283
Euchromatin, 41
Eusporangiate ferns, 89
Evolutionary potential, 35
Excision repair, 64
Excurrent growth, 153
Exine, 178
Exonuclease, 63
Expressivity, 260, 261
Extra chromosomal DNA, 66
Extrinsic selective pressures, 29

Fabaceae, 73
Fagopyrum esculentum, 282, 283
Fagus, 149, 158, 229; *sylvatica*, 202, 245, 279
False polyembryony, 229, 233
Fanconi anemia, 299
Fasciation, 171, 214
Fate map of the corn embryo, 142
Fatty acid synthesis, 76
Ferns, 33, 256; life cycle, 257
Fertilization order, 232
Fitness, 252; organism vs cell, 5
Flavanols, 178
Flavone powder, 178
Flavones, 178
Flavonones, 178
Flax, 43, 305
Flower position, 245; inbreeding effects, 234
Fraxinus, 203
Fruit abortion, 238, 240, 241; competi-
tion, 241; growth hormonal stimuli, 247
Fungal senescence, 15
Fusiform initials, 164

Gametic mutation, 305
Gametophyte mutations, 267-69
Gametophytic self-incompatibility, 284
Gamma irradiation, 48
Gamma-plantlets, 206, 207, 210
Gasteria, 300
Gene amplification, 67; conversion, 188, 190, 294
Genet, 5, 99, 102, 144
Genet age, 238
Genetic burden, 254
Genetic death, 309
Genetic drift, 295
Genetic load, 235, 250-92, 303; *Acer saccharum*, 289, 291; *Athyrium filix-femina*, 291; *Eucalyptus regnans*, 279, 280; *Osmunda regalis*, 259; *Phegopteris decursive-pinnata*, 264; *Phlox*, 278; *Picea abies*, 272; *Picea glauca*, 281; *Pinus radiata*, 273, 279; *see also* Inbreeding depression
Genetic mosaics, 166
Genetic stasis, 145
Genome size, 8-10, 43, 55, 57
Genomic DNA values, 41
Genomic flux, 16
Genomic parasites, 68
Genomic stress, 36
Genotoxic chemicals, 175
Genotrophs, 305
Germline, 2, 3, 15, 29, 36, 303
Ginkgo, 149
Gleditsia, 148
Glucosides, 177
Glycine max, 17, 172
Gnetaceae, 110
Graft stability, 203
Gramineae, 110, 139
Grasses, 139
Growth: by cell division, 209; by cell enlargement, 209
Gymnosperm apices, 93
Gymnosperms, 271

Habituation, 202

Haplopappus gracilis, 54
Hard selection, 221, 223, 224, 304
Haworthia, 300
Hedera helix, 202
Helianthus, 28, 117
Herbicide, 75
Heterochromatin, 41, 59
Heterotic load, 252, 253, 280
Histogenic specificity, 198
Histone, 60
Holcus mollis, 306
Holocentric centromeres, 46-50
Homeostasis, 216, 293
Homo sapiens, 253, 285, 290
Homoplasmic mutant, 79; sectors, 80
Hooded phenotype, 216
Hordeum vulgare, 32, 76
Hydrolytic enzyme reactions, 176
Hymenocallis calathinum, 300, 301
Hypodermis, 114

Ignorant DNA, 43
Imaginal disks, 142
Impermanent apical initials, 120
Impermanent meristem initials, 86
Inbreeding, 272; coefficient, 254, 288;
    depression, 271, 273, 274, 276, 308;
    see also Genetic load
Incest, 285
Independent assortment, 310, 311
Information modules, 9, 10; redundancy
    of, 10; storage, theoretical limits, 9
Intercalary growth, 143
Intercalary meristems, 141
Intergenic DNA sequences, 188
Intergenomic invasions, 71
Internal selection, 3
Internal selective forces, 249
Internode, 141
Interphase chromosome volume, 50, 54;
    relationship to radiosensitivity, 53
Interspecific graft chimeras, 214
Interspecific hybridization, 296
Intragametophytic mating, 256
Intragenomic recombination, 72
Intrinsic selective pressures, 29
Introgression, 296
Introns, 27, 190
Inversions, 67, 297, 298
Inverted repeat, 17, 21, 65, 73, 74

Iojap, in corn, 76
Ionizing radiation, 46, 174, 175, 203,
    204
Isoetaceae, 93

Juncaceae, 46, 48
Juniperus, 110, 113, 133; *communis*,
    235; *davurica*, 249

Kalmia: angustifolia, 278; *latifolia*,
    278; *polifolia*, 278
Karyological anatomy, 42
Kinetochore, 54; *see also* Centromere
Knotted locus (Kn1) in maize, 214, 215

Laburnocytisus adamii, 214
Larix: laricina, 271, 279, 290; *leptole-*
    pis, 30
Lathyrus latifolium, 57, 73
Leaf development, 138; hairs, 178;
    shape, 200; variegation, 113-16
Leaky lethality, 261
Leghemoglobins, 26
Legumes, 26
Lens, 248
Leptosporangiate ferns, 89
Lethal factors, 252; equivalents, 254,
    265, 270, 277, 287
Lethals, 32, 288; effect of polyem-
    bryony, 275; leaky, 262, 263; linkage
    effects, 276; segregation patterns,
    220
Libocedrus decurrens, 235
Lignin, 176
Liliaceae, 300
Lilium, 46, 48, 229; *longiflorum*, 46, 48,
    49; *pyrenaiecum*, 57
Linear chromosome, 55
Linkage, 276, 299
Linum usitatissimum, 121, 305
Liquidambar styraciflua, 278
Load levels, 287
Logarithmic growth, 37; pattern of cell
    multiplication, 87
Lolium perenne, 306
Long shoots, 148, 200
Long terminal repeats (LTRs), 22, 25
Loranthus, 229
Lotus corniculatus, 241
Luzula, 46-50; *acuminata*, 48; meiosis

Luzula (Continued)
  of, 51; *purpurea,* 46, 49; radiation
  tolerance of, 48, 49
Lycopersicon esculentum, 181, 232, 248
Lycopodiaceae, 93

M chromosomes, 286
Maize, *see Zea mays*
Marattiales, 93
Marginal meristem, 200
Marsilea, 210, 234; *vestita,* 90, 204, 205
Maternal age, 233
Maternalization effect, 65
Mating system evolution, 308
Matteuccia struthiopteris, 34, 158, 267-
  70
Mechanical forces, 211
Mechanotransductive channel, 211
Medeola virginiana, 158, 160, 161
Medicago sativa, 73, 286, 287
Megagametophyte, 226
Megaspore competition, 288
Megaspores, 226
Meiocytes, 119
Meiosis, 50
Meiotic chromosome pairing, 296
Meiotic irregularities, 297
Meiotic recombination, 314
Melilotus, 73
Membrane-bound enzymes, 191
Mendelian genes, 293, 302-8
Mendel's law of segregation, 310
Mentha arvensis, 114
Mericlinal chimeras, 86, 120
Meristematic initials, 39, 79
Méristème d'attente, 89, 117
Méristème medulaire, 117
Meristem oscillations, 109, 112
Meristems, stochastic vs. structured, 85-
  87
Meristem size increase, 122
Merophytes, 90, 91, 194
Mesophyll, 114
Message redundancy, 9, 10
Metasequoia glyptostroboides, 271
Metastasis, 134, 169
Methylated adenines, 63
Methylation, 17
Michelia fuscata, 109, 112
Microbial infections, genetic conse-
  quences, 27

Microgametophyte, 230
Micronuclei, 48
Misalignment mutagenesis, 12, 13
Mismatch repair, 60, 63
Mispairing, 12
Missence, 187
Mitochondria, 62, 72; DNA, 71; evolu-
  tion, 81; genome, 23, 70, 72
Mitosis, 53
Mitotic crossing over, 4, 168, 193, 306
Mitotic gene conversion, 4
Mitotic inhibition, 205
Mitotic quiescence, 88
Mitotic recombination, *see* Mitotic
  crossing over
Mitotic stability, 54, 56
Mobile elements, *see* Transposable
  elements
Modified roots, 116
Modulator (Mp) in corn, 44
Molecular coevolution, 188
Molecular constraint, 191
Molecular drive, 295, 296
Molecular footprint, 18, 19
Monocotyledons, 33, 117
Monoecious plants, 245
Monomeric proteins, 191
Monopodial branching, 146, 147
Muller's ratchet, 130, 309, 312
Multigene families, 28, 189
Multimeric proteins, 191
Multiple archegonia, 235
Mutagens, 176
Mutant plastids, 76
Mutation accumulation, 90
Mutational decay, 185
Mutational load, 119, 129, 132, 143,
  238, 252, 253, 267, 270, 271, 280, 283,
  308, 311
Mutation buffering, 43, 50, 106, 175,
  182, 225, 314; apical initial redun-
  dancy, 95; gene design, 186; stochas-
  tic meristem characteristics that pro-
  mote, 107
Mutation fixation, 98, 100
Mutation frequency, 3, 29-36, 167; in
  maize, 31
Mutation hot spots, 12, 185
Mutation loading per apex, 145
Mutation loss, 98
Mutation proofing, 187

Mutation rates, 3, 29-36, 60, 80, 95, 99, 269, 270, 303; in ferns, 34; a function of age, 304
Mutator (Mu) in corn, 17
Myrmecia pilosula, 54

Neoplasms, 170
Neutral mutations, 98, 100, 128, 144
Nicotiana, 27, 117, 175; *glauca*, 214; *rustica*, 27; *tabacum*, 27, 214
Nitrate reductase, 76
Nivea in *Antirrhinum majus*, 18
Noise, 8, 9, 29
Nondisjunction, 67, 314
Nonrandom orientation of chromatids, 92
Nonsense codon, 189
Nonsense mutation, 187
Normalizing selection, 225
Nucellus, 286
Nuclease activity, 62
Nucleolar dominance, 189
Nucleolus, 189
Nucleosomes, 60
Nucleotide combinations, 184
Number of apical initials, 122

Oenothera, 30, 53, 227; *hookeri*, 288; *parviflora*, 288
Onoclea sensibilis, 34, 258, 261, 265, 267-69
Ontogenetic constraints, 108
Oplismenus imbecillis var. *variegatus*, 139-41
Opportunistic growth, 152
Organelle C-value paradox, 73
Organelle DNA evolution, 72
Organelle genes, 81
Organism fitness, 133
Organogenesis, 194
Ornithogalum virens, 49
Osmunda regalis, 258, 259, 261, 265, 268, 269
Osmundaceae, 265
Outcrossing survivorship (dominant genetic load), 290
Ozone, 170, 177

P element transposition in *Drosophila*, 36
Palindromes, 12

Pallida (pal) gene promoter, 18
Palms, 159
Pandanus, 159
Parascaris equorum univalens, 54
Peach, 197
Pea pod neoplasms, 170
Pelargonium, 210; *hortorum*, 74; *zonale*, 118, 119, 199; X *hortorum*, 197, 198
Penetrance, 261
Perennials, 33
Periclinal chimeras, 110, 114, 116, 117, 124, 129, 140, 214, 250
Periclinal cytochimeras, 195
Periclinal divisions, 124, 130
Permanent initials, 120
Peroxidase reactions, 176
Peroxygenase reactions, 176
Phalaris arundinacea var. *pieta*, 140, 141
Phegopteris decursive-pinnata, 261, 264, 266
Phenolase reactions, 176
Phenotypic phase changes, 202
Phlox drummondii, 276-78, 282, 290
Photomorphogenesis, 180, 181
Photoreactivation, 64
Photorepair, 177
Photosynthesis, 76
Physiological load, 253, 260
Phytochrome, 76, 181
Picea, 172, 282; *abies*, 173, 271, 272, 281; *glauca*, 271, 281, 286, 290; *mariana*, 271, 290; *omorika*, 273, 274, 290
Pinaceae, 271
Pinus, 94, 172, 229, 237; *canariensis*, 202; *edulis*, 154; *pinea*, 202; *radiata*, 271, 273, 274; *resinosa*, 173, 235, 247, 272, 288, 290; *silvestris*, 173, 245, 271; *strobus*, 154, 173, 230, 235; *taeda*, 271, 275, 279
Pisum sativum, 57, 73, 171
Plantago lanceolata, 203
Plant architecture, 148
Plant morphogenesis, strategies, 209
Plant radiosensitivities, 54; list of factors affecting, 52
Plant tumors, 135
Plant vacuoles, 176
Plasmodesmata, 193
Plastid distribution, 81

Plastid inheritance, 81
Plastochron, 97, 109
Platanus, 148
Point mutations, 184
Polarized growth, 206
Pollen, 64; competition, 231, 232, 288; germination, 232; grain, 230; loads, 238; tube, 230; tube growth rate, 232; tube mitosis, 167
Pollination, 230
Pollination order, 232
Polyembryony, 229, 233; genetic consequences of, 228, 275
Polyploid conifer species, 272
Polyploidy, 50, 267, 296
Populus, 158; *alba*, 203
Positional information hypothesis, 210, 211
Postembryonic inbreeding depression, 271
Postembryonic load, 274, 278
Post-reductional meiosis, 50
Postreplication repair, 62
Postzygotic self-incompatibility, 285
Postzygotic somatic mutation, 276
Pregermination selection, 275
Pre-reductional meiosis, 50
Prezygotic factors, 284
Primary DNA, 43
Processed pseudogenes, 26
Promeristem, 97
Promeristematic volume, 122
Promoters/enhancers, 188, 189
Promutagen activation, 176
Proofreading, 11, 60, 62
Prophage induction, 36
Proteases, 14
Proviral DNA, 21
Proximity-dependent loads, 282
Prunus, 38; *persica*, 197; *pissardi*, 197, 199
Pseudogenes, 23-26, 184, 190
Pseudoterminal buds, 146
Pseudotsuga menziesii, 271, 279, 288, 290
Pteridium aquilinum, 261-63, 285
Pteridophytes, 88
Pyrethrum, 229
Pyrimidine dimers, 64

Quercus, 202

Quiescent center, 119, 175

Radiation resistance, 48; in seedlings, 206
Radiosensitivity, 50, 53
Ramet, 5, 100, 106, 115; competition, 132, 146, 152, 158; doubling generation, 33; generation, 143, 144, 270; longevity, 158; variation, 143
Random chromatid orientation, 92
Random ontogeny, 194
Ranunculus, 229; *adoneus*, 178
Rates of chromosomal divergence, 297
Real time and mutation, 40
Recessive embryonic lethals, 222
Reciprocal translocations, 288, 297, 298
Recombination, 308, 310
Red Queen effect, 29, 190
Reductase reactions, 176
Redundancy, 9
Regulatory element, 16
Renner complexes, 288
Repair, 11, 60
Repeated DNA, 28, 293-97
Repeated sequence polymorphisms, 296
Repeated sequences, 28
Repetitive DNA, 28
Replication errors, 86
Replication forks, *see* DNA replication fork
Replication of linear DNA, origin of 5′ gaps, 58
Replicon, 41, 65-67, 70; clusters, 66; misfiring, 67; size, 65
Reproductive cycles, 226
Restriction site polymorphism, 32
Retroid elements, generalized life cycles, 24, 25
Retroviral LTR, 22
Retroviruses, 21-26
Reverse transcriptase, 21-26, 63
Reverse transcription, 21-26, 71
Rhizobia, 26
Rhizome branching, 158; *Carex arenaria*, 162; *Medeola virginiana*, 160, 161
Ribosomal genes, 74
RNA genomes, 8
RNA polymerase I, 189
Robertsonian fusions, 298

Robinia, 148
Rosa, 229
Rosaceae, 110
rRNA genes, 187, 188, 297
Rye grass, 306

Saccharomyces cerevisiae, 22, 54, 65, 192
Saccharum officinarum, 140
Safe sites, 219
Salix, 148, 229
Sciadopitys, 229
Scilla autumnalis, 293
Scotomorphogenesis, 180, 181
Secale: cereale, 27, 59; *sylvestre*, 27, 59
Secondary DNA, 43
Secondary growth, 159-65
Secretion, 201
Sectorial chimeras, 86, 93, 98, 100, 102, 120, 133; bud placement, 101
Sectoring, 81
Seed/ovule ratio, 238, 239
Seed abortion, 231, 277; aging, 39, 40; storage, 39, 40; storage proteins, 189
Segregational loads, 253
Selaginellaceae, 93
Self-incompatibility, 284, 286
Selfish mutations, 169
Selfish or parasitic DNA, 36, 37, 43
Selfish tissue, 134
Self-sterility, 279
Senecio, 229
Senescence, 15, 145, 169
Sequence homogeneity, 293
Sequoia sempervirens, 229, 272, 275, 278
Sex, 308-15
Sexual reproduction, 219
Sexual selection, 225, 237, 244
Shoot apex, 195, 196
Shoot tip abortion, 146
Short shoots, 148, 149, 200
Siliques, 231
Simple polyembryony, 233, 235, 247, 258, 273
Sinapis alba, 75, 181
Sister chromatid exchanges (SCE), 55
Size of the apical meristem, 121
Skewed-asynchronous patterns of stem cell divisions, 89
Skewed ratios, 248

Sociobiology, 225
Soft selection, 219-50, 277, 288, 304; in pines, 247
Soft selection sieves, 225, 226, 232, 247, 274, 277
Solanum tuberosum, 180
Solidago altissima, 297
Soluble enzymes, 191
Somaclonal variation, 34
Somatic chromosome variation, 300
Somatic crossing over, 4
Somatic mutations, 3, 31, 77, 85, 108, 134, 172, 182, 305
Somatic recombination, 65
Somatic selection, 306, 307
Somatic stability, 202
Sorghum vulgare, 181
SOS repair, 36
Speciation, 296
Sphenophyllum plurifoliatum, 163
Spirea bumalda, 111
Spirodistichous phyllotaxes, 136
Spirogyra, 48
Spontaneous abortion, 279
Sporogenous tissue, 117
Sporophytic lethals, 260
Stabilizing selection, 225
Sterility, 297
Stochastic apical meristems, 96
Stochastic meristem concept, 96
Stochastic meristems, 85, 97, 98, 120, 148; mathematical model of stratified meristems, 127; mutation buffering, 107; mutation fixation, 100; selection of initials, 99
Stratified apical meristems, 109-35, 305; mathematical model, 127
Structural epigenesis, 211
Structural proteins, 191
Structured meristems, 85, 89
Structured ontogeny, 194
Substitutional loads, 253
Substrate-nonspecific enzymes, 191
Substrate-specific enzymes, 191
Sugarcane, 140
Sympodial branching, 146-48
Synonymous codons, 184; use in proteins, 183
Synonymous mutations, 184

Tam1 in *Antirrhinum*, 18, 19, 26

Tam2 in *Antirrhinum*, 18, 19, 26
Tam3 in *Antirrhinum*, 18, 19, 26, 45
Tandem repeats, 28, 187, 190
Tangential pattern of cell multiplication, 87
Taraxacum officinale, 117
Taxodiaceae, 271
Taxodium, 229
Taxus canadensis, 229
Telocentric chromosomes, 56, 298
Telomeres, 41, 56, 70; DNA, 59; DNA replication, 56; heterochromatin, 27, 59; sequences, 70
Teratological phenotypes, 170-72, 203
Terminal flowers, 245
Tetrad analysis, 50
Tgm 1 in soybeans, 26
Threshold selection, 312-14
Thuja, 235
Thymine dimers, 64
Thyrsopteris elegans, 158
Tilia, 148
Ti-plasmid, 170
Tissue culture, 34
Tomato, 17
Topoisomerase II, 65
Tradescantia, 167; *albiflora*, var. *albovittata*, 137
Transformation, 27
Transitions, 184
Translocations (chromosome), 288, 297, 298
Translocations in chimeras, 67, 124
Transorgan pleiotropic effects, 215
Transplacement in yeast, 192
Transposable elements, 16-26, 43, 44
Transposition, 16-26, 71; molecular model of *Ac* movement, 45, 46; mutational consequences, 20
Transposon movement, 18, 47
Transversions, 184
Tree architecture, 152, 154
Tribolium: castaneum, 290; *confusum*, 290
Trichimeras, 124
Trifolium, 73
Trillium grandiflorum, 167
Triticale, 59
Triticum aestivum, 181
Tropical environments, 177

True polyembryony, 233
Tsuga canadensis, 172, 173
Tumor-forming hybrids, 174, 175
Tumors, 169, 174
Tunica, 33, 97
Tunica-corpus component meristems (LI, LII, LIII), 114, 195
Tunica-corpus organization, 88, 109-35; arrangement of mutant meristems, 125, 305
Ty elements in yeast, 22
Typological apical initial definition, 96

Ulmus, 148; *americana*, 279, 290
Ultraviolet: radiation, 64, 177-81; shielding, 177-81; spectrum, 178; *see also* UV-A, UV-B, UV-C
Unequal crossing over, 26, 190
Unit biological space, 221, 223
Unstratified meristems, 96
UV-A, 177
UV-B, 177; environmental aspects, 179; photoreceptor, 181; radiation, 177
UV-C, 177
Uvularia, 229

Vaccinium, 278
Variegated grasses, 140, 141
Variegation, 16; *Chlorophytum elatum*, 140; *Commelina benghalensis*, 139; *Coprosma baueri*, 115; *Elaeagnus pungens*, 116; geranium, 198; *Juniperus*, 113; *Mentha arvensis*, 114; *Oplismenus imbecillis* var. *variegatus*, 140, 141; *Pelargonium* X *hortorum*, 198; *Pelargonium zonale*, 118, 119, 199; *Phalaris arundinacea* var. *pieta*, 140, 141; *Tradescantia albiflora* var. *albovittata*, 137
Vascular cambium, 159-65
Vegetative growth, 33, 116
Vicia faba, 46-49, 57, 73, 167, 286
Viroids, 27

Wax-layers, 178
Waxy phenotype, 18, 30
Weismann's doctrine, 2
Wheat genome, 21
Whiteshell Nuclear Establishment, 48
Wisteria, 73

Witch's broom, 172, 174; offspring, 174
Within-individual selection, 106, 305
Woodwardia japonica, 34

Xenobiotics, 175, 176; metabolism, 176; excretion, 177
Xenopus: laevis, 189; *borealis*, 189

Yeast, 54, 142, 212

Zea mays, 16-18, 26, 31, 32, 44, 57, 58, 71, 76, 140, 189, 204, 276
Zebrina pendula, 137, 138
Zein gene families, 189
Zein protein, 190
Zingeria biebersteiniana, 54